BIOGAS HANDBOOK

BY DAVID HOUSE

Illustrations by Vernon Surratt

PEACE PRESS

10 9 8 7 6 5 4 3 2 1

Peace Press, Inc.
3828 Willat Avenue
Culver City, California 90230

Printed in the United States of America by Peace Press, Inc.

An earlier edition of this book was self-published and entitled *THE COMPLEAT BIOGAS HANDBOOK*.

Other books by David House
Methane Systems
Wind and Windspinners (co-authored with Michael Hackleman)
Journey (a book of poems)

Library of Congress Cataloging in Publication Data

House, David, 1948–
The biogas handbook.

Bibliography
Includes index.
1. Gas-producers. 2. Methane. 3. Farm manure.
I. Title.
TP762.H68 1981 665.7'76 80–8998
ISBN 0–915238–47–0 AACR2

There is a great deal of satisfaction available to an author. Good books lead a life of their own, meeting people the author has never met, sharing thoughts with people in far off places, sparking ideas in others which the author may never have himself considered.

Just as a person does not exist in isolation, so too, a book is not written as the product of a solitary mind. This book is the result of interaction with many minds, as witness the Bibliography. As well, the strength to continue writing during those times when it seemed the book would never be, and its promise seemed to be empty wind, was not solely my own. The will to strive, the stamina to persevere, and the heart to have begun, all these I owe in various measures to others. Therefore, I dedicate this book to my wife, Janeth, and my baby daughter, Sarah. And I thank my parents, Bill and June House, noble and selfless, for their help. Finally, in hopes that the world will hear, I want to express gratitude to my Baha'i brothers and sisters throughout the world, for theirs is the greatest work of all.

Figures and Tables

Figures

Tables

Contents

SECTION I

OVERVIEWS

This section contains four chapters which give general information about this book, energy, math, and biology.

1: This Book

Why Here ?

Most often, the kind of information presented in this first chapter is part of the introduction. But people, most often, don't read the introduction, and since the best reason for writing one is that it be read and applied, this is called Chapter 1. But, it's the introduction.

What's Here ?

Many books of this sort, that is, books about Ancient Sources of Energy (A.S.E.), proclaim in one way or another: FREE AND ENDLESS ENERGY!!

This book will not. If you actually apply all the information in this book (and a great many will not) it is highly unlikely that your application will take place without a lot of hard work on your part.

This book makes no claim to startling originality or clever breakthroughs, and its usefulness comes mainly because here, gathered in one place, is a great deal of information on biogas: what it is, where it comes from, and how to make and use it.

There are only a few designs for biogas generators given in detail. Nevertheless, there is a sufficient variety and depth of information given so that you will be able to intelligently choose between designs or design your own generator if you wish. Several ideas are presented for generators which have never been constructed — have at it.

Text-Like

The world around us does not come convenience-packaged or predigested, and it is often not as advertised, in the sense that mistakes occur, some things are painful or difficult and so on. The design of the world seems very carefully thought out, and so this book is not predigested either. Much of this information will have to be chewed on, and you may have to reread parts of the book which have unfamiliar concepts in order to understand them.

For this reason, it has been assumed that whoever reads this book is willing to struggle where necessary, and pause and ponder if required. To help you, the book is organized like a textbook. At the end of most chapters you will find new terms used, questions, and problems. Read these, for they often contain additional information.

Terms

The new terms introduced and used in each chapter are found not only at the end of the chapter, but also at the end of the book in Appendix 16. Some of the words are strange, or strange sounding, but English was once a strange language to all of us, yet we learned to use it.

Questions and Problems

Answers to some questions (where it seems necessary) and all problems are found in Appendix 17.

General Organization

There are nine major sections in the book. Within each section are several chapters pertaining to that general subject, and within each chapter there are smaller divisions. You may have already noticed that various subsections in this chapter have their own headings. This helps put information across, since it gives a title to each new area and puts us on the same track.

Because the book is organized in this manner, you can skip something you may know, go back and find a bit of information which may have been improperly digested, or read ahead if you wish. This allows you, in essence, to reorganize the book without physically tearing it apart. When you're standing over a pile of straw waiting to put it in your generator, and you remember that you forgot something, this peculiar type of organization helps you find the right piece of your particular puzzle.

The book is not written to entertain; it is written to use. Doubtless, few people will sit down and digest the whole thing, but take time to familiarize yourself with the contents. The motion is often from the general to

the specific, so that once the bare bones of the idea are explained, you'll find more specific information on application. If you don't want to apply that idea, skip the specifics and go on to the next heading or chapter. A lot of the information presented is unnecessary for building or running a very simple biogas system. If you feel like you'll get lost in the wilderness of facts and figures, just go directly to the Plans section, where you can learn to build a very small generator — and begin to run it — for under $10.00. Then, when you want to know more, read some other sections or chapters. Use the book your own way; take charge. The information is here for you, not you for the information.

If you run across a bit of "biogasese," a technical word you don't understand in your skim-and-read search, just turn to Appendix 16 which contains the terms and there all will be revealed to your gaze.

Technical Matters

Information which it is not necessary to have in order to understand the narrative won't be in the text, though it may show up in an appendix. Sometimes an interesting but esoteric and useless bit of information is simply given, without the surrounding explanation which would render it understandable to those of us who never finished college, but if a bit of information on chemistry or physics is needed, you'll have it.

The numbers are usually given in the metric system. It's unfortunate that more of us aren't familiar with the feel of the metric system. Most of us know the heft and feel of a pound, 16 ounces, but how many could identify a kilogram or two by lifting it? For purposes of illustration and visualization of a concept, then, more familiar measures will often be given (in parentheses) after the metric measurement, unless the point is obvious without this repetition. The metric system is known generally as the Systeme Internationale, or *S.I.* system, and that is how it is identified here. The pounds, gallons, rods, and bushels which make up our own system are referred to as "American." (If this is an insult, then we have only to change to S.I. to alleviate the slur on our collective character.) Tables for translating one system to another are found in Appendix 14.

Grass Roots Research

Soon you will be prowling through the rest of these pages. One thing you'll find out after being deluged with all this information, is that very little is known about biogas generation. The process is both very simple, in that it happens more or less by itself, and very complex, in that there are a great many aspects which can be modified, eliminated, or emphasized.

Information such as that presented in this book grows by a circular process. First there is a need. Research is done to satisfy that need and the information is disseminated. The fact that the information is disseminated creates an awareness of the possibilities, and it enables a wider application of the information. The circle returns as this increased awareness and further application generate more information.

This book stands somewhere in that circle. Only a small portion of the information it contains comes directly from the author's hands-on research in the diverse fields represented. Because the area of small scale and local biogas generation is in its infancy, this is not a "how it's been done" book, it's a "how it might be done" book. The information you can generate in applying what's found herein to your own situation will further help this very real alternative energy possibility become more of a reality.

If you do actually build a biogas system, let us know. With sufficient feedback, the book can be revised to include new and better ideas, better designs, and better information. This is grass roots research, and this book is intended to stimulate such work. Write.

Philosophical Comments

Knowledge such as that herein represents a certain kind of freedom. When we understand how to gain for ourselves what we once could only buy from a faceless company (P.G. and E.?), then we are, to a greater degree, free, more self-reliant. Yes, but. We are not freed simply because we have a wider or more perfect control over matter. If it were so, then the automobile and airplane, the telephone and radio, the Aswan Dam and Empire State Building would have already given us liberty, instead they seem to be giving us death.

We will not change the world merely because we can generate biogas. Rather, we face the more difficult problem of generating hope, peace, justice — and even, outworn as the word may be, love. If there is one overwhelming reason to write this book, it is to point this out, and point further, since many people who are interested in A.S.E. are sincerely interested in the fate of mankind and the planet as well.

There is a feeling, not often expressed and yet quite pervasive, that here, in what is called appropriate technology, lies an answer to these deeply troubling problems. This is half correct.

But back to the subject at hand. If you are interested in change, and what the views which shaped this book might be, read the afterword. You'll find it interesting. Meanwhile, as promised, this chapter will end with a section, very short this time, of questions and problems, and terms used.

Terms:

A.S.E.: Usually, *A*lternative *S*ources of *E*nergy. Here, half in jest, called Ancient Sources of Energy, because, historically, it is oil, gas, and nuclear energy which are alternative, or newcomers.

Biogas: The original natural gas. The gas about which this is, and much more on this subject follows. The main components are CO_2 (carbon dioxide), and CH_4 (methane). It is methane which is the burnable part of biogas, so many people call the gas that we gather methane gas. It is not. It is biogas. Methane gas is pure CH_4, no matter where it comes from.

Grass Roots Research: Homegrown, relevant information.

Questions

Why isn't this the introduction?

Problems

None here. All the problems are out there, prowling around the world.

2: Energy

We mentioned physics in the last chapter and some of you might have cringed a little when you read the word; if you did, you may cringe again when you realize that this chapter has a lot to do with physics. But don't fret. For the curious person, each new dawn opens up a vast treasure chest, so put on your curiosity glasses and tag along.

Subject A

Energy is often defined as the ability to do work. So energy is 1% inspiration and 99% perspiration. Looking around us, and within us, we will find energy everywhere. Einstein said that mass (substance) and energy were interchangeable. That's what $E=Mc_2$ means. Does an idea have energy?

Force, Work, Power

These are terms basic to an understanding of how to convert ideas of "better living through biogas" into reality. Force is, loosely speaking, energy. In physics, *force* is the cause of motion, or the cause of a change in motion. *Work* occurs when force moves through a distance. When a pound is raised by a foot, work is done; in fact, a "foot pound" of work is done. When two pounds are raised a foot, 2 foot pounds of work have been done. *Power* is a term, in physics, which describes the amount of work done in a certain period of time. If 2 foot pounds of work is done in one minute, then 2 foot pounds per minute of power is needed for this. If 2 foot pounds of work occurs in 30 seconds, then 4 foot pounds per minute of power is responsible for this. Right?

Heat and Temperature

The most important units of energy, as far as biogas is concerned, are the units of heat. A *British thermal unit (Btu)* is the amount of heat it takes to raise the temperature of one pound of water one degree Fahrenheit. A *gram calorie (cal)* is the amount of heat it takes to raise the temperature of one gram of water by one degree Celsius. A *(kilogram) Calorie (Cal, or K-cal)* is the amount of heat it takes to raise the temperature of a kilogram (a thousand grams) of water, one degree Celsius. Evidently, 1,000 cal = 1 Cal. (Just for the record, 1 Cal is nearly 4 Btu.)

Heat is different than temperature, for if we apply the same amount of heat to a liter of water as to 100 liters, the temperature of the liter will be much higher than the temperature of the 100. On the other hand, if we heat them both to the same temperature, it will take 100 times the heat energy for the larger amount of water as for the smaller.

Kinetic vs. Potential

Kinetic energy is energy that moves and flows. Energy that rests, waits and is "in storage," is called potential energy. Examples of both are given in Table 2.1.

Kinetic	Potential
wind	wood
flame	lake
sunlight	battery
waves	gasoline
river	coal
tides	coiled spring
geysers	balloon
flying bird	methane
thought	knowledge

Table 2.1 Kinetic and Potential Energy

In the left-hand column are things which have kinetic energy. In the right-hand column are things which have potential energy.

Things with kinetic energy are working, and can cause other things to move or change. Things which have potential energy are resting, but they can release their energy so that it becomes kinetic. Wood (potential) can burn (kinetic), the battery can start a car, and so on. The energy in some of these things is easier to keep potential (stored), and consequently they are important as means of storage.

The energy in a battery discharges rather rapidly, but wood kept dry or gasoline kept bottled remains good for a long time. But to be able to use the energy to accomplish something, it must be transformed from a potential to a kinetic state. Knowledge has potential, but it must be transformed into action before it has any effect.

Transform, Transfer

Whenever energy is transformed, say from sunlight into wood, or wood into fire, or fire into steam or steam into mechanical energy, not all of the energy available in the first form can be changed into another. Something is "lost," in the sense that it does not show up as useful work or stored energy.

For an example, consider an innocent blade of grass, moving in the wind, growing in the sun, smiling in the flowers. If the blade could completely use all of the sunlight which falls on it, then we couldn't see it (!), for no light would be reflected from it into our eyes. Walking through a field of such 100% efficient plants would be a very strange experience.

If the blade of grass did not use the sunlight at all, then it would act something like a mirror, reflecting the light completely. You may not realize it, but the reason plants are green is that they don't use this part of the light as much as red or blue. To prove this, shine a red light on a green leaf — it usually looks black or dark, meaning that very little of that color is being reflected, but rather it is being absorbed and used by the plant.

So, much of the sunlight that falls on the blade of grass is simply reflected again as light, and a good portion of the light which is absorbed is not stored. To find out why, let's use another example.

A cow, munching her way thoughtfully through a field of grass, might eat our unsuspecting sample grass plant, but she would not be able to transform all of the energy in the grass plant (R.I.P.) into any one kind of energy. Rather it would probably show up as milk, manure, gas, mooing, meat, and heat among other things. Some of these we can use, others not, and so inevitably some is lost.

Whenever energy is transformed, say from light to heat, or from steam pressure to mechanical energy, and so on, it is transformed as in our grass/cow example from the first form, into several others; some potential, some kinetic, some useful and used, some unusable and so lost to us. The cosmos, it could be speculated, can still use them.

If we want to make mechanical energy out of heat, and we put 100 units of heat into the process, and get 70 units out as mechanical energy, then we got 70% of what we put in, back out again. This is called 70% conversion efficiency.

Similarly, whenever we transfer energy, we use or lose some, and we end up with less. If we transfer electricity through wires, some of it is transformed into heat because of resistance — electrical friction — and so less comes out the other end. If we drive a gasoline tanker to a gas station to deliver gas, some of the gas must be used in the engine to get us there (and back), and so, again, energy must be used to transfer energy.

Lowest Common Denominator

Energy, like water, tends to seek a common level. For example, if it's hot on one side of a barrier and cold on the other side, various things will happen which will tend to force both sides to a common temperature. Energy tends to go from higher vibrational states (light) to lower vibrational states (heat), from greater concentration to greater dilution, and from ordered states to disordered states.

In thermodynamics, (the science that deals with energy and the way energy moves), this annoying tendency of energy to become less useful is called entropy. It's important to emphasize that entropy is theoretical. In other words, we really don't know if energy always runs down, on the universal scale. How did everything get so orderly to begin with? The fact that we can make a law out of the matter of entropy is a little like saying "I never tell the truth." (Intelligence acts counter to entropy. In fact, we haven't identified any other force that is or can be "anti-entropic." That says something for how everything got so orderly to begin with...)

Intelligence can reverse entropy to a certain degree, and when we apply our intelligence, we can force energy to go from lower to higher vibrational states, from dissipation to concentration, and from disorder to order, but only at the cost of some or much of the total energy of the system.

We can take the rather dilute energy of the sun and concentrate it (with a mirror) to produce sufficient heat to make steam (solar steam) and run a steam-powered electric generator. If we use the electricity we produce to power a radio which amplifies transmitted signals into music (Beethoven's Fifth, which is *real* music), then we have used our intelligence to reverse the tendency of energy to go to lower vibrational states, dilution, and disorder.

We took the heat of the sun, and made it into the sound of beauty (lower to higher); we took the solar energy and concentrated it into a pair of wires (dilute to concentrated); and we took the random motion of hot steam molecules and set the energy to marching in an intelligent pattern out of a speaker (disorder to order).

But while we may at first have had four or five thousand equivalent units of sunlight falling on our mirror, we may only end up with two or three hundred equivalent units of sound energy (oh, but what sound!). The point is then that we cannot expect to transform energy from one kind (abundant but not entirely useful to us) to another kind, without losses — even, in many cases, as much as a 99% loss! We can hold this loss, these energy leaks, down if we realize several things.

Greater Differences, Greater Problems

Between two different levels of the same kind of kinetic energy (such as two temperatures, two voltages, or two pressures), whenever the difference is great, the leaks will be rapid.

Take a look at a difference in temperature. Heat is molecular motion. The hotter air molecules are, the faster they will move. Suppose you are an air molecule. You're warm and you're moving along at a good clip, which is fine, but everybody else (the other air molecules) is shuffling and jiving too and, frankly, it's crowded where you are (inside a warm room). If you can, you'll shuffle on out the door to the outside where everybody is moving slower. It's much more likely, in fact, that some of your crowd (the hot bunch) will go outside to cool off, than that very many of the outsiders (who are cool and slow) will be able to force their way inside.

On the other hand, if it's the same temperature outside as it is inside, then nobody wins and nobody loses. Six leave; six come in.

It doesn't matter what temperatures are involved, high or low. What matters is the amount of difference between the inside and the outside. The greater the difference, the harder we have to work to maintain that difference, because the two different temperatures have what amounts to a pressure between them, and we have to keep pumping to maintain that pressure. (This, anyway, is one way to describe it.)

Following the same logic, it is easier to extract heat from a small volume of high temperature air or water, than it is to extract heat from a larger volume at a lower temperature, even if they have the same total amount of heat.

The first way to keep leaks at a minimum (very generally speaking—since this is not true in all cases) is to keep our kinetic energy levels fairly similar—not too much more heat, not too much more voltage, not too much more pressure, and so on.

Appropriate Energy Use

Since energy is lost every time it is transferred or transformed, it makes sense that the most efficient use for a particular kind of energy is a use which does not require very many transfers or transformations.

The sun gives us heat and light. When we use it for heating and natural lighting, we have a greater potential for using it efficiently than if we use it for something that requires many transformations or transfers. Biogas can be burned for heat. As we will see, when we try to use it for other things (mechanical energy, light), a lot is lost.

Efficiency

In practice, however, an efficient use of an energy source depends on our situation and—much more intangible—our values.

If we have several (not very efficient) steam engines in good working order around, then it may be more efficient to use these in various tasks than to go out and buy a system which would lose less energy.

Up to a point anyway. We, as a society, have been doing exactly this kind of "Here–it–is–so–I'll–use–it" planning, and as a consequence, our whole system is geared around fossil fuels.

Is this efficient? Our values, pivoting around short-term dollar gain, have brought us to this point, for it was more efficient, or possibly just more expedient, to do things in the manner we have. Now we must change. A simple chart like Table 2.2 doesn't tell the whole story. There's a lot of talk about our nuclear options, but some studies have shown that we get more energy in the United States out of burning wood than we get out of nuclear reactors. Think about that.

If You've Got It, You Don't Need It...

Further, there is a tremendous difference between the

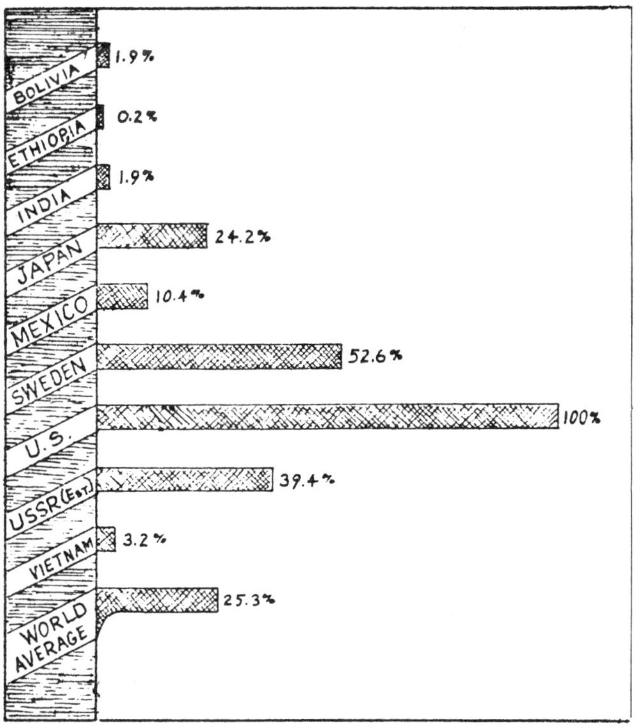

Table 2.1. Percentage of U.S. Energy Use

energy that's used and the energy that's useful. This is because in matters of energy, as in matters of money, it takes some to get some.

An acquaintance bought some chickens because he wanted to use their manure for biogas experiments. The amount of energy he used (as gasoline) going to and from the feed store for grain to feed the chickens so they would make droppings to use in a biogas generator, far exceeded the energy produced as biogas. Of course, he also obtained eggs and Kentucky fried hens, but these benefits had to be added in, in order to balance the dollar cost against the dollar benefit. But nevertheless, they still didn't gain him anything on the energy balance sheet—he was still using more energy than he gained.

If You Need It, You Can't Get It...

If this were an isolated example, it might be funny, but the whole of our society is doing the same sort of thing. The example of nuclear energy is in this category. So much energy is required to find, mine, transport, and triple-refine the uranium ore, and then to transport the uranium itself, plus the energy required to store the deadly residue... that it may well exceed the amount of energy generated as electricity. This kind of thing occurs because nuclear-generated electricity is dollar cheap, even if it's energy expensive. (Also, we bought in. We've got so much invested in it that it's no longer a question of energy—it's a question of finances and saving face.)

However, since much of the energy subsidy which goes into the nukes is from fossil fuels, the cost of this atomic electricity will go up almost as fast as the price of gasoline. Even in economic terms, the full cost is not counted, as the health hazards eventually result in medical problems for a certain portion of the population, and the energy and dollar costs of this must be borne by society, again reducing the amount of useful energy available.

But, it's not only the strange waste of nuclear energy which creates such a vast energy gap between the United States and other countries. The peculiar way of life we have developed here in the United States requires great gulping amounts of energy to produce very little, but very impressively. The result is that our standard of living, as measured in basic necessities, educational level, health care, and longevity is no greater, and in some measures less than that of other countries which use much less energy than we do. Sweden is one example.

Biogas relates to this in two ways:

1. The process converts waste into burnable gas and useful fertilizer, thus reducing, by some small portion, our energy needs, and encouraging us to grow our own food—if possible—and thus greatly increase our chances of health and reduce our reliance on an energy-intensive, nutritionally dilute, agricultural and food processing industry. (We get a product.)
2. The attempt, if we make it, to produce and use our own energy may well shock us into recognizing the truly enormous amounts of energy we use in our daily life, and the impossibility of powering our present lifestyle on ancient sources of energy. If we are ingenious and concerned, we will opt for conservation, again increasing the quality of our life by reducing the deadening and disintegrating energy fix which cushions us against natural processes and simpler living patterns. (We get—we should get—an awareness.)

Terms

Btu: British thermal unit. A measure of heat energy.

cal: Small calorie. A measure of heat energy in SI units.

Cal: 1,000 small calories. SI unit.

Conversion efficiency: Percentage of energy transformed into another useful kind.

Energy: Ability to do work.

Entropy: An annoying tendency of energy.

Force: Cause of motion.

Heat: Thermal force.

Kinetic: Energy in motion.

Potential: Energy at rest, in storage.

Power: Force through a distance in a certain time.

Temperature: Measurable result of the application of heat.

Transform: Move energy from place to place.

Work: Force through a distance.

Questions

1. If grass were red, why would it be red?
2. List three kinds of energy (excluding nuclear energy) which are not derived from the sun's energy.
3. Is gravity a force, work done, or power expended?
4. Name five kinds of energy transformation that occur in a 1978 Cadillac.
5. In the above mentioned Cadillac—which is bright baby blue—how many energy transfer systems are there, and what powers them?
6. What is entropy? (No fair giving the definition just given in Terms.)

Problems

No problems—yet.

(For answers to questions, see Appendix 17.

3: Math

Due to our lamentable educational system, many people do not like math. This is tragic, on the order of not being able to enjoy a sunset, or never having made a campfire. Mathematics can enrich our lives tremendously, since it can give us a whole new way of looking at the world. However, there's a more immediate reason for this chapter—if you don't understand the concepts it presents, you won't be able to fully use this book.

Times

Often, to indicate multiplication of numbers, the terms will be placed next to each other with parentheses like this:

$$(3.15)(4.19) = 13.20$$

Multiplication of letters (which represent variable or changeable quantities in a formula) can be indicated by simply putting them next to each other:

$$(A)(C) = AC$$

Powers of Ten

Ten is a beautiful number. It's the reason the S.I. is such a beautiful system. Everybody knows that to divide or multiply by 10, you simply move the decimal point:

$$36.1 \div 10 = 3.61$$
$$(36.1)(10) = 361.00$$

There's another way to express this kind of thing, and that's by powers of ten:

$$10^0 = 1$$
$$10^1 = 10$$
$$10^2 = 100$$
$$10^3 = 1{,}000$$
$$10^4 = 10{,}000$$
$$10^5 = 100{,}000$$

10^5 means ten to the fifth power, or one hundred thousand. And so forth. Notice that 10^2 has 2 zeros, 10^4 has 4 zeros, and so on. There are also negative powers of 10:

$$10^{-1} = 0.1$$
$$10^{-2} = 0.01$$
$$10^{-3} = 0.001$$
$$10^{-4} = 0.000\ 001$$
$$10^{-5} = 0.000\ 000\ 1$$
$$10^{-6} = 0.000\ 000\ 01$$

Ten to the minus three (10^{-3}) equals one one-thousandth. Notice that 10^{-1} is one place to the right of the decimal point, 10^{-3} is three places to the right and so on. Powers of ten are used mostly to provide a multiplier for a particular number. To find out where the decimal point should be in a number, just move the decimal point either to the right (if it is a positive power) or to the left (if it is negative) according to its magnitude:

$$0.000\ 002\ 060 = 2.06 \times 10^{-6}$$
$$12{,}000{,}000{,}000 = 1.2 \times 10^{10}$$
$$1.6 \times 10^3 = 1{,}600$$

This can save considerable paper, but it has other advantages. Powers of ten can be shuffled in interesting ways. When we want to multiply one power of ten by another, we can simply add them:

$$(100)(1{,}000) = 100{,}000$$
$$(10^2)(10^3) = 10^5$$

And to divide, simply subtract the powers:

$$1{,}000 \div 10{,}000 = 0.1$$

or

$$\frac{1{,}000}{10{,}000} = 0.1$$

These are both the same as:

$$10^3 \div 10^4 = 10^{-1}$$

or

$$\frac{10^3}{10^4} = 10^{-1}$$

Significant Digits

When we have a number, for example, a number which describes the volume of a generator, or the amount of gas it produces, information about how accurate that number is, is usually given by implication rather than outright. In the case of generator volume, and speaking scientifically and mathematically, it makes a great difference whether we say the generator has 10 cubic meters of volume (two significant digits), or 10.000 cubic meters (five significant digits). While 10 cubic meters may seem the same as 10.000 cubic meters, the difference is that in the first case, the generator is about 10 cubic meters, within a few tenths of a cubic meter, and in the second case, the generator is 10 cubic meters, within a few ten-thousandths of a cubic meter. The second figure, then is about one thousand times more accurate than the first one. Thus, significant digits have to do with accuracy. In this book most figures are only given to three significant digits, since that's about as accurate as we need to be.

Reciprocal

The reciprocal of a number is that number divided into one, or in more technical terms, if you remember your high school math, you get the reciprocal when you place the original number in the denominator, and one in the numerator. So, the reciprocal of 12 (for example), is one-twelfth. The reciprocal of one-tenth (1/10) is ten. Another way to look at this is to think of "turning a fraction over."

Sec^{-1}

We can also have reciprocal "units," where units are things like pounds, grams, ounces, cubic feet, and the like. A term like the above expression, for example, means "per second." So:

$$19 \text{ ft sec}^{-1},$$

translates as 19 feet/second, or 19feet per second.

Canceling Units

A unit, as the word is used in this book, refers to those clever abbreviations, such as *cm* or *hr*. When we multiply and divide in many of the equations used in biogasology, we may have a tremendous array of confusing units, such as:

$$9 \text{ Cal cm m}^{-2} \text{ hr}^{-1} \text{ °C}^{-1}$$

We can begin to translate such a strange string of characters if we know what they represent:

Cal = Calories, a measure of heat
cm = centimeters, a measure of distance (or, in this case, thickness)
m = meters, a measure of distance
hr = hours, a familiar friend
°C = degrees Centigrade, the measure of temperature the rest of the world uses

So, sticking all this together and taking a stab at it, we might realize that this is a description of heat loss, out of a slab of material of a certain thickness and area, during the course of an hour, for every degree of temperature difference between one side and the other. In more familiar terms, it looks like this:

$$\frac{9 \text{ Cal cm}}{\text{m}^2 \text{ hr °C}}$$

or,

$$\frac{9 \text{ (Calories heat loss)(centimeters thickness)}}{\text{(sq. meters area)(hrs elapsed)(°C difference)}}$$

When we multiply a term like this with some other numbers, and if we expect to end up with a heat loss in heat units—such as Calories—then the units such as m^{-2} will have to "drop out" somewhere. Right? We've got to end up with Calories alone. Realizing this, we should instinctively know that:

$$(\text{Cal m}^{-2})(\text{m}^2) = \text{Cal}$$

or,

$$\frac{(\text{Cal})(\text{m}^2)}{(\text{m}^2)} = \text{Cal}$$

We can subtract units such as meters, if they appear on both the top and bottom of a division problem. Just like powers of ten. This kind of logic will often tell us how to deal with some otherwise meaningless number. In our above example, once we have an area, a thickness, a time elapsed, and a temperature differential (assume all of these at 10, that is, 10 square meters, 10° C, and so on), then we should be able to plug these in the proper places and run with it:

$$(9 \text{ Cal cm m}^{-2} \text{ hr}^{-1°} \text{ C}^{-1} \text{ (10 cm}^{-1}) \text{ (10m}^{-2}) \text{ (10 hr)(10°C)} = 900 \text{ Cal (heat loss)}$$

or,

$$\frac{9 \text{ Cal cm(10 m}^2\text{)(10 hr)(10 °C)}}{10 \text{ cm m}^2 \text{ hr °C}} = 900 \text{ Cal (heat loss)}$$

Please stare at this until you understand it. Notice that unless the hours elapsed (10 hr) ends up on the top of this cumbersome expression, hours will not cancel out the whole equation. Likewise, meters and degrees (10m and 10° C) must be on top and centimeter measurements must be on the bottom, otherwise, we end up with an answer in terms of:

$$\text{Cal cm}^{-2} \text{ m}^{-4} \text{ hr}^{-2} \text{ °C}^{-2}$$

which makes no sense (m^{-4} ?? per quadrubic meter?!). If you know where you want to go with an expression, this kind of logic will tell you how to get there. If you have some number like 1.65×10^{-2} kg cm^{-3} (0.0165 kilograms per cubic centimeter) and want to find out the weight (kg) of something which has that number attached to it, then you can obviously and instantly see that you must multiply by some value in terms of cm^3 (cubic centimeters volume) to get that answer. Then, having found or decided upon that second number (for example, $6.06 \times 10^3\ cm^3$), you will know what to do:

$$(6.06 \times 10^3\ cm^3)(1.65 \times 10^{-2}\ kg\ cm^{-3}) =$$
$$(6.06)(1.65)(10^3)(10^{-2})\ kg\ cm^{-3}\ cm^3 =$$
$$(\ 10\ )(\ 10^1\ )\ kg = 100\ kg$$

While the units cancel, the abstract numbers with which they are associated, remain.

One point may need clarification: in the real world, you can't take a weight and divide by a volume. These are expressions ($kg\ cm^{-3}$, mph, etc.) which describe imaginary dimensions. Don't get hung up in trying to find a foot pound in your kitchen, or searching over hill and dale for a miles-per-hour.

The Bottom Line

Okay tiger, here are some tough ones. If you understand the foregoing (and the energy chapter), you should have no trouble with these problems. If it's still a little shaky, the grunt and sweat of the pencil pushing should solidify it. It may help to think on a piece of paper. Go to it, but be on your toes.

Terms

Power of ten: Mathematical term describing a simple way to keep track of the decimal point.

Reciprocal: One in the numerator, the number in question in the denominator.

sec^{-2}: Per second squared, $1/sec^2$.

Significant digits: A means of indirectly reporting the accuracy of a measurement.

Unit: In this book, modifiers such as kg, cm, mph, and the like.

Questions

1. Are you beginning to wonder why you bought this book?
2. What does the term 10^6 Btu ton^{-1} probably refer to?

Problems

1. When a cord of wood burns at 80% conversion efficiency, contains 16×10^6 Btu $cord^{-1}$ potential energy, and takes an hour to burn, how much heat is released each second (average)?
2. When we divide a number given as "in. sec °F", into a number given as "in. sec °F Btu", what will be the units of the final expression?
3. How could you otherwise express:

$$\frac{1}{liter^{-1}}$$

4. Think up two of your own problems. This is the hardest problem.

4: Biology

Methane—It's a Gas, Man

Many people call the gas which comes out of a generator, "methane gas." It is not. Methane is CH_4, and CH_4 is a gas at all normal temperatures and pressures, and whether a particular bit of CH_4 comes from natural gas, from a biogas generator, or from some other source, it is still methane.

Biogas, on the other hand, is that gas produced by the anaerobic (airless) biological process, and it's composed of CH_4, CO_2 (carbon dioxide), H_2O (water—as a vapor), and sometimes N_2 (nitrogen gas), H_2S (hydrogen sulfide), H_2 (hydrogen gas) and minute traces of more exotic gases. In literal terms, biogas means "gas produced by life." When we speak of methane in this book, we're talking about CH_4. Biogas will be called biogas.

Anaerobiosis

That big word means "life processes in the absence of free oxygen," and that's number one for biogas. The bacteria which produce biogas, do so only where there is little or no free (gaseous, uncombined) oxygen. Thus, they are called anaerobic (*an* means without, *aerobic* means air).

Digestion and Generation

The process of the anaerobic breakdown of organic materials is digestion. The same word is used to describe what happens, and essentially the same process occurs, in our own digestive tract. We eat food, it is digested—broken down—and we gain our energy from that process.

Whenever we talk about what happens inside a generator, then we refer to digestion. If a device, such as a huge municipal sewage plant, is designed primarily to accomplish this decomposition, it is a digestor. A generator, on the other hand, is designed with the idea of producing (or generating or evolving) biogas. The big difference is how efficient the device is. Generators are more efficient, and they give us more gas.

Culture, Seed, and Innoculation

To "start" a generator, once we've designed, built, and filled it, we need a culture, some source of anaerobic bacteria. Cultures can be found in the mud under still water, in fresh manure or excrement, under an old unturned soggy compost pile, or any place organic matter has been sitting, long away from the good fresh air.

If you make yogurt, you've probably familiar with the word culture. A yogurt culture will help make milk into yogurt. An anaerobic culture will help us make organic matter into biogas.

When we add our culture to the biogas generator, we *seed* it, and whenever fresh organic material is added to a seed (or *visa versa*), then the new material is *innoculated.*

In general, these three terms are interchangeable. They all refer to populations of bacteria, and that's the thing to remember.

Batch vs. Continuous

There are as many kinds of generators as there are generators, since each and every one will be unique in one way or another.

However, there is a distinction that is important in separating one general group of digestors from another. That separation comes about because there are two basic kinds of digestible materials, the "mix well" and the "float much."

By the way, digestible materials are also known as substrates. This word will pop up again and again, so we'll label Table 4.1 by that name.

The mix-wells form a lively slurry—the mixture of water and substrate used in filling the generator. They slosh happily around, we can pump them from place to place, and even spray our friends with them, should such a thought occur to us. These materials—and chiefly the fresh manures—can be put into a *continuous-feed* generator. We mix up some slurry, pump it into the

Mix-Well	Float-Much
Fresh manures	Dried manures
Fiber-free vegetable pulp	Plant wastes
Sewage	Leaves
Algae	Straw
	Small branches
	Grease (from animal sources)

Table 4.1 Substrates

generator, and it displaces some of the older slurry, which exits gracefully.

If your family is large enough to qualify as a city, or if you happen to have three thousand pigs in the back yard, then by all means, build a continuous-feed generator.

If, however, it's just Mom, Pop, and the Kids, and you don't have enough animals kept in barns where the manure is easy to collect, then consider building a *batch-fed* generator. Batch-fed generators, except in special circumstances, are the kind which must be designed for the float-much substrates. You fill them up, they do their stuff, you empty them: a batch at a time.

This can be, and most often is, hard work. Nevertheless, because straw, leaves, and the like behave the way they do, they are hard to pump around, unless fairly finely shredded. Batch generators are usually small; they represent the "low-technology" approach to biogas generation, because they are so simple to build, even if difficult to use. Continuous generators, operating on manure as a substrate, are more often a large-scale, high-technology venture.

We'll discuss this extensively in the Design section but hang on to this distinction between batch and continuous. Quite nearly all the research which has been done on anaerobic digestion has been done on city sewage, which is dilute and of small particle size, and generally which digests rather rapidly. Therefore, to a large degree, these studies were not directed towards the substrates most of us have in relative abundance (leaves, grass clippings, weeds) nor the kind of generator (batch-fed) which these substrates require.

In fact, all the easily available information (see the Bibliography) in book form, as of this writing, is directed almost exclusively to the digestion of sewage or manure.

Here is another reason why you won't find an abundance of blueprint drawings of large generators in this book—most of us can't use them since we don't have the required substrates.

Sludge, Supernatent, Scum, and Sand,

For some reason, all these words begin with "s." As you already know, when we mix water with our substrate, we create a slurry. When we put the slurry into

Fig. 4.1 Slurry Layers

our generator, and leave it alone, it separates into layers, as illustrated in Fig. 4.1.

Below the biogas, *scum* forms. The float-much substrates are mostly scum—that is, whatever floats to the top of the slurry is scum. The scum often forms a dense mat above the slurry, as grease, hair, straw, and assorted undesirable materials float up, stick together, and dry out, forming a hard crust. More about this later.

Below the scum is the *supernatent. Super,* meaning above, and *natent,* from *natare,* to swim. So, the supernatent swims above the sludge. Supernatent is the liquidy portion of slurry, left behind when the heavy stuff (sludge) settles out. Sludge is the goopy or mud-like portion of the whole mess.

Underneath everything is the *sand.* Well... we could call it grit, but that doesn't start with an "s." Basically, the sand layer is made up of all the heavy indigestibles. Depending on how we fill the generator, and with what, there will be a layer of sand which will accumulate, rapidly or slowly, at its bottom.

These layers appear in the slurry when it is not mixed, or only very slowly mixed. When stirred somewhat more rapidly (*agitation),* the supernatent and sludge recombine, or never separate, and the scum (depending on the substrate used) does not form as rapidly, if at all. Very rapid mixing (high-rate mixing) can generally remove the scum, or prevent it from forming, but the sand must still be otherwise dealt with. In some generators, both scum and sand remain and cannot be removed by mixing. More about these difficulties later.

After generation is completed, in other words, when digestion of the slurry is finished, what we have left is *effluent.* In some generators, the used or effluent supernatent is drawn off separately from the effluent sludge and often these are used in different ways, since they can be handled a bit differently if they have separated.

In general, city sewage effluent supernatent is pumped into the nearest innocent waterway or ocean. The sludge effluent is often dried and hauled away by truck.

Most of us, however, will use the whole effluent as it should be used—that is, not simply "disposing" of it, but using it as a fertilizer, to return it to the soil. Thus, when properly digested, whatever substrates we are using will pay us back in both fuel and fertilizer.

Because the process is a natural process, something which goes on in the real world in the absence of government grants or computer programs, we need only understand it well to use it well. This implies no dependence on a military-industrial complex, but rather greater freedom from it.

Ancient History

For more insight into the biogas process, let's look back in time. According to the best information we have, the earth appears to be about three and one-half billion years old. That's a lot of history, and such history is not written in books or on clay tablets; it's written in the bones of the earth, the rocks and strata.

The primitive atmosphere was composed principally of CO_2, water vapor, and CH_4 (methane). There was little or no atmospheric or free oxygen, and thus all life at that time lived and moved in an environment which would not allow us to survive. We are aerobic, that is, we require free, uncombined, gaseous oxygen for our life processes. Whatever primitive life existed in the dawn of prehistory was anaerobic, that is, it did not need or use free oxygen in its life processes.

An interesting question is—where was all the oxygen? Answer: It was bound up in the iron oxide deposits, bound up in carbon dioxide, bound up in hydrogen oxide (Recognize that one? H_2O!), and happily and undiscriminatingly combined with whatever was available. Another interesting question is—why is the air so full of oxygen today? Answer: Green plants. Photosynthesis means *using light* (photo), *to make the chemicals* (synthesis) which accompany life. Plants *take* in CO_2, and they *discard* O_2 and keep the C. Animals take in O_2, and they *discard* CO_2. It's a circular process, and very intelligently designed.

But on the very primitive earth, there were no animals, and there was no photosynthesis, and consequently, there was little or no free oxygen, for the only important source of O_2 is the activity of green plants. *Protect your local forest!* Gradually, however, photosynthetic organisms developed and flourished, though it took a long time for the O_2 to build up to any great degree in the atmosphere.

As conditions changed on the earth, those life forms which once could live in the open "air," could not survive the gradually increasing oxygen concentration, and they were driven into places where the ancient oxygenless, anaerobic conditions still prevailed.

Today

They are still there, and they earn their keep, for in nature, everything eventually returns or cycles, and these anaerobic organisms help to return organic matter to from whence it came.

Plants come, directly or indirectly, 95% to 98% from the air around us. They take C, O, H and (indirectly) N, from the air, to make their proteins, and carbohydrates. When they die, their remains, made up of these complex molecules, are decomposed by different organisms and returned to the soil and the air. They are, to use a recently popular word, biodegraded. In the anaerobic places (the swamps and bogs, or the lake and slow stream bottoms), the only way these plant (and animal) remains can be biodegraded is by anaerobic bacteria.

Another place where these bacteria help is in the digestive tracts of many creatures. Termites use them to help break down the wood they eat. Ruminants (cloven-hoofed, four-legged cud-chewers) have anaerobic little bitty buddies in their complex digestive tract, which help them break down their food for utilization.

So, the two main places where we find anaerobic life today are under water, and in digestive tracts.

Anaerobic metabolism (the internal life process mechanics of oxygenless bacteria) is not as efficient as aerobic metabolism. Without free oxygen, and the corresponding metabolism designed to use it, anaerobic bacteria cannot derive as much energy from the breakdown of their food molecules as aerobic bacteria derive. One illustration of this is a compost pile.

When compost is made in the open air, rapid breakdown of the organic materials results, and the temperature inside a compost pile is often 70°C (160°F) during its most active period. Similar compost materials, when placed in a biogas generator, in the necessary airless environment, produce no appreciable heat, decompose rather slowly, and leave us most of the energy which was locked up in their molecules (as much as 70%) still locked up, as CH_4, methane.

This difference between aerobic and anaerobic metabolism in regard to their ability to efficiently use (biological) energy also shows up in the fact that the process of biogas generation is easier to upset than the process inside a compost pile. Changes in conditions, compost materials, or levels of toxic (poisonous) substances which would not bother the aerobic compost process, will disrupt or stop the anaerobic process.

Understanding the reason why the process is so delicate, requires further understanding of the biology of the process itself.

Biological Energy

We've been talking about the breakdown of molecules for energy, without really explaining it. Suppose there is a coil spring between your hands. When you force your hands together and lock your fingers, the spring will try to push your hands apart. It took energy to bring your hands together, and now the spring stores that energy, locked between your hands.

In a similar way, two or more atoms are locked together—combined or bonded— and they store energy between them. When they are unlocked or broken down, energy is often released.

When atoms like carbon and oxygen are put together —one carbon atom and two oxygen atoms— we get CO_2. Two hydrogens and one oxygen gives us H_2O. These are very simple molecules, or combinations of atoms, but Mother Nature often puts together hundreds of atoms of many different kinds and comes up with very complex molecules.

If a molecule is unstable, the "locks" in it are not very good, and it may break apart easily. More stable molecules are harder to break apart, just as your pushed-together hands would be hard to break apart if you had strong fingers, or if you tied them together with string or rope.

Chemical Confusion

We've thrown a few chemical equations at you without the concept behind them. They are a very simple way of representing what happens when certain molecules get together under certain conditions—a kind of chemist's shorthand. Notice the conservation of matter in the following equation:

$$CaCO_3 + H_2O \longrightarrow Ca(HCO_3)_2$$

Six oxygens on both sides (the bicarbonate ion, HCO_3^-, is doubled, since two of them hang on to the calcium ion, Ca^{++}). Two carbons, one calcium, and two hydrogens are also found on both sides. As should be obvious, the "sub" numbers ("O_3," for example) refer to the numbers of those atoms to which they are appended. Thus, O_3 means a happy group of three oxygen atoms, $(OH)_2$ refers to a gregarious gathering of two OH^- ions.

Back to Biological Energy

Biologically, in living systems, stable molecules are broken apart (or formed), not by force, but with the help of enzymes. In our spring-hands-fingers model, a bit of grease would act as an enzyme, causing your fingers to slip apart and the stored energy to be released. If your hands were tied together with string, an enzyme would act like a pair of scissors, cutting the string, whereas without the scissors, you would have to break the string with force.

In a biogas generator, complex molecules are broken apart, step by step, into simpler molecules. The process has been compared to an assembly line — except that it's a disassembly line — where one group of workers works on a complex molecule, derives energy from it and gives the parts (less complex molecules) to another group of workers, who disassemble them further (gaining energy themselves), and so on to the final group of workers, who break the molecules into the very simplest molecules possible under the (anaerobic) circumstances — H_2O, CO_2, and CH_4.

These workers are different kinds of anaerobic bacteria, and many who have studied this process agree that it takes a great many different kinds of bacteria to accomplish the complete disassembly of a very complex molecule into CH_4, CO_2 and the like.

The workers, or bacteria, are of many many different kinds, and they operate together in ways we might not expect if we studied each one separately.

Here is one of the keys to the complexity of biogas generation — the many varieties of anaerobic bacteria and the many ways they have of operating under different conditions and in different populations.

An operating generator, then, is like a factory, filled with workers, busy manufacturing biogas from the raw materials supplied.

From the foregoing, we can see that inside the factory, things happen in stages:

Aerobic. Oxygen will inevitably enter with the raw materials put into the generator, and so aerobic bacteria use this oxygen up, meanwhile doing what they can to break the materials down. CO_2 is released, and some heat is generated.

Extracellular enzymes. In this stage, anaerobic bacteria release enzymes that attack large molecules which are still outside their own bodies (extracellular),

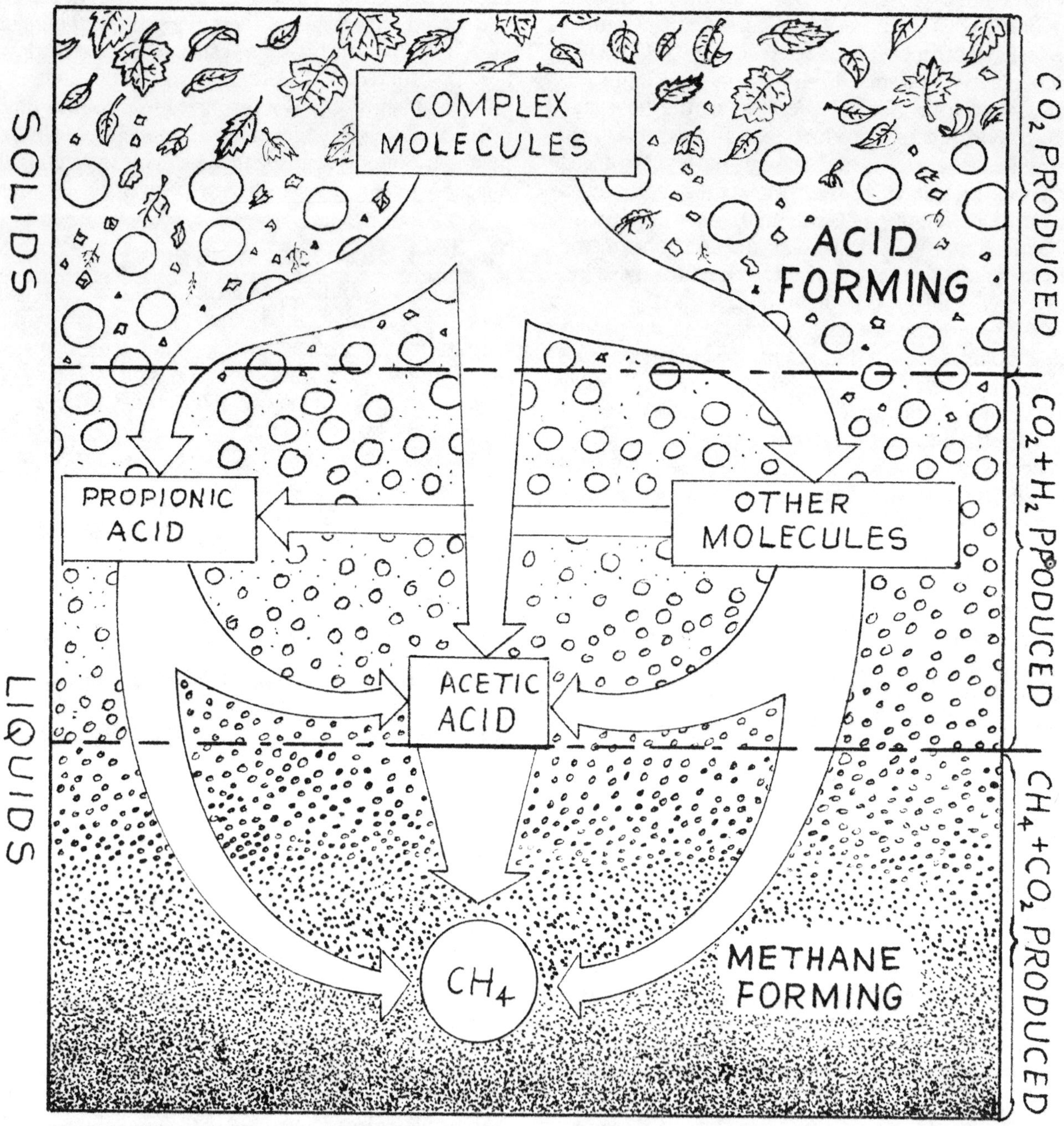

Fig. 4.2 Chemistry of Biogas Production

so these molecules will be broken down into "bite sizes."

Acid digestion. The bite size molecules (still fairly large) are absorbed by the bacteria and digested. The main byproducts of this third process are simple molecules such as the short chain fatty acids, H_2 and CO_2. (We'll get into the short chain fatty acids soon.) It's interesting to note that at this stage, hydrogen gas, an excellent fuel, is evolved. Hydrogen does not often show up in the final biogas because it is used by the anaerobic bacteria (in the next stage) in making CH_4, methane.

Gas digestion. Now comes the part we've been waiting for. The fatty acids are now gobbled up by the last group of bacteria, who turn them into H_2O, CO_2, and best of all, CH_4.

For general purposes, we will talk about biogas production as if it had only two stages (acid digestion and gas digestion), and as if there were only two groups of bacteria involved (*a*cid *f*orming and *m*ethane *f*orming — *AF* and *MF*). The second group is called methane forming even though they also produce other byproducts, because uniquely and alone, these microorganisms produce the methane component of biogas. Recent research indicates that the methane formers are not bacteria, but a whole new kind of creature, as different from bacteria, as plants are from animals. Nevertheless, we'll still call them bacteria for convenience.

The biogas process, unlike many others, leaves no residues which are poisonous to the process itself. As an example of a common biological process which does not follow this pattern, consider alcoholic fermentation. In this process, yeast metabolizes (eats up) sugars and one of the byproducts is alcohol. As the process continues, the percentage of alcohol increases in the sugar-yeast broth, and eventually (at around 17%) kills the yeast.

In the biogas process, however, it cannot be said that any major residual byproduct is poisonous to the bacteria. Sunlight, even in the absence of oxygen, can damage or kill the bacteria.

Peer Amid the Pyramids

One last point, previously referred to briefly, can now be more fully explained. When we were discussing the easily upset process of biogas generation, and the low efficiency of anaerobic metabolism, another factor, it was stated, contributes to this delicate character. It has to do with pyramids. Whenever the soil is unhealthy, plants grown in that soil will be unhealthy. Animals eating the plants will be unhealthy, and people eating the plants, animal products, and animals will be unhealthy. This is the food pyramid, sometimes called the food chain, greatly simplified.

At the bottom is the soil; standing on that are the plants, above that, the animals, and at the top is man. The only reason we're on top is that we're standing on everybody else. If anyone below us slips, we slip too. The snail darter may be more important than we realize.

There are a lot of pyramids which we can find around us. For example, our society rests on our economy, and our economy rests on our agriculture, and everything is balanced on the natural world and the things we are given freely each day. Sunlight. Oxygen. Water. Change the weather a little, agriculture suffers, the economy self-destructs, and the social animal turns savage. The point here however, is not social, but biological.

The biogas bacteria are at the top of their little pyramid as well. If anything goes wrong somewhere else, it affects them, and biogas production slows down or stops. Anybody else's trouble automatically becomes their trouble. So, we have to remember not to rock the

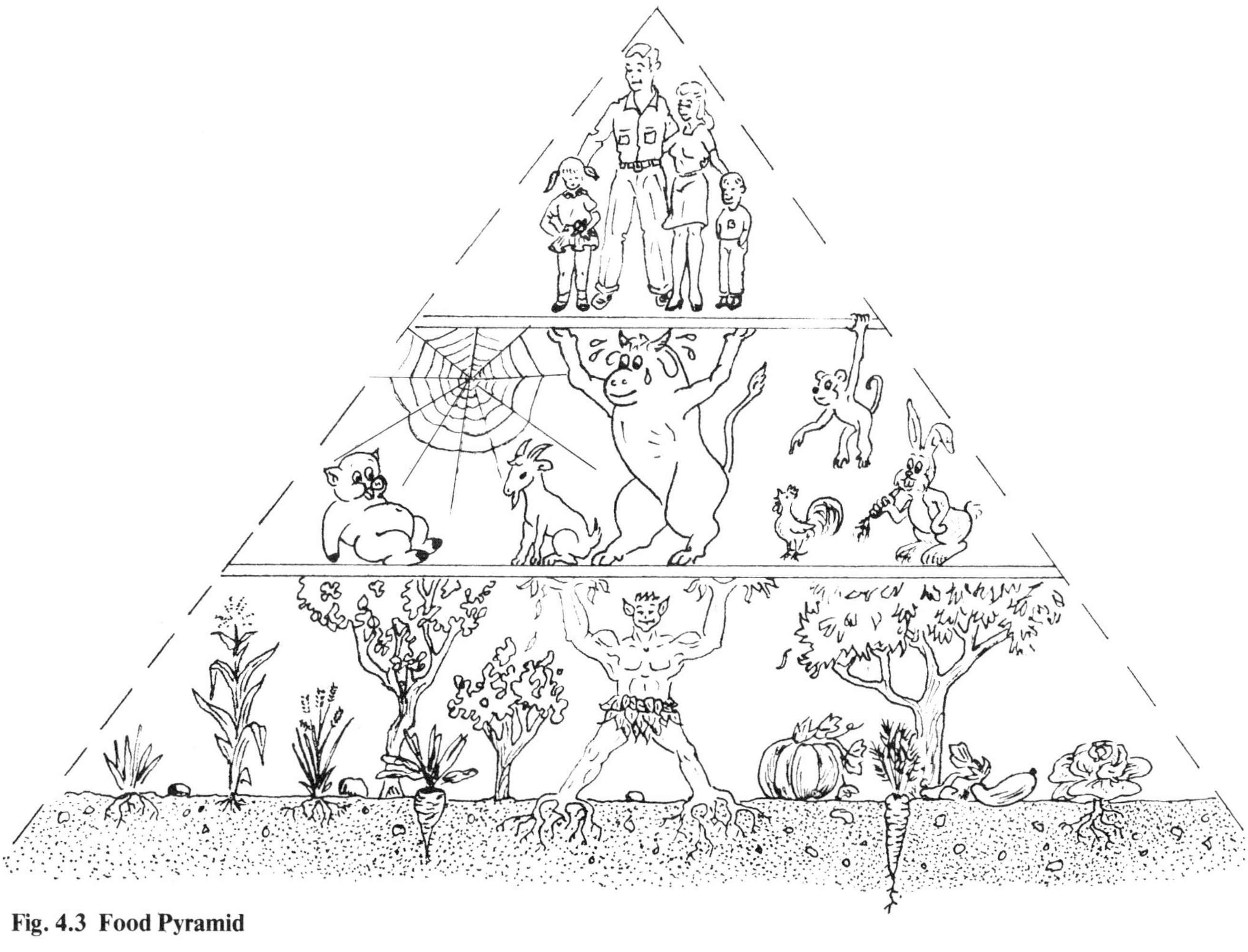

Fig. 4.3 Food Pyramid

boat. Gradual changes can be more easily tolerated by the ecosystem in the biogas generator than can rapid changes in temperature, pH, and so on. Next up, more on these conditions.

Terms

Acid digestion: The "first part" of biogas generation, where complex molecules are broken down into simpler molecules, such as the fatty acids, CO_2 and H_2.

Aerobic: Needing free gaseous oxygen to survive, or being able to tolerate it.

AF: Acid forming.

Agitation: Mixing.

AKA: Also known as (alias).

Anaerobic: Requiring an oxygenless atmospere. Poisoned by oxygen.

Anaerobiosis: Life processes carried on in the absence of free oxygen.

Batch feed: A load, generate, clean out type of generator.

Biogas: That combination of gases which is produced by anaerobic decomposition.

Biological energy: Energy available to life, most generally gained from, or with the assistance of, other life forms.

CH_4: Methane.

Continuous feed: Generators into which slurries are daily or more constantly put.

Culture: Populations of anaerobic bacteria.

Digestion: The biological breakdown of organic materials.

Digestor: A device designed to break down organic materials.

Ecosystem: The interactive web of life which covers the whole planet.

Effluent: The used slurry, sludge, supernatent or scum.

Enzymes: Chemicals which help form and break down molecules.

Extracellular enzymes: Enzymes which operate outside the bitty bodies of biogas buddies.

Fatty acids: More about this is found found later. We'll come to it.

Gas digestion: The "second stage" of biogas generation, during which the CH_4 of the biogas is produced.

Generation: Here, the production of biogas.

Generator: A device designed to produce biogas.

Food pyramid: The food chain, another reason to believe that all flesh is grass.

Float much: Those substrates which refuse to make easily pumpable slurries.

Innoculation: Populations of anaerobic bacteria.

Low Technology (or low tech): A technology available to mom, pop, and the kids. AKA "kitchen sink technology".

Metabolism: The cellular mechanics of life; the process of using biological energy.

Methane: CH_4.

MF: Methane forming.

Mix well: Those substrates which make pumpable slurries.

Molecules: CH_4, CO_2, H_2O and their brethren and sisteren.

Parameters: conditions or factors.

Photosynthesis: Another miracle.

Ruminants: Cud-chewing, four-legged, cloven-hooved animals.

Sand or grit: Sand or grit.

Scum: The floating mass of material above the supernatent.

Seed: Populations of anaerobic bacteria.

Sludge: The settled portion of the slurry; a mudlike, semi-solid mass.

Slurry: The mixture of a substrate and water which sits in the biogas generator.

Substrates: Those materials, once alive, that are mixed with water to form slurry and fed to the generator to produce biogas.

Supernatent: The liquid portion of the slurry which "floats" above the sludge.

Questions

1. Where did all the O_2 in the air come from?
2. Did you ever notice that science is good at the "how" of things, and very poor with the "why" of things?
3. When something goes wrong with our generator, we should change it as fast as possible, right?

Problems

None.

SECTION II
PARAMETERS

Alternatively, this section could be called how to be good to your little bitty buddies. Herein, we'll take each of the parameters or conditions affecting biogas production and examine them in some detail. First, however, some general points ought to be made...

The most important point is that everybody disagrees about everything, or so it seems when the many articles, research papers, theses, and books about or pertaining to biogas generation are studied. People report widely different temperatures as optimum. Some say that this material or substance is digestible, others swear on a stack of chemistry and physics handbooks that it's not. And so on. The differences arise for very important reasons.

Besides the obvious errors (mathematical and mechanical), people tend to make errors in assumption. By far the largest portion of work done on anaerobic digestion has been done with sewage, and results gained in a study on sewage do not necessarily show what the results will be for plant wastes (as one example).

Many researchers are not aware of the inaccuracies in the language they use. "Gas production" is not a useful phrase, unless we establish that we are speaking of the rate of gas production (per day) or possibly the total gas production over a period of time as defined in some manner. In the same way, we could talk about which city is "largest." This seems clear enough at first glance, but when we really get down to cases then we have to ask whether we are speaking of population, or area (square miles), or some other parameter we can speak of as being "large." Often things are reported in the literature which, because of this kind of inaccuracy, are impossible to understand and apply. In any case, these differences and the possible reasons behind them will be mentioned when the information could be useful to you.

5: The Airless World

As you already know, the environment inside the generator is not really airless. Rather, it's oxygenless, or anaerobic. Even so, it has a nice ring to it—airless world. Is Buck Rogers in your biogas generator?

Anaerobic bacteria can be either *facultative* or *obligate*. Facultative anaerobic bacteria are switch-hitters. They can change from a metabolism that uses oxygen to one that does not. These facultative anaerobes then, can survive in environments which have a low concentration of oxygen, or none.

Obligate anaerobes, however, are less adaptable. They can only survive, and thrive, in an environment where free oxygen is, for all practical purposes, nonexistent.

The methane-forming (MF) bacteria are obligate anaerobes. As a result, when there is an appreciable amount of oxygen in a generator, they will not do their stuff. So, in a generator contaminated with oxygen, the MF bacteria cannot produce CH_4 and the biology of the generator is stuck in the acid-digestion stage. Some biogas may be produced, but it's mostly CO_2.

For this reason, often when a generator is starting up and the materials inside still trap pockets of free oxygen, the gas produced will not burn. Once the oxygen has been used up, however, the composition of the biogas will change until it is high in methane—60% to 70% CH_4. This change usually occurs within two weeks of the time when gas production becomes noticeable, assuming the temperature of the generator is in a range comfortable to the MF bacteria.

Terms

Facultative: Switch-hitters.

Obligate: 0_2, won't do.

Questions

None.

Problems

None.

6: TS, VS, FS

Each substrate, to a greater or lesser degree, contains water. When we merely weigh it, we have wet weight, or total weight. If we're weighing leaves from a pile that's been in the weather, then what they weigh will depend on whether it rained last night, or it has been hot and dry for months. Yet, in each case they might be the same leaves, identical in their ability to act as a substrate.

For purposes of accurate comparison then, substrates are dried at a temperature just above the boiling point of water (105° C or 220° F). The weight left after an hour of drying, or unchanging after several dryings, is called dry weight, dry matter, or *t*otal *s*olids (*TS*).

Eventually, if our leaves were digested for a long enough time, all the elements composing them would return to whence they came. Nearly all the C, O, H, and N would return to the air or water, and the rest would remain earthbound. Since the biological decomposition of these substrates requires so much time and since it is hard to judge when it is complete, scientists instead burn the substrates at 600° C (1,112 °F), and assume that everything which departs is volatile, and considered available for biological decomposition. This missing-after-burning portion is called organic matter, available matter, or *v*olatile *s*olids (*VS*).

The ashes are called ashes, or *f*ixed *s*olids (*FS*), and it is assumed that this portion is not available for biological decomposition.

In fact, some of the substances which end up as ashes (FS) might have been used in a biological process, and there are substances which go up in smoke (VS) which would not have been used in a biological process in any reasonable (say, one year) period.

Chief among the substances which are unused biologically but which do go up in smoke anyway, is *lignin*, which is discussed further in Chapter 9, C/N. Lignin is a real stick-in-the-mud when it comes to being decomposed. It simply refuses. We expect that a substrate with a high percentage of VS will give us a lot of biogas, but if it also has a lot of lignin, it will not decompose as happily and completely as other substrates which do not contain such stubborn compounds. Thus, this method of burning substrates to determine VS is not an accurate measure of how much material is biologically available to the bacteria. Nevertheless, because it is widely used, this narrowly useful measure will be used here as well.

VS and FS usually are given as a percentage of TS, dry weight. If they were measured as a percentage of wet weight, they would decrease every time the air got humid and raised the water content and therefore the wet or total weight of the substrate.

TS, of course, and percentage of H_2O are both a percentage of the total wet weight, and so they do change as the substrate gets wet or dries out. Because of the importance of these measures, it is wise to determine TS, and then, protect the substrate from changes in moisture content as well as you can.

Further information on making these determinations is found in Appendix 1.

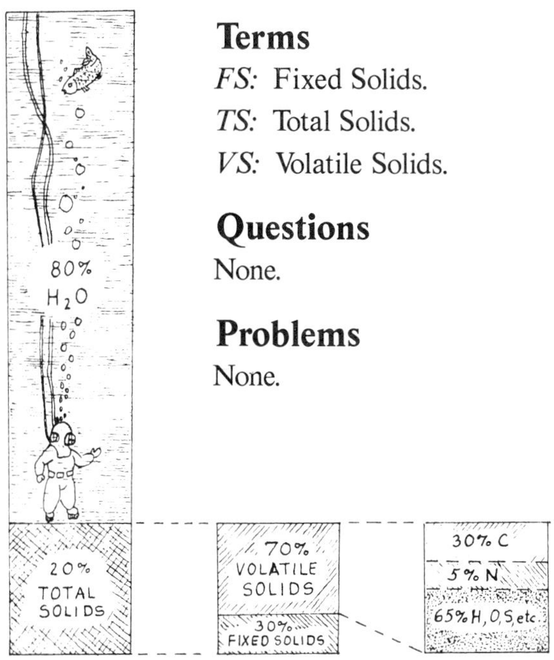

Fig. 6.1 Analysis of Cow Manure

Terms

FS: Fixed Solids.

TS: Total Solids.

VS: Volatile Solids.

Questions

None.

Problems

None.

7: pH

Next to the all-important anaerobic environment, the MF bacteria depend absolutely on several other conditions. Přimary among them is proper pH. Even to many chemists, pH is something of a mystery. Actually, you don't need to understand what pH is to use it in biogas production, so the technical material is just for the curious and the brave. What you will need to understand is how pH reacts: what makes it go up and down.

What is pH?

Water is H_2O, or (as it can also be written), HOH. A very minute number of water molecules in any given amount of water will *disassociate* (break apart) into H^+ and OH^-. These are called *ions*, atoms or molecules with a positive or negative charge. In pure water, H^+ and OH^- are equal in number. pH is a measure of the abundance of H^+ ions in a solution. The pH of pure water is 7.

This number describes a negative power of 10 (10^{-7}, in the case of pH 7), and it is a measure of the relative number of H^+ ions. Relative, since pH actually indicates the number of H^+ ions in a standard volume of water, whereas we may be finding the pH of either a drop or an ocean. (Consult a chemistry text for more information, if required.)

If we were to add more H^+ ions to the water, the pH would change. When an acid, such as HCl (hydrochloric acid, stomach acid) is added to pure water, it disassociates into ions, H^+ and Cl^-. This adds more H^+ ions to the solution, and the pH is changed.

Pure HCl has a pH of 2, and depending on how much HCl we add to our water, we can change the pH of the water from 7 down toward 2. As backwards as it may seem, when we add more H^+ ions, the pH falls, and the solution becomes more acidic (pH below 7). But consider this in terms of powers of 10. Is not 10^{-2} larger than 10^{-7}?

If we add OH^- ions to the solution, or any alkaline chemical (those which consume H^+ ions), then the pH will rise—that is, it will come up above 7. Substances which have a high pH (above 7) are called alkaline, or sodium bicarbonate ($Na(HCO)_3$), and ammonia (NH_3) are all alkaline or basic. When we add these chemicals to an acidic solution, they gobble up the free H^+ and the pH rises. (Still here?)

pH is important to all life, and the life inside a biogas generator is no exception. The AF (acid-forming) bacteria, which produce fatty acids, can tolerate a rather low pH and still keep on slugging, but the MF (methane-forming) bacteria are completely knocked out of the game around pH 5.5. They dislike going below 6.8 or above about 8.5.

Part of our job when we want to produce biogas, then, is to make sure the pH stays in a comfortable range for the MF bacteria. Another part is to make sure that the pH doesn't change rapidly, even within the range indicated above, since the MF bacteria don't like that either.

Now, we'll go on to discuss different ways to regulate pH, but the first question we should ask is: How does the generator's ecosystem regulate its own pH?

Natural pH Regulation

Since the MF bacteria use fatty acids as their main food source, they tend to keep the pH at neutral or above by taking the excess H and shoving it out the gas pipe combined with carbon (as CH_4), and by getting rid of CO_2.

The concentration of CO_2 dissolved in the slurry affects the pH:

$$CO_2 + H_2O \longrightarrow H_2CO_3$$

(Carbon dioxide plus water yields carbonic acid.)

The more CO_2 we have dissolved in the slurry, the more carbonic acid (H_2CO_3) we will have, and the more acidic the slurry will be. So the generator ecosystem can regulate its pH by giving off CO_2. This can give us an idea, if we understand what's happening: recirculation of biogas back into the generator can sometimes strip the slurry of enough CO_2 to lower the pH.

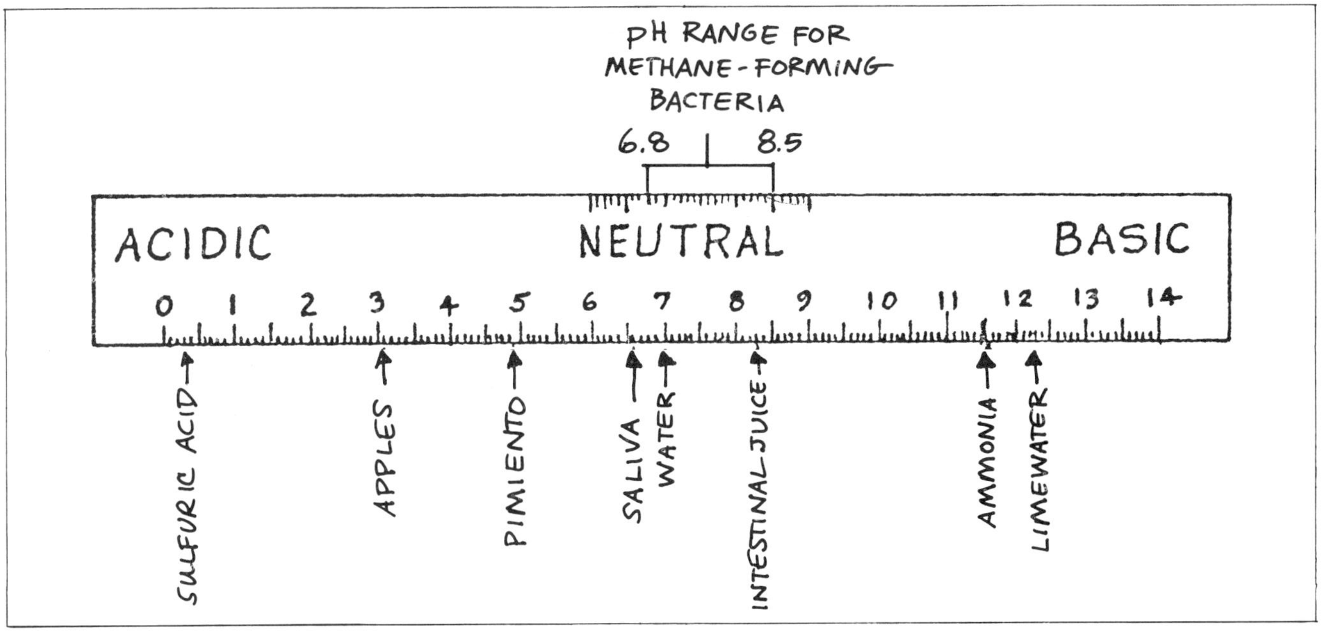

Fig. 7.1 pH Ranges

Buffer System

The pH is also regulated in the generator by a buffer system. A buffer is like a chemical shock absorber. A buffered generator will change pH less, and at a slower pace, than an unbuffered generator. So, the buffering capacity of a generator is a phrase that describes the ability of a generator to hang in there at a certain pH (usually moderate, often slightly basic) in spite of forces which might otherwise cause it to change pH. A buffer system is, in a sense, a storage depot of alkaline chemicals, which moderates the effect of any acidic chemicals in the generator.

New slurry has a weak buffer system, and so when acid digestion begins, the pH may fall, since the fatty acids produced will have no balancing alkaline chemicals. Older slurry has had time to develop these buffers, and the populations of MF bacteria which can "use up" the acidic molecules via the production of biogas. New slurry, weak in buffering capacity, can be helped by the addition of certain chemicals (discussed below) to strengthen its buffer system, or it can be mixed with old sludge, which produces a similar effect.

It's important to remember that pH can be an excellent indicator of the health of the biogas process.

Drama in the Generator

As an illustration of this, consider the following example: your thermostat decides to take a vacation. Night falls and so does the temperature in your generator, resulting in the demise of many of the delicate MF bacteria. A week goes by before you can fix the gizmo that broke. The acid-formers, less susceptible to life's occasional hard knocks, come through the experience a little bruised but still turning out those fatty acids, but the MF bacteria fall behind in using them up because they don't enjoy the cold.

So what happens? The acids accumulate and the pH falls. And now, the first problem (fall in temperature) is compounded by the second (fall in pH). The low pH further damages the MF bacteria, and eventually damages the AF bacteria which are literally beginning to swim in their own wastes. The resulting slowdown among the AF bacteria allows the suffering but silent MF bacteria to begin to make headway in using up the fatty acids and eventually the situation will rebalance itself—unless the original problem still exists.

The most visible outside indicator of this drama in the generator will be the decrease or even cessation of gas production. Eventually there will be a spurt of gas production as the MF bacteria use up the accumulated fatty acids. The pH of the generator will follow the same schedule, falling with the gas production, gradually stabilizing at some lower value, and then rising with or just before the rise in gas production.

Problem: pH is Falling, pH is Falling, All Over the Town

Instead of sitting and watching all this, what might we do? If the primary problem isn't obvious, we have to investigate. What is causing this? Some possibilities are temperature fluctuations, toxic materials, the addition of unwilted green material to the generator, and feeding the generator too fast. Air in the generator will not usually cause a dramatic change in pH.

These possible causes of falling pH will be discussed further on, but the point is that something has probably caused the MF bacteria to lose their appetite. Find it and fix it.

Then too, don't let the pH drop too far, since even if gas production is temporarily not up to standard due to some shock to the MF bacteria, an uncontrolled drop in pH will only cause more serious setbacks. There are several things which can be done.

Solutions: pHs of the World, Arise!

The first solution is to add lime. Some people say yes, add lime, some say no, add no lime. This, in fact, is one of those controversies mentioned earlier, in which opinions differ sharply and radically. As far as we're concerned, Sawyer, Howard and Pershe (1954) settled the question in favor of lime with their experiments and expertise. Lime is then recommended here. "Lime" is one of those words which is not well defined. Generally speaking, it refers to a group of calcium (Ca) compounds. Limestone, the common natural rock, is largely $Ca\ CO_3$, calcium carbonate. In this book, the phrase limewater will refer to $Ca(OH)_2$, calcium hydroxide or slaked lime. There is also CaO, calcium oxide. Lime will be used here to describe $CaCO_3$, calcium carbonate. If you read other books or articles, though, you're on your own, since any one of the above three calcium compounds is sometimes referred to as lime. It's not a big thing, since either limewater $Ca(OH)_2$ or lime $Ca\ CO_3$ will work in the jobs described below. CaO, calcium oxide, however, will not. Don't buy something labeled "lime" unless you know its chemical formula and it is indeed, what you want.

Responses to lime will also vary depending on the material being digested. Many are the unsung heroes of biogas research, and two of them, the venerable Messrs. Boruff and Buswell, did research on generation of biogas using cornstalks (1929). They found that gas production was markedly increased when cornstalks were soaked for four days in limewater before digestion.

Lime or limewater also has the interesting characteristic of combining (above pH5) with CO_2, removing it, in effect, from the biogas, and thereby increasing the percentage of CH_4 in the biogas. The addition of relatively large amounts of lime to the slurry can even cause CO_2 to be removed from the biogas atmosphere above the slurry, (rather than just removing it from the gas in solution in the slurry), thereby creating a partial vacuum above the slurry. The danger in that situation is that if there are leaks in the system, air will be drawn in, creating the possibility of explosion. The vacuum created may also be powerful enough to damage the digestor. Above pH5 then:

$$Ca(CO)_3 + H_2O + CO_2 \longrightarrow Ca(HCO_3)_2$$

or:

$$Ca(OH)_2 + CO_2 \longrightarrow Ca(HCO_3)_2$$

($Ca(HCO_3)_2$ is calcium bicarbonate.)

Not much lime is needed. Lime can be added to the tune of 0.2 to 0.3 grams per liter of slurry (0.2 to 0.3 ounces for every cubic foot of slurry), or it can be added slowly (over a period of several hours or days) while the pH is checked, until the pH rises to 7 or 8.

If lime is added after the generator becomes stuck (when it becomes acidic and digestion stops), it may cause foaming to occur. If a generator has been in serious trouble, add lime slowly, wait for the pH to rise a little, and then wait for a day or so for things to calm down. Repeat the treatment until gas production starts, and then wait and see if the pH won't begin to correct itself. If it does not, continue cautious additions of lime until all is well.

Lime has one main disadvantage. It does not dissolve easily, and thus it often does not mix uniformly throughout the digesting mass of materials. A measurement of pH, therefore, may not give an accurate picture of the true pH (average throughout the whole generator). This stubborn quality of lime is a problem because you may have added enough lime and not be able to tell—it may take awhile for the pH to reflect the quantity of lime available to the generator. For most of us, however, it's not critical. Limewater dissolves more easily than lime.

Ammonia

If you can't get lime, or don't want to use it, add a little ammonia. Ammon*ia* (NH_3) is toxic to the biogas process in too great a concentration, but a little served up in the right way has been found to help correct an acid condition. Ammon*ium*, an ion, (NH_4^+) is not toxic in ordinary amounts. One part by volume of ammonia, mixed with 40 parts by volume of warm water for every 10,000 parts by volume of generator will help correct an acid pH.

For a 55 gallon drum generator, that's 4.25 teaspoons of ammonia mixed in 3.5 cups warm water—Americanese. Or, in metric, that's 21 cubic centimeters of ammonia, 0.832 liters of water. Round it off, if you wish. Pure ammonia is almost never found in drug stores or supermarkets, not even pure ammonia diluted with water. In all but the rarest case, what is sold as ammonia has other chemicals added which are toxic to our microscopic friends. Go to a chemical supply house for a bottle of ammonia. Don't buy or use something which contains anything other than ammonia and water, unless your experiments show that the other chemicals do no harm. The 1:40 (ammonia:water) dilution is not as important, since it's done to dilute the ammonia so that the ammonia "medicine" doesn't shock the system when it's added to the generator. When you add it to the generator, be sure and maintain the 1:10,000 (ammonia volume: generator volume) relationship.

$$NH_3 + CO_2 + H_2O = NH_4HCO_3$$

pH	Indicators	Solution	Gas Composition
Low	Biogas production is low; biogas will not burn; effluent is not well digested.	Add buffer, remove CO_2 from biogas and recirculate.	High in CO_2 and H_2S.
Good	All is well.	Smile.	High in CH_4.
High	Rare situation. No common indicators.	Wait, or recirculate CO_2-containing gas. Raise C/N ratio.	Varies.

Table 7.1 pH in Brief

The third solution is to add bicarbonates. The following chemicals are not really a third choice, since they are as good or better than ammonia, but they are third here since they have to be someplace. Since the natural buffering capacity of a generator is largely due to bicarbonates of calcium, magnesium, and to a lesser degree, ammonium (Sawyer, Howard, Pershe, 1954), we would expect that these could serve quite well as buffers.

$$Ca(HCO_3)_2 = \text{Calcium bicarbonate}$$
$$Mg(HCO_3)_2 = \text{Magnesium bicarbonate}$$
$$NH_4HCO_3 = \text{Ammonium bicarbonate}$$

The last may serve, if needed, as a source of N. (Another chemical which could serve in the same way is ammonium hydroxide—NH_4OH). NH_4CO_3 is, among the three, the best bet, and may even be a shade better than lime because it adds N to the process. Adding calcium and magnesium ions (Ca^{++} and Mg^{++}) may contribute to scum problems.

The fourth solution is soda. Let's not forget good old Arm and Hammer—baking soda, sodium bicarbonate, $NaHCO_3$.

$$NaHCO_3 + HCl = NaCl + CO_2 + H_20$$

Sodium bicarb plus hydrochloric acid yields salt plus carbon dioxide plus water. Notice that the addition of sodium bicarb will cause CO_2 to be produced. Add *slowly*.

One common element to many of these chemicals is the HCO_3^- ion. It consumes H^+, as in the equation below, which shows ammonium bicarbonate gobbling up H^+ from acetic acid:

$$HCOOH + NH_4HCO_3 \longrightarrow NH_4COOH + CO_2 + H_20$$

(Acetic acid + ammonium bicarbonate yields ammonium acetate + carbon dioxide + water.)

The fifth solution is CO_2. In systems which use gas recirculation, the CO_2 can be scrubbed—removed from the biogas gathered, and the gas recirculated. This will pull more CO_2 out of the solution in the generator (similar to the process which happens when soda pop fizzes) and the pH will rise (e.g. the slurry will become more alkaline). We mentioned this awhile ago.

And finally, sixth is C/N. Lower the original C/N. See Chapter 9 for more information.

Generally, pH is measured with litmus paper, which can be found at pool supply stores or chemical supply houses. Litmus paper changes color in response to pH, so that when the litmus is dipped into the solution being tested, the resulting color can be read by comparing it with the color chart that comes with the litmus paper. For large-scale operations, electronic pH testers can be used.

Terms:

Acidic: Low pH.

Alkaline: High pH, "basic.".

Ammonia: The molecule, NH_3.

Ammonium: The ion, NH_4^+.

Buffer: A pH "shock absorber".

Ions: Loosely speaking, atoms or molecules with a positive or negative charge.

Lime: Here, $CaCO_3$, calcium carbonate.

Limewater: Here, $Ca(OH)_2$, calcium hydroxide.

Litmus paper: pH test paper.

pH: A measure of the acidity of a solution.

Questions

We have a generator with falling pH. What, in general terms, can we do?

Problems

None

8: Temperature

Quite nearly as critical as pH to the process of biogas production is temperature. There are three generally recognized ranges of temperature in which different groups of anaerobic bacteria operate, each of them named according to the temperature range it "loves." *Philos* means love, in Greek, and *psycro-* (or *cryo*), *meso-*, and *thermo-* mean cold, middle, and heat, respectively. So these bacteria are *psycrophilic* (cold lovers), *mesophilic* (middle lovers), and *thermophilic* (heat lovers).

Psycrophilic

Psycrophilic or cryophilic bacteria seem to be able to survive the range of 0° to 5°C (32° to 40°F). That's not exactly shirt sleeve weather. Specific studies on these bacteria and their (brrr!) temperature ranges are as scarce as hen's teeth. It seems fairly certain that at these temperatures we are dealing solely with digestion and *not* with biogas generation. There is some reason to believe that the effect of temperature on gas production varies with the material (substrate) being digested. Manures and sewages, on which 80% to 90% of the work in the area of biogas generation has been done, may respond less favorably to low temperatures than plant wastes.

It seems peculiar that there should be such a tremendous gap between psycrophilic bacteria at 0° to 5°C, and mesophilic bacteria at 21° to 40°C (70° to 105°F). It may well be there are other groups or single strains of bacteria which operate satisfactorily at lower temperatures, but they are not widespread, for otherwise they would now be in use, since one of the disadvantages of biogas production is the fact that, for optimum gas production in most climates, the generator must be heated; and if it were possible to obtain good production without heat, this would have been done. Or would it?

Further research on this question may show that there is a point or stage in the whole process which is more temperature-dependent than any other. In other words, going back to our disassembly line image of biogas production, there may be one place in the disassembly line where the workers suffer more quickly from cold. When they slow down in their work, they hold up the whole process, which cannot move faster than they do. Indications are that every stage of the process up to the methane-forming stage will occur at lower temperatures; the MF bacteria, however, seem to need heat. (For a report of a case where this is not true, see *Newsweek*, December 12, 1977, page 13.)

The effect of lower temperatures is lessened when some of the effluent supernatent or sludge from one batch or cycle of the generator is reused in the next batch or cycle. Besides adding nitrogen and water to the process, this will recycle some of the useful enzymes previously manufactured by the bacteria. Recycling the effluent may, however, add to the complexity of the generator.

The author found, when working with the modular batch-fed generator described in the Plans section (using unshredded leaves, which had been digested for several weeks at varying temperatures from 20° to 30° C), that when the generator was no longer heated, it continued to produce gas above what would have been expected.

Boshoff (1963, 1967) seems to agree that this is possible when he says: "Thus, when utilizing the residues of a previous fermentation... only some 15% less gas can be expected at 22° C as compared to 32° C. On the other hand, when using dung as a starter, with low bacterial activity, the yield for fermentations carried out at 32°C can be up to double that achievable at 22°C. It is clear from the context that he is speaking of the rate of gas production.

The questions involved in low or ambient (surrounding, environmental) temperature biogas production are many and complex, but this is a very worthwhile field for grass roots experimentation. See the Plans section for one exciting idea (the hybrid generator).

Thermophilic

Thermophilic bacteria have some disadvantages which do not commend them for a biogas generator which uses only simple technology. It is not that they cannot be used, but rather that they are more difficult to use.

The thermophilic temperature range is 40° to 60° C (104° to 140° F). Optimum is 50° to 55° C. These temperatures require, of course, more energy to maintain in the generator. Further, it has been reported that these bacteria are more sensitive to temperature changes, so that whatever heating arrangement is made should be able to maintain a certain temperature with only small (2° to 3° C) changes.

This may be another popular myth of biogas production, since experiments done by Heukelekian and Kaplovsky (1948) showed that thermophilic bacteria are well able to tolerate temperature changes and respond well when the temperature is brought back to normal after having dropped to a lower (colder) value, even if the colder value was held for some time. This study also pointed out that while thermophilic organisms *do* survive at mesophilic (20° to 40°) temperatures, the opposite is not true. That is, mesophilic organisms cannot generally survive thermophilic temperatures.

Another reported disadvantage to thermophilic digestion (see, for example Merril and Fry, 1973) is that the sludge from thermophilically digested substrates does not have as great a fertilizing value as mesophilically digested sludge. While this is indeed true, it is somewhat misleading. Thermophilic sludge, the settled portion of the effluent, does not have as much nitrogen in it as does mesophilic sludge, but this is because most of the nitrogen that was in the substrate will be found in the supernatent liquor, above the sludge. The thermophilic effluent (that is, sludge plus supernatent) has as much nitrogen as the whole effluent from a mesophilic generator, but in a different form—as ammonia and related soluble compounds.

There is evidence that at higher temperatures, more H_2S is produced from the slurry. This increased H_2S production can be a drawback in biogas which is used for certain purposes (see the Uses section). Further, because of this, thermophilic effluent may smell worse than mesophilic effluent, which may be offensive to you, or possibly to your neighbors.

The biggest advantage of thermophilic biogas generation is that the gas which is potential in the substrate is generated more rapidly. This is an advantage in rate of gas production, but it does not result in increased total gas production. Given enough time, approximately the same total amount of gas will be produced at any temperature from 21° to 60°C (70° to 140°F).

However, the fact that the rate of gas production is higher, means that the generator can be smaller and still produce the same given amount of gas (per day for each unit of generator), as compared with a larger mesophilic generator. Another way of saying the same thing is that, at a higher temperature, a smaller generator volume will handle the same amount of substrate that a larger volume generator would handle at a lower temperature. One study (Fong, 1973) showed a 200% increase in the rate of gas production at 42°C (107°F), as compared with the rate at 32°C (90°F). Another study (Savery and Cauzen, 1972), which was very shortlived (8 to 9 days), showed 130 liters of biogas per kilogram of wet manure at 50°C (122°F). In this study, the high rate (thermophilic) biogas was 60% CH_4, the low rate (mesophilic) biogas was only 50% CH_4.

After we punch our calculator, we find that this means that the high temperature produced about 90 liters (3.2 cubic feet) of CH_4, but only 46 liters (1.6 cubic feet) of CH_4 was produced with the lower temperatures. The dramatic difference is probably due to the fact that, in both cases (we may assume) oxygen was introduced with the wet manure, and the thermophilic generator more quickly used this oxygen up, and began to produce less CO_2 and more CH_4. Over a longer run, the mesophilic generator would probably have done better, after it had stabilized. Most studies of this kind do not show this much of a difference, possibly due to the fact that they are longer lived. Whether or not the slurry is well-seeded will also make a great deal of difference.

Notice carefully that these studies, and the figures we gained from them, do not tell us how much more energy it took to heat the thermophilic generator. In some cases, where cold weather and poor insulation combine to conspire against us, it could take more energy, per day, to heat the thermophilic digestor than we would gain by the increased rate of production. This is a hidden disadvantage of thermophilic digestion. The use of a source of heat other than biogas, such as solar or wood, may make the equation balance more favorably.

Your Basic Biogas Production Curve

The balance here is still time. Under any temperature conditions in the two ranges, biogas production accelerates rapidly, peaks, and gradually falls. Fig. 8.1 shows batch digestion, started without a seed, and charted from the time of perceptible gas production.

If we stay near points B or C, then we get 40% to 50% of the possible total gas production in a relatively short time. If we move on out toward D or E, we may double the time needed, and only increase our total gas output by 20%. If we have the time and the space, then it matters not; we can shoot for whatever amount of time we wish. The last 15% of possible total gas production, however, may require 3 times the number of days it took to get the first 85% of the gas, and there comes a point in anyone's figures where it's no longer worthwhile to continue.

So, the logic behind thermophilic biogas production is: Punch it through. Balance the cost of space (e.g., size of generator) against the value of the gas produced,

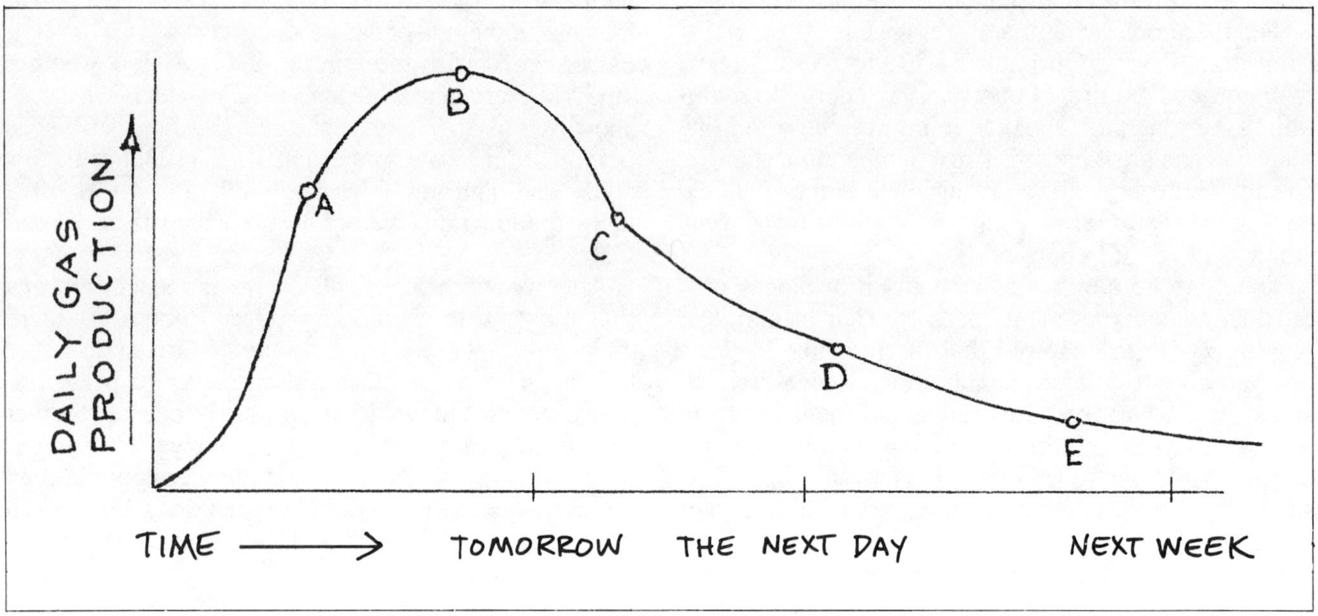

Fig. 8.1 Gas Production Curve

and punch it through, staying just on the right side of the production apex. For longer runs, thermophilic digestion is not worthwhile, since as we mentioned, the energy cost of heating the digestor will soon equal or exceed the energy gained in biogas, unless the heat requirements, due to climate and insulation, are low.

For these reasons, thermophilic digestion is not generally recommended for batch generators, for small-scale generators, and for long runs. Whether or not thermophilic digestion will prove economical in another situation should be demonstrated experimentally.

Artificial Thermophilia

There is some reason to believe that additives could stimulate thermophilic type (very rapid) digestion at lower temperatures.

Human excrement or cow manure (to name only two) are sources of thermophilic bacteria, yet neither the human intestine nor the cow's intestine ever reaches thermophilic temperatures. How then can we explain these heat loving bacteria? Does it make sense that they would not have adapted their metabolism to the low temperature conditions in which they can apparently survive and compete?

The conclusion seems to be that these bacteria prefer, not heat, but rather biochemicals manufactured by the intestine or digestive system, to stimulate their metabolism. Several studies have been done which show that the fluids from cow rumen (part of the cow's digestive tract) can help stimulate thermophilic-type digestion at mesophilic temperatures. This is another subject worthy of research, for the same biochemicals may stimulate gas production at still lower temperatures.

Mesophilic

For most applications, mesophilic digestion is the best bet. The temperatures involved (21° to 40°C, 70° to 104°F) are low enough to be more easily maintained, not as much H_2S is produced, and temperature fluctuations are better tolerated in stable digestors. Optimum temperature is 30° to 35°C.

And Below

Biogas production is very slow below 15°C (60°F), although digestion of the substrate continues well below that temperature. This fact is fairly obvious if we consider the general design of septic tanks, which are, one could say, household-sized, anaerobic, continuous-feed, unheated digestors.

Since septic tanks are unheated, their temperature is essentially the same as that of the surrounding soil, which stays in the neighborhood of 10°C (50°F) at a depth of several feet, for most soils and climates. Septic tanks are nearly never designed with provisions for releasing excess biogas, since so little is generated due to their low internal temperature. (Chiefly for this reason—low temperature—septic tanks cannot be converted into biogas generators, unless an adaptation of the hybrid design idea is used.)

Table 8.1 adapted from Fair and Moore (1934) gives a general idea of the difference temperature makes in the time it takes to achieve 90% of the possible biogas production. Because of the complexity of the biology of the biogas process, this table may not give you a very accurate estimate of the time required for you to achieve 90% of the possible biogas latent in your substrate, under your specific conditions. The chart was derived from many different studies of sewage sludge, a substrate which is bound to be different than your own material because of its relatively high dilution, its generally small particle size and the fact that it digests rapidly. For information on mathematical formulation of gas production, see Appendix 4.

Terms

Ambient: Surroundings, environment, usually referring to temperature.

Mesophilic: Middle lovers.

Psycrophilic: Cold lovers.

Thermophilic: Heat lovers.

Questions

None

Problems

None

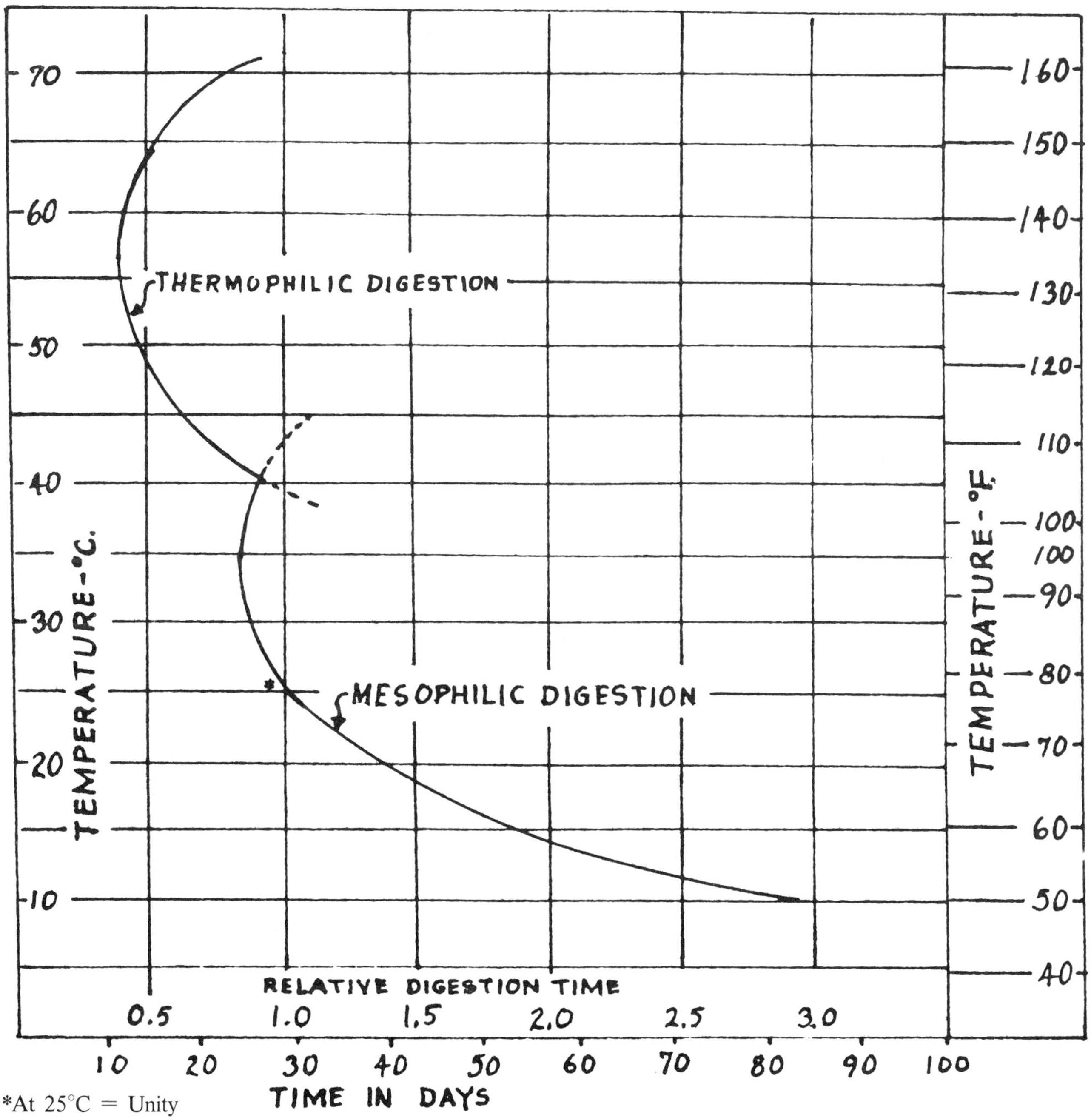

*At 25°C = Unity

Table 8.1 Time Required to Achieve 90% Digestion

9: C/N

The carbon-nitrogen ratio of a substrate is a measure of the number of carbon atoms in a substance divided by the number of nitrogen atoms. For example, casein (pronounced "case-een") is a protein molecule with the general formula $C_{148}H_{233}O_{48}N_{37}S$. (That's a long way from CO_2.) Casein has 148 carbon atoms, 233 hydrogen atoms, 48 oxygen atoms, 37 nitrogen atoms and one sulfur atom in every molecule. Its carbon-nitrogen ratio is ($148 \div 37$) four carbons to one nitrogen.

A carbon-nitrogen ratio is generally greater than one, (more carbon than nitrogen) and so it is often written as a number (casein = 4) rather than as a ratio (4:1). "Carbon-nitrogen ratio" is abbreviated to *C/N* or *C:N*.

Some of the carbon atoms in any substrate are given off as CO_2, which is a product of the respiration of the microorganisms. In other words, both the acid formers and the MF bacteria "breathe," and one of the by-products is CO_2, as it is with our own respiration. (The C/N of the effluent will, for this reason, be lower than the C/N of the fresh slurry.)

Carbon is also incorporated into the bodies of the little bitty buddies, which have a C/N of between 4 and 9.

All in all, bacteria both in aerobic and anaerobic digestion use about 30 carbon atoms for respiration and body building for every one nitrogen atom they use. This also means that recipes for compost piles are good recipes for biogas generators, as far as the C/N is concerned.

Carbon: Available vs Present

However, some kinds of carbon-containing substances resist breakdown and so they don't easily or immediately release their carbon atoms for use. Lignin, for example, is a carbon-containing compound which holds on to its carbon atoms.

In fact, lignin is the most common stubborn carbon compound, for all land plants use it to help stiffen and support themselves. As you may suspect then, there is more lignin in old grass than in new grass and more in wood than in leaves.

Recognizing this, sometimes the C/N of the substrate is given as "non-lignin," and in this case, 25 or 30 is most likely the best C/N, since then the bacteria will get 25 or 30 available carbon atoms to one available nitrogen atom, and they will be happy and busy.

However, C/N is more often *not* given as non-lignin, and when this is the case for paper, straw, and other woody and fibrous plant materials, we will have to give it our best guess and up the ante. A C/N of 40 or 50 might be advisable in this case, so that the bacteria can find 30 available C atoms for each available N atom.

Some studies have shown that digestion of cellulose can take place quite well at a C/N of between 50 and 20 (Sanitary Engineering Research Laboratory, College of Engineering and School of Public Health, University of California, Berkeley, hereinafter referred to as SERL, 70-2 and Hill, 1939) but nearly everyone aims for 30 when possible, for best digestion of any substrate.

Of course, when we can select and weigh the materials we put in our generator, we can mix and match to come out near our desired C/N. Sewage treatment plants, for one example, cannot be so choosy.

Table 9.1 lists some reported C/Ns (if available) and other information about many different kinds of materials. Some of the columns may be unfamiliar, but they will be explained.

Comments on the C/N Tables

Before we launch into a discussion of how to use the C/N table, some points may need mentioning. The table is divided according to the kinds of substrates represented:

Animal Wastes contains manures and other associated wastes, including human wastes.

Plant Wastes has everything which could be called by that term, except wood wastes.

Home Wastes has some duplication, because therein are found such substrates as are found around the home.

Purified Substances is a short list which may be useful if all you know is the proximate analysis of a substrate in terms of carbohydrates, protein, etc.

Wood is a fairly short list, containing some information on wood substrates. No C/Ns are given for wood and paper. For all practical purposes, wood and paper can be considered a source of carbon only.

"Median" in the *Comments* column, is based on the "middle" values (*e.g.,* halfway between the highest and the lowest). Note that this may not be average, in the ordinary sense, where several values are added together and divided by the number of values. Where there is a question mark in the comments column, preceded by a word or two, this means that the comment is an educated guess.

Wide variations in the listed values have been reported, and your own substrate will probably differ from what has been reported. For example, under Plant Wastes, take a look at leaves. With great certainty, these figures have been given by different authors, yet the variations in reported C/Ns and percentages of nitrogen (%N) is startling. The difference is probably explained in the comments column, but the point is still valid. These values may look very reassuring and accurate, complete with decimal points; nevertheless, they are approximations.

	C+N number	H_2O number	cmb number	C/N	%N	%C	% H_2O	%VS	Comments
Animal Wastes									
Animal tankage				4.1					
Blood	2	16.7	33	3.25	12.0	39	94		median
Cow (India)	2.3	5	11.3	25.0	1.7	43	80		low protein feed (?)
Cow (beef)	2.3	5	11.3	16.0	2.6	42	78		alfalfa fed
Cow (beef)	2.8	2.2	6.2	28.0	1.2	34	46	69	dried, packaged
Cow (dairy; plus bedding)	2.3	4.6	10.3	21.0	2.0	42	78		grain fed; plus rice hulls
Farm yard manure	3.1			14.0	2.2	30			
Fat scraps				76.0					
Fish scraps	2			5.1	8.3	4.2			median
Horse manure	1.7	4	6.7	25.0	2.3	58	75		
Human, feces				6–10	4–6		73–75		reported variations
feces	2.2	3.9	8.6	8.0	5	40	74	85	median
mixed	2.7	12.5	34	2.4	10.9	26	92		median
urine				0.8	15–18		95–96		reported variations
urine	3.4	20	67	0.8	16.5	13	95	69	median
Meat scraps				5.1					
Mixed slaughter Wastes	3.9			2	8.5	17			median
Pig manure	1.8	5.6	9.9	14	3.75	53	82		
Poultry	1.7	2.9	5	9.3	5.6	52	65		general average
chicken					3.6	25–38			reported variations
chicken	2.8			7	4.5	31.5			median
turkey	2.6	4	10	8.3	4.2	35	75		
Rabbit					1.7				
Sheep	1.3	3.1	4	20	3.75	75	68		
PLANT WASTES									
Alfalfa				16–20	2.4–3				reported variations
Alfalfa	2			18	2.7	49			median
Algae	1.9			4.7	9.1	43		93	
Amaranthus	2.3			11	3.6	40			pigweed family

	C+N number	H_2O number	cmb number	C/N	%N	%C	% H_2O	%VS	Comments
Bagaase	2.5			113	0.35	40			
Beet	7.1	3.1	22.2	27	0.5	14	68		root (?)
Buttercup	1.9			23	2.2	51			ranunculus
Cabbage	2.1			12	3.6	43			
Carrot, whole	2.2			27	1.6	43			
Clover, hay	5	1.7	8.4	24	0.8	19	40		
Clover, red	2			27	1.8	49			
Cooksfoot	1.9			19	2.6	49			
Corn, stalks & leaves	2.2			53	0.84	45			
Cottonseed meal				5					
Fern	2			43	1.15	50			
Flax waste	1.8			58	0.95	55			phormium
Grass,	2.1	2.9	6	19	2.4	46	65		lawn cuttings
hay, young	1.9	5	9.6	12	4	48	80		fresh (?)
johnson		4			0.19		75	98.6	low N (?)
Kentucky blue	2.1			19	2.4	46			
timothy	2			58	0.89	49			
Kelp	6	9.1	55	7.3	2.0	14.5	89	50.5	macrocystis pyrifera (fresh)
Lavatera trimestris		5.3		30			81		gardens, South Europe
Leaves, fresh				40–80	0.5–1				reported variations
fresh	2.2			60	0.75	45			median
rotted	2.5	2	4.9	203	0.2	41	50		and leached (?)
Lettuce					3.7				
Malope grandiflora		5.3		17			81		gardens, South Europe
Mustard	2.5			26	1.5	39			
Onion	2.4			15	2.65	40			
Oats, straw	1.9			48	1.05	50			cut for hay (?)
straw	2.3			83	0.52	43			cut for grain (?)
Peanut hull				36					non-lignin
Peat	1.6			67	0.9	60			
Pepper	2.4			15	2.6	39			
Potato top	2.5			25	1.5	39			
Purslane	2.5	2.2	5.4	8	4.5	36	55		fresh
Ragwort	2.1			21	2.15	45			
Rape shoots	2.7			8	4.05	33			
Seaweed	2.4	3	7.2	17	2.4	40	67	78.7	ascaphyllum nodosum (fresh)
Seaweed	2.4	1.1	2.6	17	2.4	40	11	78.7	(air-dried)
Sunflower, older		5.9		30			83		striped, 1.8–2.4m
Sunflower, younger		10		14			90		striped, 1.2–1.5m
Tobacco	2.4			13	3	39			
Tomato	2.3			12	3.3	40			fruit waste (?)
Turnip, tops	2.2			19	2.3	44			
Turnip, whole	2.2	14.3	32	44	1.0	40	93		
Water hyacinth	3.3	1.4	4.6	21	1.4	29	30	91	partly dried (?)
Vegetables, non-leguminous				11–19	2.5–4				reported variations
Vegetables, non-leguminous	1.9			15	3.25	49			median

	C+N number	H_2O number	cmb number	C/N	%N	%C	% H_2O	%VS	Comments
HOME WASTES									
Bread					2.1				
Dog food	2.4	1.1	2.6	11	3.4	38	6.4	91	Purina dry chow
Grass mowings	2.1	2.9	6	19	2.4	46	65	90	young grass
Kitchen garbage	1.8	3.6	6.3	16	33	52	72	90	no paper
Kraft paper				—		44	6	99+	
Newspaper		1.7		—	0.11	41	6	99	average
Potato tops	2.6			25	1.5	38			
Rags	1.7			25	2.15	55			
Raw garbage	1.7			25	2.15	55			
Refuse, combined	2.7			34	1.05	36			
Turnips, tops	2.2			19	2.3	44			
Turnips, whole	2.2	14.3	32	44	1.0	40	93		
Wheat flour					1.7				
PURIFIED SUBSTANCES									
Cellulose	—	—	—	—	0	44	—	100	approximate
Protein	1.5			3.5	16	53+	—	100	approximate
Carbohydrates						40			approximate
WOOD									
Chestnut	2			—	0.1	50.3		99+	
Larch	2			—	0.14	49.7		99+	
Sawdust, fresh	1.8			—	0.11	56		99+	douglas fir (?)
Sawdust, rotted	1.9			—	0.25	52		99+	douglas fir (?)

Table 9.1 C/N

Using the C/N Table

Let's say you are harvesting wheat in a field by hand. You stack up the bundles you have cut, and later, when they are drier, you return to the field, gather up the bundles, and take them back to the place where you will separate the wheat grain from the chaff and straw. Clearly, if you weigh the bundles before they are dry, a large part of the measure you get will be water. In the same manner, if you weigh the dry bundle, this will not tell you very much about the weight of the grain you have. There may be a lot of grain on each stalk, or not very much. If a grain mixture for animal food or for cooking calls for a certain portion of wheat and a certain portion of oats, it wouldn't make sense to use the weight of either the wet or the dry bundles of wheat and oats, for we would be concerned with the weight of the grains and not the weight of the grain plus stalks. the weights of either the wet or the dry bundles of wheat and oats, for we would be concerned with the weights of the grains and not the weights of the grain plus stalks.

In a very similar way, when we mix substrates to balance the C/N, we should remember to use only that portion of the substrate which is made up of the elements of C and N. In many books about biogas, this point is not made, or not made well enough

Organic materials have not only carbon and nitrogen, they are also composed of oxygen, hydrogen, sulfur,

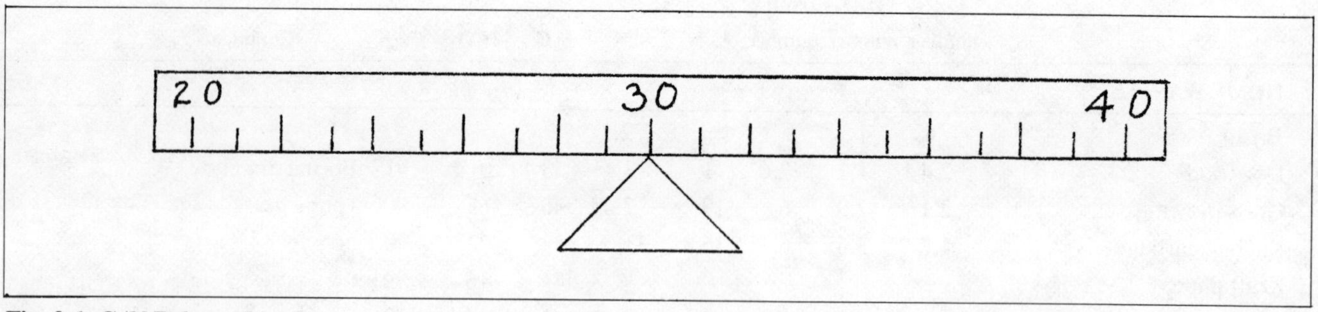

Fig. 9.1 C/N Balance Line

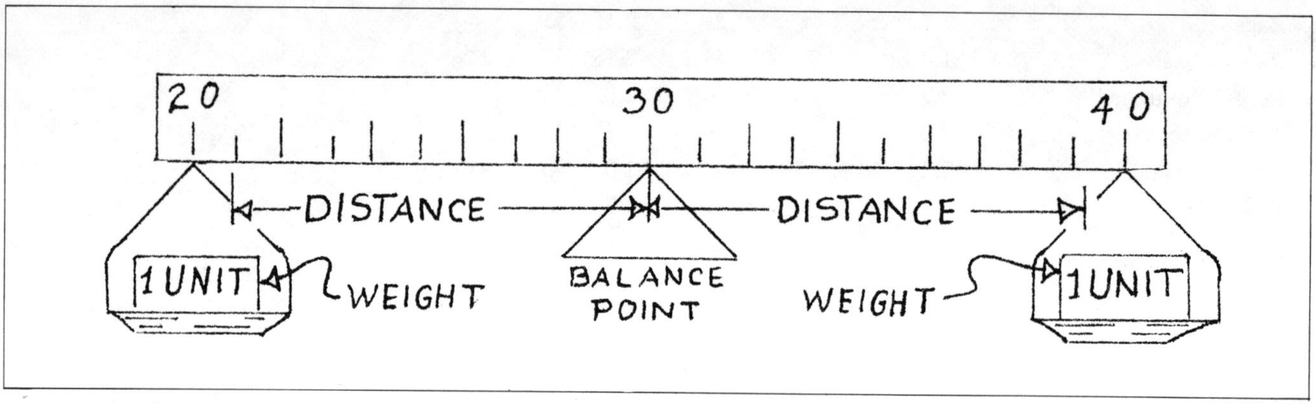

Fig. 9.2 Weight x Distance

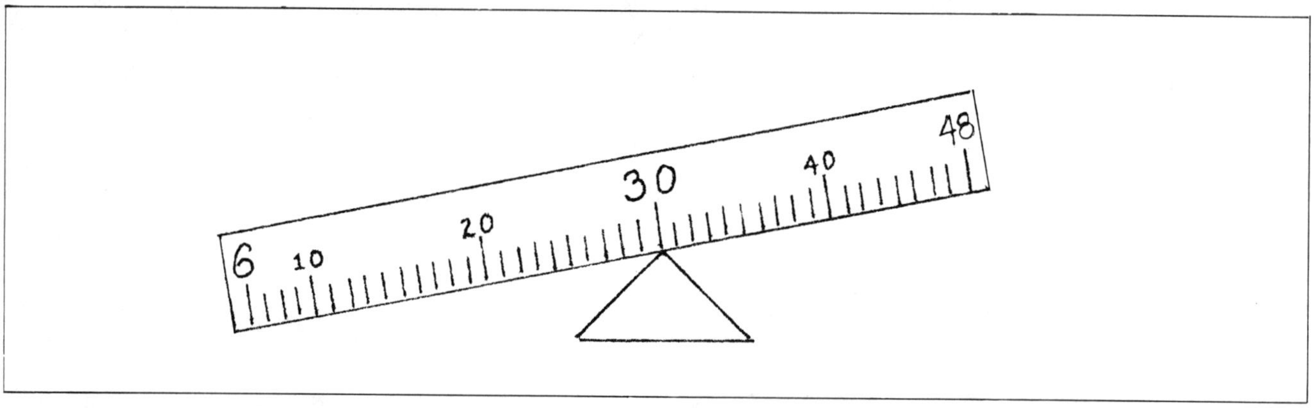

Fig. 9.3 Unbalanced C/N

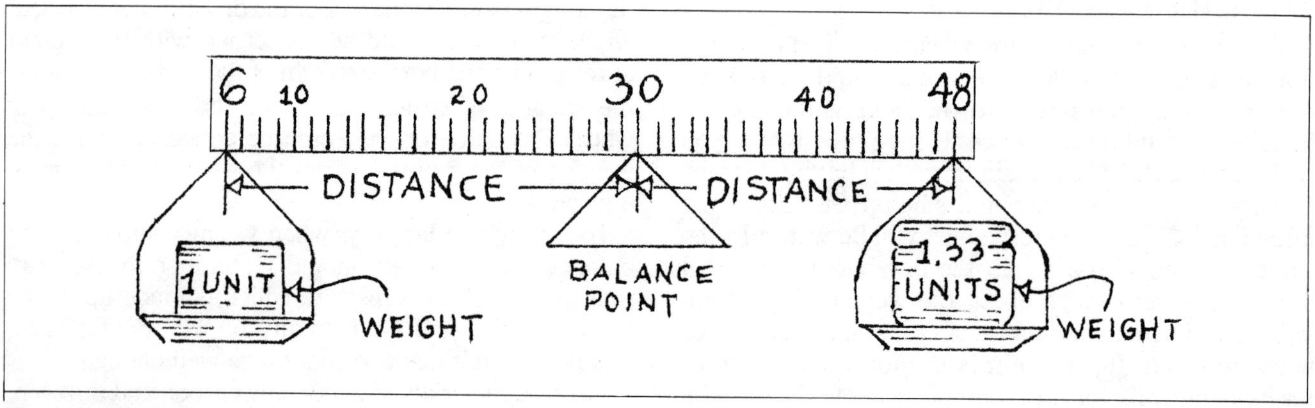

Fig. 9.4 Balanced C/N

phosphorus, potassium, and so on. As far as C/N is concerned, these other elements are the chaff and straw. (This is not to say they are not important in the biogas process—we're just not interested in them as far as the C/N is concerned.)

So, let's assume for the moment we are only dealing with the grain; just that portion of the material which is only C and N.

In order to balance two different C/Ns so that the mixture we end up with has the target C/N (the C/N we want for our generator), we can construct a simple balance line. (See Figs. 9.1 through 9.4.) This line is simply a way of visualizing the problem. The line is marked off like a ruler, but instead of inches, the units are the C/Ns of the substances we are working with. Suppose, for example, we have a C/N of 40, and one of 20, and we want a C/N of 30.

Both C/Ns are the same distance away from our target C/N of 30; that is, 10 units. We can see clearly that if we put equal weights on each end of the line, we should come out with the right balance. So, in this case, we can mix equal weights of the "grain" portion—that portion of the substrate which is only C + N—of each substrate to balance our C/N at 30.

In Fig. 9.3 we have different distances. We have substrates with a C/N of 48 and of 6, and we want a C/N of 30. This again is easy to solve, for the line will be balanced when the weight times the distance equals the weight times the distance. So, if we hang one unit of weight on the 6 mark (Fig. 9.4), then:

$$(\text{weight of x})(\text{distance } 18) = (\text{weight of } 1)(\text{distance } 24)$$
$$(x)(18) = (1)(24)$$
$$x = 24 \div 18$$
$$x = 1.33$$

We need 1.33 units of weight hung on the 48 mark.

In other words, when we mix 1 unit of that portion (of the C/N = 6 substrate) which is C + N, and 1.33 units of that portion of the C/N = 48 substrate which is C + N, then the resulting mixture will have a C/N of 30.

Right? Just the grain portion, so far. To figure out what the dry weight will be, we turn to the column labeled "C+N number" and multiply by this number.

For you math buffs: the C+N number is the approximate reciprocal of the percentage of C + N in each substrate times 100, or 100÷(%C+%N).

Let's suppose in our example above that the C+N number for the first (48) substrate is 2.2 and of the second, 2.5. In this case:

(1.33 units of weight) (2.2 C+N number) = 2.9
and
(1 unit of weight) (2.5 C+N number) = 2.5

These numbers can be taken either as direct weights (2.9 kg., 2.9 pounds, 2.9 ounces, 2.9 tons), or as ratios: 2.9 parts by weight to 2.5 parts by weight. However, these are dry weights.

To get from the dry weight (TS) to the wet weight—the ordinary weight of the substrate—we multiply by the H_2O number for that particular substrate.

Again, just for you math buffs: The H_2O number is the approximate reciprocal of the percentage TS, times 100, or 100÷%TS.

This (finally!) will give us the ordinary weights of the substrates (or the ratio of weights) we need to have to get our target C/N.

In the example given above, and again assuming arbitrary H_2O numbers (of 2.5 and 1.7):

$$(2.9)(2.5) = 7.3$$
$$(2.5)(1.7) = 4.2$$

So, 7.3 units wet weight of the C/N 48 substrate and 4.2 units wet weight of the C/N 6 substrate will give us (7.3 + 4.2) 11.5 units wet weight of a C/N 30 substrate. The process is a little cumbersome, but quite simple and accurate.

You may wonder why the H_2O number and the C+N number are considered separately—why not just multiply them together and deal with only one number? The answer is that while the actual C/N of different samples of a substrate will vary somewhat, the percentage moisture in them will vary a great deal, and thus the C/N chart is set up to allow you to determine your own H_2O numbers for greater accuracy. If the percentage of water in your materials changes, then you can change your math accordingly.

For anyone who does not want to find his/her own H_2O numbers, use the "cmb (*n*)"—the combined number—it's just the C+N number times the H_2O number.

The sequence is: Determine the proper weights or ratio of weights of the "grain" (C + N only) by the balance line method.

$$C_a - C_t = D_a$$
$$C_t - C_b = D_b$$

and:

$$(D_a)(1) = (D_b)(W_b)$$

or,

$$\frac{D_a}{D_b} = W_b$$

Where:

C_a = the C/N of substrate A

C_b = the C/N of substrate B

C_t = the target C/N

D_a = the distance C_a must go to get to C_t (when C_a is larger than C_t)

D_b = the distance C_b must go to get to C_t (when C_b is smaller than C_t)

W_b = the weight ratio of substrate B (C + N only) needed to balance 1 unit weight of substrate A (C + N only)

Then, (if you have determined your own H_2O numbers) multiply by the C/N numbers of substrates A and B to get the dry weights needed:

$$(D_a)(CN(n)_a) = DW_a$$
$$(W_b)(CN(n)_b) = DW_b$$

Finally, multiply by the H_2O numbers of A and B to get the wet weights needed:

$$(DW_a)(H(n)_a) = WW_a$$
$$(DW_b)(H(n)_b) = WW_b$$

Where:

$CN(n)_a$ = the C+N number of substrate A

$CN(n)_b$ = the C+N number of substrate B

DW_a = the dry weight of substrate A

DW_b = the dry weight of substrate B

$H(n)_a$ = the H_2O number of substrate A

$H(n)_b$ = the H_2O number of substrate B

WW_a = the wet weight of substrate A required for the target C/N

WW_b = the wet weight of substrate B required for the target C/N

Or, if you have not found the H_2O numbers of your substrates, then multiply D_a and W_b by the combined numbers of substrates A and B to find the wet weights required:

$$(D_a)(CMB_a) = WW_a$$
$$(W_b)(CMB_b) = WW_b$$

Where:

CMB_a = the combined number of substrate A

CMB_b = the combined number of substrate B

For practice, we'll run through an example, and some problems are found at the end of the chapter which will give you additional practice.

Rummaging around in your backyard, you find young grass hay and oat straw. First, choose a target C/N. We'll say 30 in this case, but the oat straw is bound to have lignin in it, and that will mean that the available carbon will be less than the measured carbon. Checking the chart, you find:

C/N of oat straw (cut for hay) = 48 (substrate A)
C/N of oat straw hay = 48 (substrate A)
C/N of young grass hay = 12 (substrate B)

So:

$$48 - 30 = 18 \quad C_a - C_t = D_a$$
$$30 - 12 = 18 \quad C_t - C_b = D_b$$

Very convenient. This means (in terms of ratios):

$$D_a = 1$$
$$W_b = 1$$

Glancing back at the chart, you find that oat straw has no listed H_2O number, nor combined number. Cursing quietly, you find out (from Appendix 1.) how to find this information for yourself, and having done this, we'll assume you discovered that your batch of oat straw has an H_2O number of 1.7. The listed H_2O number of young grass hay is 5, and both of them have a C+N number of 1.9:

$$(1)(1.7)(1.9) = 3.2 \; (D_a)(H(n)_a)(CN(n)_a) = WW_a$$
$$(1)(5)(1.9) = 9.5 \; (W_b)(H(n)_b)(CN(n)_b) = WW_b$$

So, for every 3.2 pounds or kilograms or tons or units of oat straw, we will need 9.5 pounds or kilograms or tons or units of grass hay for a $C_t = 30$.

To determine the amounts required for more than two substrates, simply use the process two or more times. For example, for a four substrate mixture, pair them off in twos and run the process. For a three substrate mixture, run the process twice using A + B and A + C combinations, but remember that you must add the amounts of A required to balance B as well as the full amount of A required to balance C.

Obviously you cannot produce a target C/N of 30 from a C_a and C_b which are both either above or below that target.

More on C/N

The C/N is not an extremely critical parameter, since such a wide range of C/N values can support the process of biogas generation. Further, it tends to be self-regulating, since if there is too much N, NH_3 is produced. When this passes off as a gas, the amount of N in the slurry drops. Too much NH_3 (ammonia) poisons the bacteria, so the process may slow down if the C/N is too low and a great deal of NH_3 is produced.

So, you don't want to go too far in this direction, but lowering the C/N will increase the initial buffer capacity of the slurry, if this is needed.

If the C/N is too high, more CO_2 will show up in the gas, lowering its heat content (because of a lower percentage of CH_4), and the pH of the slurry will be low (acidic). Again, the process will slow down or stop.

The slurry left after all the shouting's over (the effluent, remember?) will have dropped from a C/N of around 30 to a C/N of around 15, because the N stays put, and the C leaves, tied up in CO_2 and CH_4. This fact is useful to know, since the liquid portion of the effluent (supernatent liquid) has a great deal of N in it, and it can be recycled into another generator to: (1) add water; and (2) lower a C/N which may be too high; and possibly (3) to "seed" the next batch of substrate with bacteria. Possibly because these bacteria adhere to surfaces and they are not fond of floating about, pound for pound, the solid portion(sludge) of what's left has a great many more bacteria than the supernatent.

A Last and Trivial Word

According to Golueke (1972), if the percentage ash of a sample of garbage (or possibly paper) is known, its carbon content can be approximated:

$$\% \text{ carbon} = (100\% - \% \text{ ash}) \div 1.8$$

This formula has limited usefulness.

Terms

C/N: Carbon to nitrogen ratio.

Lignin: Stubborn substance.

Median: The middle value.

Non-lignin C/N: The C/N based on the available, rather than the total carbon.

Target C/N: The C/N we want to achieve in mixing substrates.

Questions

None

Problems

1. Target C/N is 35. Using rotted (leached) leaves and purslane, what is the wet weight ratio?
2. Target C/N is 38. Using beets and whole turnips, timothy grass, and clover hay, give the wet weight ratios. Assume timothy grass has a 50% moisture content, and therefore an H_2O number of 2.
3. Give the wet weight ratio you would use for timothy grass and whole turnips, for the same target C/N as above (careful now).

10: Percentage of H_2O

All living things require water to survive, and it is interesting to note that many biological fluids, like blood, resemble sea water. The sea, of course, is the mother of life, and so in more ways than one, we all have the ocean in our blood.

In an anaerobic generator, the water is used primarily as a transport medium. Although it has other uses in the generator, only a small portion of it is involved biologically and chemically in the biogas process. The generator, like a pond, or the coastal sea water, or a forest, is an ecosystem, supporting many different kinds of life, all of which interact in a manner that tends to produce the greatest good for all. In several senses, the generator, like a pond or any other ecosystem, can be regarded as a single organism, and this organism needs not only food, but also water.

Less Is More

But how much water? This is an important question, since the water itself, for the most part, does not add anything to biogas production. So, if we can cut down on the amount of water needed in the generator, we increase its effective volume, since we can cram more of the essential substrate into our generator, and, theoretically, it should then produce more biogas, all else being equal. This is advantageous.

There is one other major advantage to a drier anaerobic digestion and that is that the leftover material is easier, in most cases, to handle and transport. If needed, water can be added to the spent slurry, but it is very difficult—energy plus dollar costly—to remove excess water. But, as usual, there are trade-offs; not only advantages but disadvantages to using less water. Some of the disadvantages are mechanical. Mixing or stirring municipal sewage or manure slurries with paddles or other devices becomes difficult above 6% solids. Pumps for mud and concrete which accept up to 4 centimeter (about one and a quarter inch) rocks are in common use but they require quite a bit of energy to operate, and of course they are expensive, which would limit their use to large scale operations. G.M. Wong-Chong (1976) suggested that a low-speed Archimedes screw could be used for transporting a very dry slurry into and out of a generator. A modified version of his idea is shown below, in Fig. 10.1.

This might save considerable energy.

By the way, all the percentages given here refer to a percentage of weight or mass. Ten percent solids, then, means that 10% by weight is dry matter, and 90% by weight is liquid that can evaporate, or water.

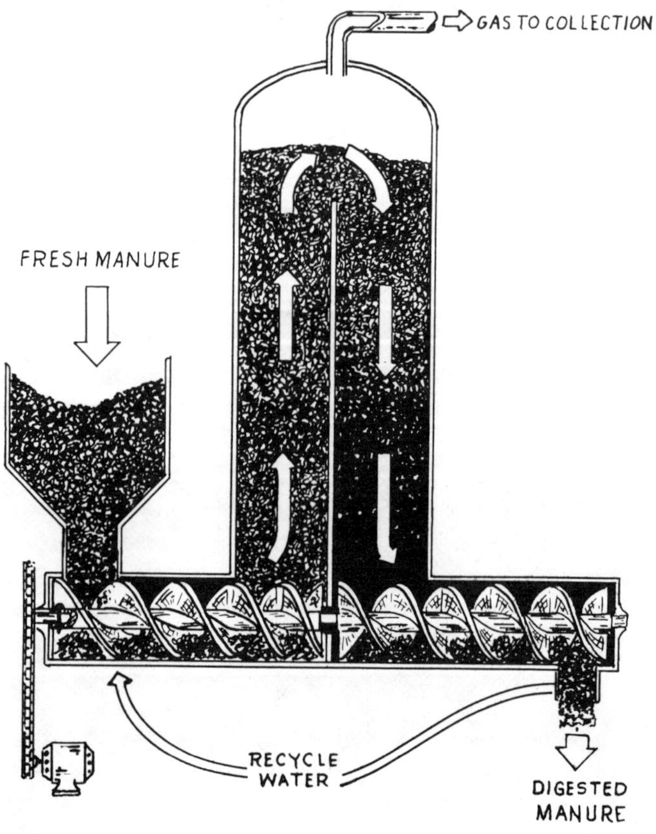

Fig. 10.1 Dry Digester
(Adapted from Wong-Chong)

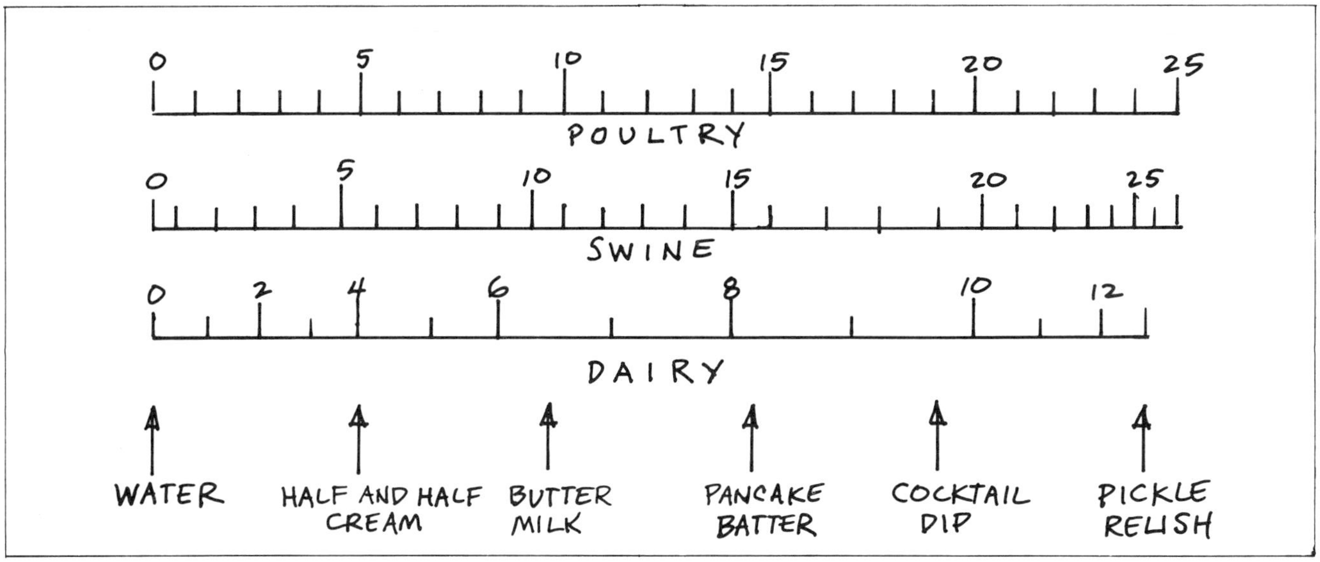

Fig. 10.2 Percent Solids and Consistency

More Stirring Information

The mechanical difficulty involved in stirring a certain slurry depends not only on the percentage of solids, but also on the percentage of VS. A higher percentage of VS tends to make the slurry more difficult to stir. This may be the reason that, for example, different manures, when measured at the same percentage of solids, nevertheless present different degrees of difficulty. For example, the consistency of cow manure at 8% solids, according to Hart, Moore and Hale (1966), is about the same as the consistency of pig manure at 15% solids.

These researchers tested poultry, pig, and cow manure, and, in a novel and even humorous chart, showed how consistency differed among these three. It should be remembered that feed probably has a great deal to do with this particular quality of manure, so that different rations will affect to some degree the consistency of the manure produced.

In any case, you can see their work in Fig. 10.2.

Dry digestion

We are going to refer over and over again to "dry digestion," or "dry slurry." Don't think that this means bone dry. It simply means less than 90% water. This is dry, because ordinary slurry is more than 90% water.

Studies (G.M. Wong-Chong, 1976, Schultze, 1958) show that dry anaerobic digestion (digestion above 10% solids) is entirely feasible, although most information on biogas generation continues to state that 7% to 9% solids is the necessary, indeed the only, range of concentration that can be used. This is at least partly because, at higher percentages of solids, not as much biogas is produced per pound of substrate as it is at lower percentages of solids. But, the question of what amount of biogas (per pound of substrate) is optimum is relative to other factors.

For example, some studies have shown that the optimum concentration, in terms of the total amount of gas produced from a certain weight of sewage, is about 2% solids. However, when 98% of the volume of the generator is given over to water, and only 2% is biogas-producing solids, that particular waterlogged generator will need to be about 5 times as large as a generator operating at 10% solids. At 10% solids, the "loss" in total gas production over a short run may be 6% to 10%, granted, but does it make sense to increase the generator volume by a factor of 5 to try to capture that extra bubble?

Because municipal sewage comes the way it comes (diluted below 10% solids), not much work has been done on more concentrated slurries. The work which has been done, however, suggests that successful anaerobic digestion at increasing solids concentrations can only take place when possible problems are recognized and preventative measures taken. There are four such problems: start-up, innoculation, concentration of toxins, and pH.

Start-up

The biogas process, as you remember, is, in very simplified terms, a two-stage process; acid digestion, and gas digestion. If something happens to disrupt the MF (methane-forming) bacteria, then the organic acids produced in the first stage begin to accumulate. When they reach a concentration of 3,000 to 4,000 *p*arts *p*er *m*illion (*ppm*), they begin to be toxic to the MF bacteria, whereas at 300 to 1,500 ppm, they are a necessary nutrient. In a similar way, we must drink water, but too much will drown us.

Too great a concentration of volatile acids can occur in dry anaerobic digestion when either a batch or a continuous-fed generator is started with inadequate seeding—without enough methane-forming bacteria to rapidly use up the acids.

When first starting a dry generator, there are three ways to avoid an overly acid condition. One is the use of a heavy seed—the use of 10 or more percent actively digesting dry slurry, mixed completely with the fresh dry slurry. Another approach is to gradually increase the solids content of the generator's slurry after an active healthy wet digestion (below 10% solids) is in process. Last, we can add a buffering material, such as lime, as described in Chapter 7 on pH. One, two, or all three of these solutions in any combination may be tried to help in the start-up of a dry generator.

Innoculation

In continuous-fed dry generators, where new material is being added daily, there is a problem with innoculation. With a 5% solids manure slurry in a well-stirred generator, the anaerobic bacteria will be moved from place to place, and chances are excellent that fresh material will soon be home base for a numerous and well-balanced population. In drier slurries, however, establishing a well balanced population of bacteria (innoculation) within entering or fresh slurry is more difficult. A great many of the AF (acid-forming) bacteria are facultative anaerobes, and, since they are thus able to survive in many environments, AF bacteria begin to thrive soon after the fresh material is introduced to the dry generator. The MF bacteria, on the other hand, are obligate anaerobes (so they are killed by the air) and they multiply more slowly in any case. So fresh material, unless innoculated, will soon go sour (turn acidic) when added to the generator.

This may seem like the same problem mentioned earlier, yet they are a bit different. Both have to do with establishing active digestion, but different solutions may be practical. During start-up, it may be inconvenient to add a large amount of seed material, but it can be done. The problems involved in adding active slurry to every loading of fresh slurry can be quite a bit more complicated, since any means of accomplishing this must be mechanically incorporated into the generator. On the other hand, if a heavy seed is used only in start-up, it need not affect the generator's design.

Stirring is also possible, but, for the reasons pointed out earlier, this is not an excellent alternative. The buffers mentioned previously can be added to absorb the shock of the initial abundant production of acids.

G.M. Wong-Chong found an interesting phenomena which might, happily, solve the problem of innoculation in a dry generator. The gas escaping from the dry mass of slurry apparently fluffed it up, and moisture percolated into the bottom of the generator (through the porous substrate), washing the needed bacteria down into the fresh dry slurry, which was added at the bottom. When Wong-Chong observed water on top of the dry slurry, he took this as an indication that the process was not well established. Low gas production accompanied this and was another indicator of unfavorable conditions. After stirring, however, the process generally resumed and the fluffing/percolation started again, stabilizing the innoculation. This indicates that a dry generator should be designed with bottom loading. He suggested a non-confined screw auger for stirring (since such stirring was uniformly successful at solving the problem), and found that the digested dry slurry had a very stable nitrogen content, which is important for fertilization, as will be discussed in Appendix 3.

Toxins

A third problem of dry digestion is concentration of toxins (poisons), such as ammonia, antibiotics (from medicated feed or from veterinary treatment), or heavy metals, all of which are discussed in Chapter 13 in more detail. But, briefly, the solutions to the problem of toxins are three. (1) Don't add toxins; (2) change the form of the toxins; or (3) dilute the slurry. Of course, the last will bring us back to stodgy old wet (below 10% solids) anaerobic digestion, but then, dry digestion is not a cure-all.

Both Wong-Chong and Schultze found, in their studies of dry digestion, that a lower percentage of volatile solids were decomposed than the percentage that is usual for wet digestion (but neither apparently attempted pH control). In essence, this means that less of the potential biogas was produced in these studies under dry conditions than would be normal under wet conditions.

Nevertheless, the two areas of low temperature generation and dry digestion hold great promise, but much work needs to be done to show methods of realizing these promises. It may be that plant wastes substrates respond better to dry digestion than animal wastes substrates.

Once Again

Just to recap, then, 7% to 9% solids is generally recommended. Six percent solids and above makes for some difficulty in paddle stirring. Above 10%, pumping becomes increasingly difficult. With pH control, it appears that up to 20% solids in the slurry can be handled biologically, though use of this dry a slurry may require design modification in the generator. Over 20% solids results in decreasing the rate and total amount of biogas evolved for any given dry weight of manure substrates and is difficult at best.

Terms

Dry slurry: Above 10% solids.

Effective volume: That portion of the generator which is used, and/or how well it is used.

Fluffing/Percolation: The process described by one researcher which indicated healthy dry digestion.

Wet slurry: 10% solids or more dilute.

Questions

None

Problems

None

11: HRT and Loading Rate

These parameters are used exclusively in continuous-fed generators.

Loading Rate

Loading rate refers to the number we get when we divide the weight of volatile solids loaded each day into the generator, by the volume of the generator. A typical loading rate then, might be 2 grams of volatile solids per liter of generator, or 0.15 pound per cubic foot. (These are not equal quantities.) Loading rate is an important parameter, since it tells us the amount of food—raw substrate volatile solids—which enters the generator each day. The AF bacteria can break this food down rather rapidly, forming volatile acids, but unless the MF bacteria can use up these acids at the same rate they are formed, then the generator's pH falls (it goes acid), and it shuts down.

Clearly, wherever we haven't got a well-established population of biogas bacteria, our generator will not be able to handle a high loading rate (or, in other words, a lot of food). If we have a stable, well-established population in our generator, it can handle a larger loading rate—more food each day. At very high loading rates, the feeding has to be more nearly continuous (hourly). At lower loading rates, once a day is okay. Specific loading rates recommended for various manure substrates are found at the end of Chapter 16.

HRT

In a continuous-fed generator, the idea usually is to squeeze as much biogas out of the substrate as we can, in as short a time as possible. For example, when sewage is being digested, about half the biogas that can be generated in 90 days, shows up in the first 15 days (at 90°F, 32°C). The longer we leave the slurry in the generator (meanwhile adding more each day), the larger the volume of the generator has to be, and so the more it will cost. If we load a liter of slurry into a generator every day, and if the slurry stays in there for 40 days, our generator will have to have a capacity of 40 liters. If we only retain the slurry 20 days, the generator will only need a volume of 20 liters, and although it's half as large, it may still generate as much as 75% (three-quarters) of the gas the 40 liter generator will (at the same temperature and so on).

HRT, then, is Hydraulic Retention Time, or the average number of days a unit volume of slurry stays in the generator. When we're dealing with small-scale generators, the difference in cost between one hydraulic retention time and another in terms of generator size and cost won't be very great, but in large-scale generation, costs add up fast, since everything—tank size, heating equipment, agitation systems—has to be scaled up.

So, for small-scale, continuous-fed generators, we may decide on a hydraulic retention time of almost any length up to 30 or more days. For larger scale systems, we have to find the balance between several factors. When using HRTs of below 7.5 days at 35°C (95°F) or 12.5 days at 25°C (77°F), we begin to experience washout, a condition where the biogas bacteria are washed out of the generator about as fast as they can multiply, and so the population of bacteria is unstable. The lower limit then, of the HRT is the washout time.

The upper limit is a question of the economics, and this varies in each situation. Refer to the charts and formulas in the section on temperature, or refer to Appendix 4 for information helpful in determining when, under the conditions which apply to your proposed generator, you have reached the maximum gas production for the minimum possible investment. Municipal sewage treatment plants are generally designed for HRTs of 30 days for extra process stability, but we must remember that they are not designed with the economics of biogas in mind. Large-scale, high-rate mixed, continuous-fed, animal manure substrate generators generally have HRTs from 12 to 20 days (at 35° C).

One final factor must be considered; at a given loading rate and for a fixed size of generator, the only way to vary the HRT is to change the amount of water loaded into the generator along with the substrate. If we load some fixed weight of VS into each liter of generator each day, then enough water will need to be added to the substrate so that the final volume of slurry (substrate volume plus water volume) added per day equals the volume of the generator divided by the HRT. So, if we have a 100 liter generator, and an HRT of 10 days, we will load 10 liters of slurry every day. If, in the above situation, we change the HRT to 20 days, then we can only load 5 liters of slurry each day; but if we keep the same volatile solids loading rate as before, the only way to produce that change in HRT is to double the solids concentration of the slurry. (For high-rate mixed continuous-fed generators, this solids concentration generally does not exceed 10%.) Now, of course, since HRT and loading rate are design parameters, they will generally not change after a generator is operational, so the above example is used mainly to illustrate the relationships involved.

Review the chapters on Agitation, Percentage of H_2O, the Design section and Appendix 4 for further ideas and information related to HRT and loading rate.

Terms

HRT: Hydraulic retention time; the number of days an average unit volume of slurry stays in the generator.

Loading rate: A number derived by dividing the total weight of VS fed to the generator each day, by the volume of the generator.

Washout Time: The minimum HRT, based on the time required for the MF bacteria to replenish their numbers at a certain temperature.

Questions

None

Problems

Assume a generator of 100 liters:

1. If you use an HRT of 15 days, how much slurry do you load each 24 hours?
2. Using a substrate of 80% VS and a loading rate of 2.5 grams VS per day per liter generator space, what is the HRT? (Assume that there are 20 grams of substrate solids in each liter of slurry at 10% solids.)

12: Agitation

Tests have shown that when the slurry is well mixed, an increase of 10%-15% may be expected in gas production (Sierp, 1931). However, it should always be remembered that energy invested must be exceeded by energy gained, just as dollars spent must be exceeded by dollars earned. Otherwise, as in the case of an electrically powered agitator which costs as much or more in electric Btu than we gain in methane Btu, we are simply buying electricity and turning it into biogas, at a loss.

Sierp pointed out that the CO_2 component of biogas is more soluble in the slurry than the CH_4 component, and that greater pressure will cause even more CO_2 to be dissolved in the slurry. Thus, at the bottom of the generator, where the pressure is greatest, a good deal of CO_2 will be dissolved in the slurry, and not much CH_4. Sludge at a depth of 33 feet contains dissolved gas which is 96% CO_2, and 4% CH_4 and according to Sierp this is a major source of the increase in biogas in generators where all the slurry is active. (Whether or not the slurry can be active in the absence of agitation is another question.) As we stir the slurry, this dissolved gas is released, and the result is apparently that we will often get more biogas, but not as often will we get a great deal more methane.

Agitation also increases the complexity and operating cost of the generator because of the added equipment and energy required.

Why Agitate?

Well, there are several advantages to agitation beyond the possible increase in biogas production:

1. Agitation is one good way to control scum.
2. It tends to maintain a more uniform temperature throughout the generator. Because slurry is sluggish, and because heat moves through slurry only very slowly, agitation is one good way to spread heat evenly through slurry.
3. In spite of this chapter's opening statements, it is still true that agitation can promote a more rapid digestion, and a more regular gas production in larger generators. These generators can develop pockets where the slurry isn't doing anything. Agitation eliminates these pockets. Agitation helps keep the entire contents of large generators in active digestion through continued re-innoculation, dilution of localized volatile acid buildups, and the maintenance of uniform temperatures.
4. Agitation prevents some of the caking or depositing of sludge on internal heating pipes as compared to what happens when unmixed slurry is allowed to stand near heating pipes at the same temperature.
5. As a result of the release of CO_2 from the slurry, it can more easily maintain a proper (alkaline) pH. (Refer to Chapter 7 on pH for the reasons.)
6. Violent mixing can break sludge particles apart, increasing the effective surface area. This, of course, will depend entirely on what substrate is being digested.

None of these items will be of overriding advantage in small generators, since they do not have the same problems as larger generators. Again, it will depend on your unique situation if agitation will prove beneficial. Design questions related to agitation are discussed in the section of that name.

Agitation that is truly continuous—24 hours a day—has, for some reason, a depressing effect on the biogas buddies. Intermittent agitation, for example agitation for 15 minutes each hour, is much better tolerated.

Terms

None

Questions

Why agitate?

Problems

None

13: Toxins

Toxins are poisons, and, as with any other living creatures, anaerobic bacteria can be slowed down or killed by certain toxins.

NH_3

The most common of these toxins is ammonia, NH_3. Some points about ammonia were discussed earlier in Chapter 7 on pH.

We learned there that a little ammonia may not harm, but rather can help the biogas process by helping to control pH, and that too much ammonia is toxic to the process. The line cannot be sharply drawn between a little and too much, since the bacteria can be conditioned to accept different levels of ammonia, by a gradual increase in its abundance. The general level of ammonia toxicity is around 2 grams per liter of slurry (2.67 ounces per gallon).

Any change in the conditions of the generator—temperature, pH, type of substance—within certain broad limits, is less disruptive to the biogas process when it occurs gradually. A gradually increasing dose of ammonia will result in a change in the metabolism or the nature of the population within the generator, so this kind of change is better tolerated than a shock introduction of a large dose of ammonia.

"Intensive animal production" is the flesh factory kind of farming—*e.g.*, put as many units as you can in a small space and pump them with food and chemicals to get the greatest weight gain in the shortest time. This is what happens when economics controls (overwhelms?) biology. Manure collected from intensive production often contains a lot of ammonia, since both feces and urine are generally collected in these operations. (It often contains antibiotics as well, since disease is a fact of life under such conditions.)

Pastured animals do not give a manure as high in ammonia, not only because of their different physical situation, but as well because they are not often fed the kind of high-protein rations which confined animals are, and these high-protein (processed) food rations also contribute to the high ammonia level of intensively produced animal manure.

Poultry do not urinate; they deliver both feces and urine in one load, and this means that even non-confined poultry manure is generally higher in ammonia than other manures (horse manure, for example). Besides poultry, pigs also have a manure high in ammonia, although collecting their urine and feces separately will decrease this to some degree.

Finally, if you wish to use your own excrement for biogas production, the addition of your urine may greatly add to the toxic stress of the generator. Urine from any source is fairly high in ammonia (or urea, a related chemical). Anytime we collect urine, we collect ammonia.

Toxic Mechanism

Toxins operate by what is called a toxic mechanism. This is the mechanical, chemical, or biochemical process by which a toxin acts.

Diatomaceous earth, used for filtration in swimming pools, is toxic to insects. Apparently what happens when insect larvae come in contact with it is that the extremely fine abrasive (like sandpaper) particles in the diatomaceous earth draw the water out of their bodies. They dry out and die. This is an example of a toxic mechanism which is largely mechanical in its action, as far as we know.

The toxic mechanism of ammonia may have a great deal to do with pH, since the pH of undiluted ammonia is 11. As the concentration of ammonia increases, the pH rises, unbalancing the biogas process. This is a chemical toxic mechanism.

However, there is evidence that the bacteria involved in biogas production can survive fairly high short-term pH values without great harm, as long as these pH values are not held high by the addition of alkaline materials. While some of the toxic effect of ammonia may be due to changes in the pH or buffering capacity of the slurry, it is more likely the toxic effect of ammonia is due to a biochemical toxic mechanism.

In many biological processes, or at least those involving only a single organism or type of organism, the byproducts of the biological process act as toxins to the process. The creatures which carry out the process, in other words, begin to be poisoned by their own wastes.

Alcohol, for example, is produced by the fermentation of sugar by the yeast. The concentration of alcohol in the sugar-yeast brew begins to poison the yeast at about 13%, and kills the yeast before it reaches 17%.

In biogas production, the ecosystem of the generator prevents the accumulation of toxic byproducts. An ecosystem is a living system which supports not one, but many different kinds of life. In a biogas generator, this means that one bacteria's waste is another bacteria's food. The AF (acid-forming) bacteria of many different kinds, with many different food preferences, break very complex molecules into the simpler molecules of organic acids, alcohols, CO_2, H_2 and other chemicals with strange sounding names.

If the C/N is low, meaning that there is an overabundance of N, then ammonia will also be formed. Otherwise, much of the N is used up in the formation of protein. (Protein is 16% N, generally.)

The organic or volatile acids, the alcohol, and other chemicals, and the H_2 are nearly all used by the MF bacteria to make biogas. Some of the CO_2 passes off in the biogas, some stays in solution, some is used as a source of carbon for the CH_4.

The ammonia, however, which doesn't show up in great abundance in a generator which has a balanced C/N, apparently acts as a toxic byproduct rather than a food when it is more concentrated. In a similar manner, sugar is an excellent food for bacteria. However, honey, with its very concentrated sugar content, will not support bacterial life. As with other toxins, we can do three things to modify the toxic effect of ammonia: dilute it, change its chemical form, or avoid adding it to the slurry to begin with.

Sometimes a higher ammonia concentration will be tolerated if the generator has gas recirculation (without removal of CO_2) since the added CO_2 can help buffer the high concentration of ammonia.

Urea, another molecule common in urine, is also toxic to bacteria. However, with sufficient CO_2, and in the presence of enough carbon-containing materials, urea (H_2NCONH_2) can be broken down into ammonium bicarbonate, an alkaline chemical, which can be buffered by more CO_2:

$$H_2NCONH_2 + 3\ H_2O + CO_2 \rightarrow 2\ NH_4HCO_3$$

If strong, positively charged ions are introduced into the generator (such as those naturally occurring in brackish water), these will reduce the toxic effect of ammonia, and possibly of urea. Ordinary table salt (NaCl) has been used (Cassell and Anthonisen, 1966) at the rate of 0.41 grams per liter of slurry, to combat ammonia toxicity. The sodium (Na), however, has an adverse effect on soils at these concentrations when the effluent is used for fertilizer, unless abundant water is used for irrigation and the soil is of light texture. Other chemicals which might work—due to their similar molecular nature—are potassium chloride (KCl), or other molecules with ionic bonds, or calcium or magnesium compounds, and these might not so easily damage soils.

Other Toxins

Other toxins to the biogas process are obvious and probably expected. Pesticides harm the bacteria. Dilution or avoidance are the only cures. Herbicides have the same effects as pesticides.

Antibiotics used in animal food or injected into the animals, can cause a bad reaction in a generator using that manure as a substrate. While the *antibiotics* (meaning: against life) themselves may soon disappear in the animal's digestive tract or bloodstream, they apparently leave more lasting byproducts which harm the biogas bacteria. Avoid antibiotics, since many experiences indicate that when a generator fails because of antibiotic toxicity, it must be cleaned out and restarted.

Heavy metals, which are often the consequence of industrial pollution, harm the biogas bacteria. The principal culprits are Cr (chromium), Cu (copper), Ni (nickel), Zn (zinc), and Hg (mercury). Of these, copper is commonly used in compounds to combat fungus, such as fungus on grapes. Zinc is used in galvanizing buckets, pipes, and other metal devices exposed to weather or water. Probably the zinc would not appear in the generator unless a galvanized surface were directly exposed to organic matter, as with galvanized slurry pipes. Use plastic or clay pipe. Anything acidic next to a galvanized surface, will leach the zinc into itself. Dilution or avoidance are the only cures.

Synthetic detergents have also proven harmful. If you plan to use a biogas generator as your sewage disposal system, there are several ways to approach this. If the house to which the biogas septic tank is to be attached is as yet unbuilt, consider installing several drainage systems for various water-using appliances. The hot water from the bath or shower, which, as water, adds nothing to the actual production of biogas except the increased cost which results from the larger volume needed, may serve to help heat the generator when used in a heat exchange unit. On the other hand, toilet water should go into the generator. (However, see Chapter 16, Manure Substrates for more detailed information on this.) Some sinks have a disposal separate from the main sink, and this kind could easily be hooked up to drain dish water to the heating system, but shunt disposal wastes to the generator. Many good publications are available on these and related ideas. See, for example, McGill University's, *Stop the Five Gallon Flush*, or publications and magazine articles on grey water. In any case, with reference to toxins, make sure to watch your use of toxic cleansers (such as Comet with chlorinal), chemicalized soaps, detergents, shampoos, toilet bowl cleaners, caustic drain cleaners, and so on. If codes will not allow aboveground disposal of sewage, consider installing the generator where it can be easily heated and serviced (in a basement?) and run the effluent out through an ordinary leach line system or into your septic tank.

Terms

Intensive animal production: Flesh factory or feedlot animal production.

Toxic mechanism: The process by which a poison, poisons.

Urea: A N(itrogen)-containing chemical, related to ammonia.

Questions

None

Problems

None

14: Pressure and Surface Area

Pressure

There is little or no information available about the effect of increasing gas pressure on the biogas process. For low-technology, simple applications of the biogas process, self-generated gas pressure is an obvious plus, since the generator itself would then provide sufficient pressure (as a result of gas evolution) to push the biogas through pipes from source to use, and give the rate of gas flow needed in whatever application (stove, lights) it is being used.

It is undoubtedly true that increasing pressure above the slurry will retard the formation of gas, but to what degree, why, and how this affects the composition of the gas all remain experimentally unknown. Much more information about pressure in general, its measurements, and the pressures needed for various application, appears in the Uses section. See also Chapter 38.

Since increased pressure will increase the amount of CO_2 dissolved in the slurry, it will affect the pH by producing carbonic acid. The effect of increased pressure, we may then presume, can be partially overcome by the addition of a buffering agent, possibly lime.

Meanwhile, one reference (Babbit and Bauman, 1958) states that pressures allowed above the slurry should not exceed 15 to 18 centimeters (6-7 inches) of water. There is also a report which states that one plant was operating, with apparent success, at a pressure of 25 centimeters (10 inches) of water (Sewage Works Journal, June 1954). Pressure is often measured in terms of centimeters of water, or inches of water. This conventional way of stating pressure comes from the most common low-cost method of measuring pressure: the open-tube manometer.

As you can see in Figs. 14.1 and 14.2, the fluid inside the tube is being forced up toward the atmospheric or air pressure by the greater pressure of the gas coming in from the left-hand side of the tube.

As the water rises in the right-hand side of the tube, gravity pulls it down, and so its weight, and the force of ordinary air pressure, balance the force of pressure from the gas. If the gas exerts too much pressure, it will blow the water out of the tube. If the gas pressure is equal to atmospheric pressure, the water will be level across both sides of the tube.

In Fig. 14.2 you can see that the gas pressure is greater than the air pressure, and the difference is measured in centimeters or inches of water. It is important to understand and remember that with an open-tube manometer, the measures are of biogas pressure over and above atmospheric pressure.

Table 14.1 relates centimeters (and inches) of water to other kinds of pressure measurements (millimeters of mercury and pounds per square inch).

Surface Area

The finer a substrate is shredded, ground, or pulped, the easier the digestion process will be. Any given substrate will digest more rapidly, and possibly even more completely, when taken apart into little bitty bits. Of course, the energy it takes to shred a substrate will generally need to be subtracted from the energy gained in biogas. Ingenuity may come into play here. A wind-powered shredder is a possibility. (Use a high-torque, low-speed machine such as a water-pumper or Savonius rotor). Another possibility is to put a stubborn substrate (cornstalks?) in a place where cars, tractors, or animals will travel over it. Straw that has been wet and then sun-dried is more brittle and easier to crush.

For larger scale attempts, possibly a silage machine would produce the desired results, if the trade-off seems worthwhile.

Terms

Centimeters of water:
Dynes per square centimeter:
Inches of water:
Millimeters of mercury:
} Different Ways of expressing pressure measurements.

Open-tube manometer: A low-cost device for measuring pressure.

Pounds per square inch: Pressure measurement.

Questions

None

Problems

None

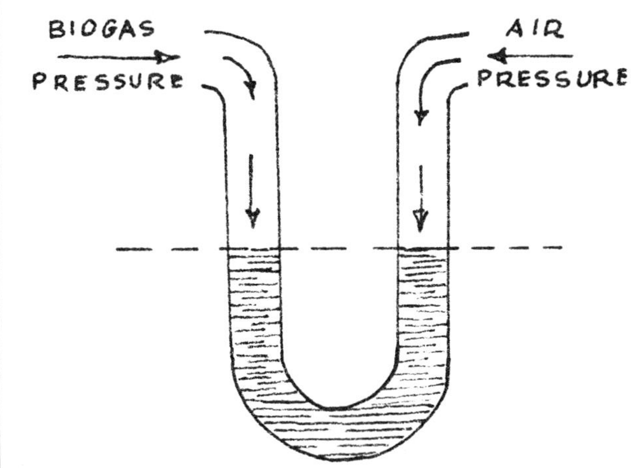

Fig. 14.1 Open-tube Manometer

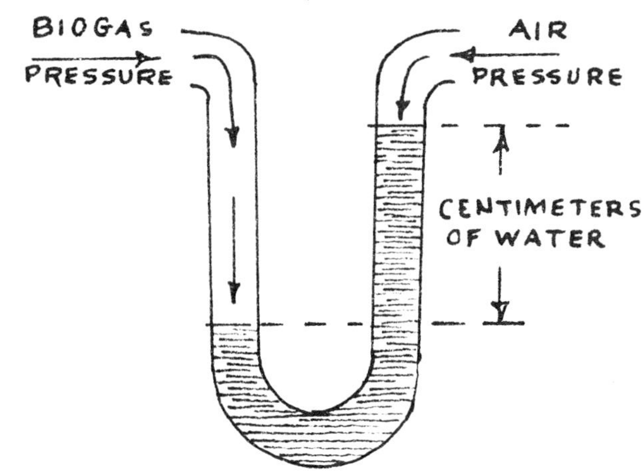

Fig. 14.2 Gas Pressure Greater than Air Pressure

cm H_2O	in. H_2O	mm Hg*	psi
1	0.39	0.74	.014
1.27	0.5	0.93	.018
2	0.79	1.47	.028
2.54	1.0	1.87	.036
3	1.18	2.21	.043
3.81	1.5	2.80	.054
4	1.57	2.94	.057
5	1.97	3.68	.071
5.08	2.0	3.74	.072
6	2.36	4.41	.085
6.35	2.5	4.67	.090
7	3.0	5.15	.100
8	3.15	5.88	.114
10	3.94	7.36	.142
10.16	4.0	7.47	.145
12	4.72	8.83	.171
12.70	5.0	9.34	.181
15	5.91	11.03	.213
15.24	6.0	11.21	.217
20	7.87	14.71	.284
20.32	8.0	14.95	.289
25	9.84	18.39	.356
25.40	10.0	18.68	.361

*Hg is the chemical symbol for mercury.

Table 14.1 Pressure Measurements Correlated

Section III

Substrates

In most instances when people keep livestock, they feed them according to the kind of animals they are. In the biogas process, it is just the reverse. Our livestock, biogas bacteria, are developed depending (among other factors) on what we feed them. We're not concerned with pedigrees here.

If we have leaves, we feed the generator leaves, and the bacteria we get may be different than those we would get if we had manure and fed that to the generator.

In this section, we'll talk about different materials, how they respond differently to the biogas process, and the different problems with each. To start with, we'll discuss purified substances, such as cellulose, protein, sugar, and so on. These substances don't occur naturally in isolated forms. They must be produced by physical or chemical means from other naturally occurring materials. Sugar, for example, comes from a refining process in which raw sugar beets or sugarcane are cut up, boiled down, bleached, and processed until they come out as white sugar, molasses, and pulp. Generally, unless you live in a very special situation, or are interested in digesting the by-products of such industrial processes, this sort of information will not be directly useful to you. It may be indirectly useful, however. Paper, for example, has a great deal of cellulose in it, and so information on how cellulose digests can be helpful if you are working with paper. Sugarcane has a lot of sugar in it, and the same sort of comment applies.

Remember, it's not possible to accurately predict how a particular material will respond to digestion merely because we know its chemical composition. The biogas process is a biological process, and biological processes tend to be rather individualistic and often unpredictable. Following Chapter 15 on chemically purified substances are Chapter 16 Manure Substrates, Chapter 17 Plant Substrates, and Chapter 18 On Growing Substrates.

Chapter 16 and 17 contain tables that provide very specific information on the digestion of particular kinds of manure and plants.

15: Purified Substances

Carbon

Carbon is an element, a substance familiar to most of us as the building block of the material in charcoal, lead pencils, and diamonds. Carbon is also the basic element of all molecules associated with life, such as methane, acetic acid, glucose (ordinary sugar), and so on.

Many of the purified substances which we can decompose anaerobically yield a greater weight of gas than the original weight of substrate. This is because water, H_2O, enters into the reaction and adds hydrogen to it. This is the major source of the hydrogen atoms that show up in CH_4. The theoretical chemical equation for the transformation of carbon into methane shows us an example of this. In words, the equation runs: Two carbon atoms plus two water molecules yield one methane molecule plus one carbon dioxide molecule.

In terms of those chemical equations which you have come to know and love:

$$2C + 2H_2O \longrightarrow CH_4 + CO_2$$

$$24 + 36 \longrightarrow 16 + 44 = 60$$

Below the chemical equation are some numbers. These represent the relative weights of the molecules, and show that, even when the atoms which make up the molecules change partners, the weight on either side (60) remains the same. Each atom of carbon has a relative weight of 12 units, each atom of hydrogen has a relative weight of 1 unit, and each atom of oxygen has a relative weight of 16 units. Thus, each molecule of water has a relative weight of 18 units (1 +1 +16), and each molecule of methane has a relative weight of 16 units (12 +1 +1 +1 +1), and each molecule of carbon dioxide has a relative weight of 44 (12 +16 +16). The end products have a total relative weight of 60 units, or 2.5 times the weight of the original carbon.

In this example, which is very unlikely to occur in real life (impossible?), we put one gram or pound of carbon into an anaerobic and biologically active environment, and get 2.5 grams or pounds of biogas after the carbon has been completely digested. This translates into 1,870 cubic centimeters of biogas *volume* per gram of carbon substrate fed into the generator. In Americanese, this is 30 cubic feet of biogas per pound weight of substrate. These figures (as unlikely as it may be that we will ever find ourselves feeding carbon to our generator), are important, since they represent a maximum gas weight and gas output for any substrate. Ordinarily, our figures will be much lower for two reasons. (1) Only a particular percentage of any substrate is even theoretically available for decomposition, and that is the VS, volatile solids. (2) Only a certain percentage of the VS will decompose in any reasonable amount of time.

Thus, on the average, we can expect only 1,250 cubic centimeters per gram or 20 cubic feet per pound of biogas volume per unit weight of sewage sludge (VS) decomposed. If we measure on the basis of TS, then, our "volume per pound" measure drops even further, to around 1,060 cubic centimeters per gram or—cubic feet of biogas per pound (TS) of sewage sludge. And, remember, sewage sludge is easy to digest.

That's the picture then. The maximum possible biogas evolved per gram of VS decomposed is 1,900 cubic centimeters. Since it is a bit more than highly unlikely that 100% of the VS will be decomposed, then per gram of VS added, anything from 350 to 1,500 is ordinarily to be derived from VS.

It's important to notice the great difference between reporting biogas production per unit weight decomposed and per unit weight added. Suppose that each gram of VS decomposed produces 1,000 cubic centimeters of biogas. The big question then becomes: what percentage of VS decomposed? If only 50% of the VS is decomposed by the time the slurry leaves the generator, then we would only get 500 cubic centimeters per gram VS added. If only 10% decomposed, the figure would be only 100 cubic centimeters.

Carbohydrates

Carbohydrates are described by their name, since they contain *carb*on; *o*xygen, and *hydr*ogen. (Although, technically speaking, they need only contain carbon and oxygen.) Sugar, starch, cellulose, alcohol, and lignin are all in the class of chemicals known as carbohydrates.

In plants, cellulose, starch, hemicelluloses, and that part of the crude fiber which will decompose give about 110% of their weight in biogas, at 50% CO_2, 50% CH_4 (Buswell and Boruff, 1932). The most important of these carbohydrates, cellulose, is discussed next.

Cellulose. Cellulose is probably the abundant "kind" of organic molecule. It is here called a *kind* of molecule since, to some degree, the particular process which is used to extract this purified substance from its original home (such as a cotton fiber) determines how much will be gained. For this and other reasons, it cannot be said to be a specific molecule.

Both green garbage and garden debris contain about 80% cellulose dry weight. Much paper, for example Kraft paper, the kind of paper from which grocery bags are made, is also pure cellulose. Cotton is 91%, flax is 82%, ramie is 85%, jute is 65% to 75%, kapok is 55% to 65%, and wood is 40% to 50% cellulose dry weight. Cotton is the purest natural source of cellulose. Cellulose is 44% carbon dry weight, and it contains no nitrogen, two general facts which may help determine C/N ratios when we know something of the chemical analysis of a particular substrate.

When cellulose is anaerobically decomposed, 890 cubic centimeters of biogas per gram of cellulose decomposed are produced (14 cubic feet per pound). For pure cellulose, this evolved biogas is 50% CH_4, 50% CO_2, but since some of the CO_2 remains dissolved in the slurry, the percentage of methane may be higher in the gas we actually collect. (When lime or limewater is added to the slurry, this further increases the amount of CO_2 dissolved in it, and thus the percentage of methane given off by a lime-buffered slurry is higher than in a non-buffered slurry.) As with carbon, the weight of gas given off by a certain weight of cellulose exceeds the weight of the cellulose. One gram of cellulose, completely digested, will give us 1.11 grams of biogas. This happens because for each molecule of cellulose decomposed, one molecule of water gets into the act and adds to the weight of the biogas. When reduced to a formula, it looks like this:

$$C_6H_{10}O_5 + H_2O \longrightarrow 3CO_2 + 3CH_4$$

cellulose + water ——> carbon dioxide + methane

162 + 18 ——> 132 + 48 = 180

Pure cellulose takes 35 days at 21°C (70°F) to completely (90%) decompose. Natural celluloses, those which have not been altered chemically and are still found where the plant put them, often take a good while longer to decompose, and they may give not one, but two peaks of gas production, the second coming as many as 120 days after the start of anaerobic decomposition. When either natural or chemically purified cellulose decomposes, it tends to produce an acid environment. This is not because cellulose produces more acids as by-products than some other materials, but rather because the bacteria which accomplish this breakdown have a more difficult time producing an adequate buffer system. That is, they cannot produce alkaline chemicals to balance the acids as easily with cellulose as with, for example, proteins. The difference is largely due to the fact that cellulose has no nitrogen. When using cellulosic materials, such as paper, leaves, and other plant and wood products, watch the pH carefully, and, as insurance, add a buffer.

Not only does cellulose completely lack N, but it also lacks phosphorus (P). Both of these elements, plus others, are necessary for the bitty buddies which do the work, so we must add them in some form, either chemically or organically (with substrates which contain them) to substrates high in cellulose. For example, in the digestion of paper, the addition of lime and P (phosphorus) has been shown to be helpful. B vitamins, and minor amounts of molybdenum (Mo), cobalt (Co), or magnesium (Mg) seem to stimulate the decomposition of cellulose, or cellulosic materials such as paper.

Whereas any C/N from 20 to 50 (depending on the substrates used) seems to be all right with the anaerobic bacteria, any C/P (carbon-phosphorus ratio) from 160 to 420 (160 up to 420 C atoms for each P atom) will provide enough phosphorus (SERL 70-2).

Cellu*lose* is broken down by a group of enzymes, known as cellu*lases*. These enzymes are found in the guts of termites, snails, silverfish, and ruminants (cows, goats, buffalo). They are also produced by molds, fungi (of which mushrooms are part), lichens (those peculiar growths found on rocks, very hard, often very colorful), and seedlings, as for example the barley seedling. It may be possible to use this knowledge in helping to stimulate the breakdown of cellulose and cellulose-containing substrates within the generator, by adding sources of these enzymes.

When using rumen (cow gut) extract to decompose cellulose, Hall and Cheng (1955) found that many of the B vitamin group stimulated this decomposition. Biotin and B_{12} (two of the B vitamins) were synergistic. That is, they seemed to stimulate the process best of all when they were both present, as compared with the stimulatory effect of either one alone. These researchers also found that yeast extract stimulated the process better than any mixture of B vitamins with the same analysis. This, they concluded, might have been due to some factors found in the natural B vitamins (e.g., those in the yeast), which are not found in purified, or chemically produced B vitamins. Natural B vitamins are produced when legumes (alfalfa, pea) sprout, or when lactose (milk sugar) is fermented by lactobacillus (milk bacteria). Yeast is widely used by the brewing industry, and sometimes the yeast cakes which are left after the brew has been filtered are available from breweries.

While it doesn't make sense for us, either economically or in terms of energy, to go out and buy chemical additives, it may be that, being aware of the capabilities of these bacteria, we can find some natural, low-dollar-cost, low-energy-cost way to reproduce more favorable conditions in our generator.

It's well known that there are enzymes in papaya which promote digestion. Would these be useful in biogas production? Are there other fruits, herbs, or vegetable enzymes available in different climates which could be used, with a minimum of preparation, that might stimulate biogas production?

These are questions of appropriate technology, and as such, they haven't been answered by work that has been done in the past by people focused on the problems of large-scale biogas production. Maybe you'd like to work on these problems.

Lignin

We've talked about lignin before, particularly in the Chapter on C/N. As you may remember, lignin is the material which gives support to land plants. The stronger and sturdier the plant, the more lignin it has in it. Logs have more than leaves, stalks have more than blades.

Lignin not only holds the plant upright, but also helps keep the plant from decaying, and when it finally has broken down some, it forms compounds which hold the soil together and keep soil nutrients from being released too soon into the environment. Altogether, it's a tough customer. Lignin seems to exert some biological or chemical effect on the plant molecules with which it is associated to stop them from decomposing like they would in its absence. This effect has been hard to study since when lignin is chemically extracted from the plant, it seems to have an especially depressing effect on anaerobic decomposition.

For example, when we chemically tear an innocent cornstalk down into its constituents—cellulose, lignin, pentosans, fats, and miscellaneous juices—and then we put all these things back together again in a generator and sit down to wait... nothing happens. When we leave the poor cornstalk alone, it can decompose nicely, and on analysis, we find that some (about 20%) of the lignin has decomposed. Chemical lignin will not decompose, but natural lignin, left alone just where nature put it, will, to some degree. However (even if we don't manipulate the lignin chemically) if our substrate has a high analysis of lignin (e.g., mature bracken fern), then we can assume that it will not perform well in a biogas generator. This may be one of the reasons that fir sawdust does poorly in biogas generators.

Sugars, Starch, Alcohol

Laura and Idnani (1971) did research in which they tried to stimulate the anaerobic decomposition of cow manure by the addition of common materials. The purpose of this was to allow farmers in India to utilize these materials, if possible, to obtain more biogas from their biogas plants. They obtained results which seem to indicate that while canesugar was readily decomposed (in 8 days at room temperature), it did not, in the manner used, stimulate better or more complete decomposition of the cow dung.

Starches are digested into sugars, and sugar most often into alcohol. This is the process by which grain, potatoes, and various other starchy plants, or honey, fruit, and similar sources of sugars, are turned into alcohol. Alcohol can be made easily into biogas, and if the above results reveal a general pattern, it can be very rapidly made into biogas. Substrates high in sugar can be expected to have an acid tendency, so a buffer should be added.

Fats

Fats form an interesting group of molecules, giving "the best quality and the greatest quantity of gas" of any of the purified substances (Buswell and Boruff, 1932). They are very important, since they form the major food source for the MF bacteria; therefore, we will discuss them in some detail.

Fats are expressed in chemical analysis as "ether soluable matter" (that's diethyl ether ($C_2H_{52}O$), and include fatty acids, grease, oil, soaps, and so on.

The very simplest fats produce slightly less biogas weight (90% plus) than the original weight of the fat when completely decomposed, but the more complex fat molecules, of higher molecular weight, produce more than 100% of their weight in biogas.

The maximum weight gain of biogas (as a percentage of the weight of the original molecule) which is theoretically possible from fatty acids is 164%. The maximum percentage of CH_4 is 75%, or a ratio of CH_4:CO_2 of 3:1. Since CO_2 is more soluable than CH_4, the biogas finally evolved from some particular slurry actually composed of fatty acids, will show a higher percentage of CH_4 than these theoretical figures.

The general equation for the anaerobic decomposition of fatty acids, which can give us the above information (if we know the chemical formula) is:

$$C_nH_{2n}O_2 + \frac{n-2}{2}H_2O = \frac{n+2}{4}CO_2 + \frac{3n-2}{4}CH_4$$

(Buswell and Boruff, 1932)

Name of acid	Chemical Formula	Weight of biogas, % of original weight	% CH_4 in biogas	AKA
Formic	CH_2O_2	80	25	Methanoic
Acetic	$C_2H_4O_2$	100	50	Ethanoic
Propanoic	$C_3H_6O_2$	112	58	Propionic
Butanoic	$C_4H_8O_2$	121	63	Butyric
Pentanoic	$C_5H_{10}O_2$	127	65	Valeric
Hexanoic	$C_6H_{12}O_2$	131	67	Caproic
Heptanoic	$C_7H_{14}O_2$	135	68	
Octanoic	$C_8H_{16}O_2$	138	69	Caprylic
Nonanoic	$C_9H_{18}O_2$	140	69	Pelaigonic
Decanoic	$C_{10}H_{20}O_2$	142	70	Capric
Octadecanoic	$C_{18}H_{36}O_2$	151	72	Stearic

Table 15.1 Fatty Acids *(Adapted from Buswell and Boruff, 1932)*

More technically, the formula works only for saturated monocarboxylic acids.)

Therefore, in the case of octadecanoic acid;

$$C_{18}H_{36}O_2,\ \text{“}n\text{”} = 18$$

and the specific equation is:

$$(C_{18}H_{36}O_2 + 8\ H_2O = 5\ CO_2 + 13\ CH_4)$$

Notice that there are ten oxygen atoms on one side, ten on the other. Also, carbon atoms balance carbon atoms (18), and hydrogen balances hydrogen (52 atoms on both sides). This means the molecular weight on both sides is the same, and so the weight of the biogas evolved must exceed the weight of the octadecanoic acid by exactly the weight of the water molecules added.

The molecular weight of octadecanoic acid (1 atom of C = 12, 1 atom of H = 1, 1 atom of O = 16) is 284. The molecular weight of the water molecules is (8 x (1 +1 +16)) 144. To prove that the weight on both sides is the same, we can find the weight of the CO_2 molecules, which is (44 x 5) 220, and the weight of the CH_4 molecules, which is (13 x 16) 208. Then we have:

$$284 + 144 = 220 + 208$$

$$428 = 428$$

This means that the weight of the biogas generated is approximately (428 ÷ 284 x 100) 150% greater than the weight of the octadecanoic acid. If you got through all that, you are now a certified chemist.

Soap and Scum

The fatty acids, when exposed to alkaline materials like sodium hydroxide (NaOH), calcium hydroxide (CaOH), or magnesium hydroxide (MgOH), react to form soaps, or in chemical terms, "the alkali salts of fatty acids."

Although sodium soaps are soluble (they dissolve in water), the calcium and magnesium soaps are much less soluble, and so they precipitate (they come out of solution) and they form an important component of scum. Scum, the portion of the slurry which floats on the liquid portion, forms a dense hard mat if left alone, which will eventually completely stop gas production in a biogas generator. These insoluble soaps, and various greases and oils, bind the other materials in the scum together and make it more difficult to break up the scum. In a similar way, wet hair, caught on a screen or in a trap in a bathtub drain, gathers oils, greases, and soaps, and begins to stick together in a way that clean wet hair would not.

These insoluble soaps, and their friends, the greases and oils, if allowed to dry, repel water, and so they are difficult to wet again and decompose, after having been dried.

Although fatty acids, in the presence of Ca^{++} ions, or Mg^{++} ions, form insoluble soaps, these acids are nevertheless extremely important in biogas production. Even if we did not want them to be present in the slurry, they would still be formed in the process of disassembling more complex molecules, for the volatile acids passed on from the AF bacteria to the MF bacteria are mostly fatty acids. Further, some researchers, working with unbuffered sewage sludge, have found that 70% of the CH_4 evolved comes from the acetic ion, the ion which is half of acetic acid. So, whatever disadvantages the fatty acids have, we are stuck with them.

In fact, in any generator, one of the surest methods of charting the health and progress of the biogas process is to follow the changes in the concentration of volatile or fatty acids. Some details on how to do this are found in the Appendix on Analysis. Because the concentration of these acids is so critical in the biogas process, some books will tell you never to add acid to a generator. It's true that the so-called mineral acids (HCl, hydrocloric or stomach acid, and H_2SO_4, sulfuric acid, to name two) can indeed disrupt the biology of the process very quickly. However, the fatty acids, such as acetic acid (found abundantly in vinegar) when added with great care, can be used by the generator just as well as any other substrate. If we buy these acids as chemicals, it is an expensive way to make biogas, but it will work.

The loading rate for fatty acids depends on three things: (1) the rate at which the MF bacteria use up the volatile acids; (2) the rate at which volatile acids are being produced in the generator; and (3) the overall level of volatile acid concentration, which should not exceed three grams of acid per liter of slurry. If it begins to rise up to and above this level, then trouble is brewing for the small biogas bacteria.

If you wish to experiment with fatty acids as substrate, use an established culture from a generator that was fed (preferably) garbage or manure, feed the acids continuously in small (drip...drip...drip...) amounts, and use high-rate or constant mixing.

Oils and Greases

Oils and greases are decomposed only if they come from animal or vegetable sources. Petroleum or other mineral oils and greases, slow or stop the digestion process, depending on their quantity and type. They interfere with digestion either by coating slurry particles so that enzymes manufactured by the bacteria cannot penetrate the particles and do their work, or by poisoning the bacteria directly.

Due at least partly to the proverbial difficulty of mixing oil and water, digestible oils and greases tend to take longer to decompose than most purified carbohydrates.

Protein

The last group of purified substances which needs exploration is proteins. Nearly everyone has heard of protein, but very few have any idea of what it is. Proteins are composed of simple building block molecules, called amino acids, and these simple building blocks are used to make what are sometimes very complex molecules.

Proteins generally have more nitrogen than other kinds of molecules, so that where cellulose, a carbohydrate, may not contain any significant amount of nitrogen, a protein molecule is generally 16% nitrogen by weight.

An average analysis for a particular type of protein is:

53% C
23% O
16% N
7% H
1% S ("S" is sulphur.)

From this, we can see that the C/N of pure protein often is very low, at 3 or 4.

When added to a biogas generator, protein begins a rapid breakdown and produces abundant volatile acids; more acids, in fact, than either the carbohydrates or the fats, but due possibly to the excess of nitrogen, the bacteria which feed on the protein are able to produce an equalizing abundance of alkaline molecules, and so the pH in a protein-fed generator generally remains high (alkaline).

Gas production of proteins is comparable with that of cellulose, either per gram added, or per gram decomposed. When mixed, protein and carbohydrates may have a stimulating effect on one another, for the former has a low C/N, and produces a high pH, while the latter tends to have a high C/N and to produce a low pH.

Terms

Amino acids: The building blocks of protein.

Carbohydrates: Compounds containing carbon, oxygen, and hydrogen.

Cellulases: Enzymes which help decompose cellulose.

Cellulose: Main compound in wood and paper.

Cubic centimeters per gram VS added: A measure of biogas production.

Cubic centimeters per gram VS decomposed: A measure of biogas production.
hydrogen.

Fatty acids: Major food source of the methane bacteria.

Soaps: The alkali soaps of the fatty acids.

Questions

Which of the general categories of substances described would probably need added buffers?

Problems

None

16: Manure Substrates

Cow Manure

Cows are ruminants, a general designation for a group of four-legged, fur-bearing, cloven-hoofed, milk-giving, cud-chewing, horn-possessing animals. The most interesting characteristic of ruminants is the fact that they have such an unusual digestive system.

Briefly, a cow will graze a pasture, swallowing most of what she collects largely unchewed. Later, when the animal is resting, small portions of her previous meal will be elevated to her mouth and chewed. This is why a ruminant, resting in the shade, can be seen to repeatedly chew and swallow. When swallowed the second time, the meal is further digested and passes into her stomachs (she has four) and intestine. Once there, as in other digestive systems, anaerobic bacteria help the cow make the most out of her meal.

Ruminant digestion allows the animals two great advantages: (1) they can stay for shorter periods of time in open meadows (where there is greater risk of attack) gathering a meal, and can leisurely chew and digest it in more protected surroundings; (2) ruminants are able to make use of foods which other animals cannot, due to the low-protein content of these foods. Cows transform low-protein straw and grass into high-protein milk and meat. We might expect from this that the manure of a cow would be high in carbon (because the original food was high in carbohydrates like cellulose) and low in nitrogen (because the cow extracted much of the nitrogen in the form of protein). However, cows generate a tremendous amount of gas (yes, biogas) thus dissipating much of the carbon, and much of the nitrogen collected by the cow is eventually returned to the soil either in the manure or the urine. (Some nitrogen, of course, is in milk, meat, hide, and hair, as protein.)

The result of this is that a cow eating from a pasture with an overall mixed plant C/N of between 40 and 60, gives a manure with a C/N of about 25 or 30.

Obviously, some of the potential of the plant material for biogas production must have been used by the cow, since she has digested the material anaerobically and aerobically, before we get to use it in biogas production. In fact, cows generate an average of 155 liters of CH_4 (or a greater volume, in total biogas production) per day. Mrs. O'Leary's cow—the one that started the great Chicago fire—didn't really need to tip the kerosene lamp over. All she needed to do was get close to it.

At any rate, because the cow's food has been digested already, a dry kilogram of manure will usually generate less gas than a dry kilogram of whatever the cow was eating. The difference is not as striking over the short haul as it might be however, since the cow has helped to make some components of the plant matter more available, even while she used up other components. The cow manure is generally digested more rapidly than plant matter, probably because the cell walls of the plant material in the manure have been broken down, and thus bacteria can more easily attack what is there. Also, as we mentioned previously, the cow adds certain biochemicals that the anaerobic bacteria apparently like.

In a very interesting experiment (see Table 16.1) done by Laura and Idnani (1971), these workers tried adding different materials to cow manure to stimulate its ability to produce biogas. The idea was that the cow manure, (which under ordinary conditions will only lose about 30% of its volatile solids in the process of biogas production), could produce more biogas per kilogram dry material from the volatile solids which are usually not decomposed, and yet the spent or finished slurry would not as a result lose much of its worth as a fertilizer. More gas, no loss in fertilizer. Much of the value of the experiment came from the fact that common materials (of the sort considered available to Indian farmers) were used. The manure used (they refer to it as dung) was without urine, and the three most successful additives in terms of increased volume of biogas were canesugar plus urea (a simple nitrogen-containing molecule), canesugar plus lime, and urine.

Oddly, the addition of either leguminous leaves alone (peas, alfalfa) or non-leguminous leaves alone did not stimulate biogas production very much, although there was some result. The addition of canesugar alone, or what they refer to as "sarson oil cake" alone, or filter paper (essentially pure cellulose) alone, had no effect on the total amount of gas produced. Ashes and charcoal both reduced gas production, charcoal rather dramatically. (Although some researchers claim that activated charcoal helps city sewage digestion and gas production.)

These treatments also had an effect on the composition of the gas. The sugar plus urea and the non-leguminous leaves had a great effect on increasing the percentage of CH_4, followed closely by urine, and then by sarson oil cake. But the highest percentage of CH_4 in the biogas was produced by canesugar plus lime. Cellulose (filter paper) reduced the percentage of CH_4 in the biogas. (This may have been due to adverse pH changes).

Based on these results, we can assume that lime, leaves, urihe, sugar (or substrates high in sugar, such as sugar beets, if well shredded), or the residue from an oil-extraction process, such as safflower oil cake, will stimulate the digestion of cow manure.

If the manure was collected with urine, and lime added to the generator, this would probably be the least expensive and least problematic method of stimulating decomposition. Well-shredded materials, such as leaves, sugar-containing substrates, or oil cake, added to the generator, might increase scum problems, but would further stimulate decomposition and therefore biogas production.

Material	Proportions (material : manure)
Sun-dried leaves	1 : 20 to 1 : 5
Air-dried sugar – containing substrates	1 : 20 to 1 : 5
Fresh oil cake residue	1 : 250 to 1 : 50

Table 16.1 Stimulating the Digestion of Cow Manure

The proportions in Table 16.1 are ball park suggestions, and could well vary in any particular situation depending on the availability of materials, but they include the proportions upon which Laura and Idnani's experiments were based.

In discussing biogas production from animal manures, many authors use the term, "animal unit," which, translated, is 1,000 pounds of live flesh. If a cow weighs 1,400 pounds, she is 1.4 animal units. (One hundred ducks is one animal unit, assuming each duck weighs about 10 pounds.) In metric, one animal unit is around 450 kilograms.

The amount of biogas which can be evolved from each animal unit of cow (or any other creature) depends on many factors. The most important of these are:

1. Amount of manure produced per animal unit.
2. Feed composition.
3. Manner and completeness of collection (Dung plus urine? Bedding as well? Is all or only some of the manure collected?)
4. Percentage of volatile solids in manure.
5. Percentage of volatile solids destroyed by the generator under normal operating conditions.
6. The amount of biogas produced per each unit weight of volatile solids destroyed.

Since each of these factors can vary widely, widely varying amounts and qualities of biogas can be produced from one animal unit of cow (or any other creature). Much of the available literature fails to take these factors into account, especially the most important factors of feed—kind, quality, and quantity—and collection.

These two factors are often related, as can be easily seen, for the cow keeper who pastures his animals will not be able to collect as much manure as the cow keeper who confines them, although the former may have healthier cows. Cattle kept to be slaughtered for beef are often fed high-protein feed supplements which lower their total manure output, but increase the percentage of VS in the manure available for decomposition. Dairy cattle are more often pastured, or fed hay and silage in the winter, which increases manure production, but lowers the percentage VS available.

Assuming, however, 100% manure collection and an average feed situation, whatever that may be, one cow unit will produce enough manure to make from 0.85 to 1.70 cubic meters (850 to 1700 liters, 30 to 60 cubic feet) of biogas per day. An average value hovers around 1.1 cubic meters (1,100 liters, 39 cubic feet) per day. In continuous-fed generators, production will hover around 900 cubic centimeters per gram VS decomposed. However, not more than 60%, and often not more than 10%-15% of the VS added will decompose, resulting in low gas production. For example, for 20% VS destruction, gas production would only average 180 cubic centimeters per gram VS added. Notice the difference between this figure and 900 cubic centimeters.

Human Wastes

Excrement is, for many people, a squeamish subject. Many swearwords and bad jokes make reference to this material. But this is a cultural phenomenon, since other peoples have virtuous regard for human excrement as a soil amendment (fertilizer). F. H. King, a truly visionary professor of agriculture, noticed in the early 1900s that our agricultural practices were robbing the soil of fertility, and, being also somewhat philosophical, he noted a connection between civilizations and the way they cared for their farmlands.

Some civilizations are the Fourth of July variety, rocketing upwards to a political or technological peak, and then just as suddenly fading away, a whisper on the wind, sand blowing around some colossal ruin. Other civilizations outlast the centuries, and Professor King felt that at least one reason they did so was that they practiced an ecologically sound agriculture. Translated, this means that whatever originates from the soil must return to it.

So he traveled to China, which was much further away in the days before ICBMs, and there he found many fascinating things which he describes in his book, *Farmers of Forty Centuries.* Farmers in China regarded excrement as an extremely valuable fertilizer. They had no sewers; chamber pots filled with that day's excrement were put outside the doors, and men pushing carts collected it at night. The Chinese equivalent of the city fathers would accept bids from businessmen for the collection rights in various sectors of the city, and those businessmen who made successful bids hired the cart pushers and sold the wastes to farmers outside the city.

Because of such systems, some Chinese farm soil had borne healthy crops for four thousand years, and because of the lack of such systems, American soil has become depleted in one hundred years. New and ever more marvelous and expensive technology has delayed by a few years the final death of much of this country's soil. (Soil is alive, and it can be killed. Then it becomes dust and concrete.) Of course, the problem is more complex than this. We need to change a lot more than our manner of sewage disposal. City sewage in America is not always a safe fertilizer as it has often been contaminated with industrial chemicals and heavy metals. Further, home re-use of human wastes brings with it—if improperly done—the risk of parasites and disease.

Small-scale (1 to 20 people) use of excrement in a biogas generator can be an excellent way to complete the circle of life. That is, by returning our own wastes to the soil via the anaerobic compost process, we can return fertility to the soil. It may, however, be a relatively poor source of energy (biogas).

While the danger is not great in most (not all) areas of the United States or Britain, it must be remembered that improper composting of human excrement, and careless use of the effluent can be a factor in the spread of diseases. The list of diseases and parasites (such as tape worm) which can be thus spread is impressive. Some of the more common diseases and parasites which can be spread are:

baccilary and amoebic dysentary	coccidiosis
cholera	swine erysipelas
vibrosis	ascariasis
leptospirosis	cysticercosis
	tape worm infection
infectious hepatitis	facioliasis
tuberculosis	schistosomiasis
brucellosis (undulant fever)	typhoid
	listeriosis

A review of the literature shows that there is rather extensive use of anaerobically digested city sewage on agricultural lands, with few or no reports of disease or infection. This might be due to the fact that in all, or nearly all of the cases reported, those people who were using sewage effluent were aware of its hazards, and thus took adequate precautions.

Agriculturally speaking, sewage effluent disease hazards, if they exist, can be reduced or eliminated by: (1) not using sewage sludge around or upslope from a well; (2) not spraying or overhead irrigating with the diluted sludge or supernatent liquid; and (3) not using the effluent on root crops or crops bearing food near the ground.

If effluent is to be used at all on human foods, the safest use is furrow irrigation of diluted effluent under fruit or nut trees. If effluent is used on pasture land, there is sometimes a risk of cross-infection from parasites which can inhabit either animls or man.

In the small-scale situation, the dangers are not as great, if and only if, the use of sludge for fertilizer is not general in that area and foods grown on effluent are not widely traded, sold, or otherwise available. These diseases and parasites do not arise spontaneously, and if you and your family are disease- and parasite-free, and if no outside infection occurs, then use of effluent for agriculture should be safe.

Probably the safest agricultural practice for suspect effluent is to grow a crop for aerobic compost, or to use it strictly for ornamental plants. You may have a hard time, however, explaining to your neighbors why your flowers are so especially brightly colored.

Biologically speaking, these diseases can be controlled by two agencies—heat and other organisms. In the aerobic compost environment, the temperatures generated and the organisms present kill all parasites and pathogens within several hours of the time when maximum temperatures are reached. If the compost pile is improperly made (for example, if some materials are too dry or too wet) or if the outside of the pile is not put on the inside when the pile is turned two weeks or more after having been built, then infected materials may still be infected when the compost is used.

In the anaerobic compost environment, the heat is not generally great enough to kill the parasites and pathogens, and so disinfection occurs mainly by biological antagonisms over a period of time—the bitty buddies slug it out to the finish.

Complete safety is had after six months (!) of anaerobic composting which, in a small-scale situation, need not represent a tremendous volume of storage.

We are not recommending the agricultural use of home sewage effluent, but at the same time, we recognize that many people want to know about this. (Our main worry is not sewage, but the fact that this is a "sue age".)

Besides the agricultural and biological precautions which can be taken, there are medical precautions as well. These include vaccinations and practices similar to those necessary for a trip to the Far East. Consult your physician or a travel agent about these.

One adult on an ordinary diet will produce from 100 to 250 grams of feces per day. On a vegetable diet, an adult will produce from 300 to 400 grams per day. (Respectively, 0.22 to 0.55, and 0.66 to 0.88 pounds per day.) Feces are usually neutral to slightly alkaline in pH, 24% to 27% TS (dry weight), with a C/N of 6 to 10, nitrogen 4% to 6% of TS, VS is 85% of TS. Normal values for urine are 1 to 1.6 liters volume per day, average pH 6.0, 4% to 6% TS, with a C/N of 0.8, nitrogen 15% to 18% of TS, VS is 72% of TS. (That's 1.06 to 1.69 quarts volume produced daily). Every liter of urine weighs about 1,020 grams. Every quart of urine weighs about 2.9 pounds.

Studies (such as Snell, 1943) suggest strongly that while feces introduced into an active generator will digest readily, urine and feces, undiluted, will not digest. The problems are several. Ammonia toxicity for human urine is not as significant as urea (H_2NCONH_2) toxicity. Ammonia nitrogen only accounts for 4.6% of the total N in urine, while the nitrogen found in the urea molecule accounts for 84% of the total N in human urine. But urea breaks down into ammonia, and so more often than not, only ammonia toxicity is spoken of, and the precautions and remedies for either urea or ammonia toxicity are the same in any case.

One possible solution to this problem is gas recirculation to increase the amount of CO_2 available to the slurry (see Chapter 13). Another solution is the addition of water to dilute the toxins. Biogas production is completely inhibited by a concentration of 2350 to 3500 parts per million (by weight) of urine nitrogen (Snell, 1943). Ordinary undiluted urine must be mixed with about 4 times its own volume in water to bring the urine nitrogen concentration below this point. Therefore, if someone plans to use a high solids (undiluted) excrement slurry in a home sewage biogas generator, urine cannot be added in a very great amount.

For further discussion of such questions and their economic importance, see Chapter 50.

Biogas production from human feces is good, averaging 500 cubic centimeters per gram VS added, at 70% CH_4. Notice, however, that one or two people don't produce a great whopping volume of excrement. For this reason not much biogas can be expected out of a generator fed solely on home sewage (about 1 cubic foot per day per person).

Pig Manure

Due to the fact that the pig is such a useful animal, providing bacon, sausage, ham, and pickled pig's knuckles, there are a great many pigs in both the United States and Britain. Biogas research has occurred using pig manure as a substrate because of the economic importance and the large population of pigs. Therefore, information specifically about pig manure is available whereas information about llamas, peacocks, gerbils, elephants, gnus, and the like is totally absent.

The problem with the numbers generated by this research, however, remains the same. Since type of feed and confinement, etc., vary so much from one situation to another, variations in any particular characteristic of the swine manure will arise. Most authorities agree, however, that with mesophilic (35°C), continuously-mixed, continuous-feed digestors, a loading rate of 4.0 grams of volatile solids per liter of generator volume (0.25 pounds per cubic foot), as a maximum, with 3.2 grams per liter (0.20 pounds per cubic foot) as a relatively easily obtainable average value. (Continuous-fed generators should be started on lower values, and the loading rate gradually increased as the generator stabilizes, or becomes more able to accept higher loading rates.)

Hydraulic retention times of 10 to 20 days under the above conditions could mean a variation in the solids content of the slurry, of from 8% to 3%, in a fixed size of generator.

Increasing the solids concentration while maintaining the same volatile solids loading rate should allow an increased retention time, and thus more biogas per gram of manure. This, of course, is what we want. The main problem with this approach, though, is the generally high ammonia concentration of pig manure. When the ammonia concentration rises above 1,800 parts per million (about 1.8 grams of ammonia per liter of slurry, or 0.11 pounds per cubic foot), then the biogas bacteria suffer, although they wil be more successful in producing biogas at a high ammonia concentration if they are introduced to gradually increasing concentrations. This is done by slowly increasing the solids concentration of the slurry which is fed to the generator over a period of time. In the absence of the equipment needed to determine the PPM (parts per million) concentration of ammonia in the slurry, the percentage CH_4 in the biogas or possibly the total amount of biogas evolved can indicate the general health of the biogas bacteria.

In a well edited and highly informative book, *Energy, Agriculture and Waste Management*, Fisher, Sievers and Fulhage, in their contribution to the book (pp 307-316), give some figures on an experiment with a 100 gallon generator (13.2 cubic feet, 374 liters) using pig manure (feces plus urine) from hogs fed "a typical "finishing ration" of 14% corn/soybeans". Their figures should not be taken as applicable in all situations, as the biogas process seems to vary so much from situation to situation. However, they indicate the ball park. Under various conditions (changes in loading rate, etc.) they found an average of 636 cubic centimeters per gram VS added, and an average of 58% VS destruction.

Poultry Manure

Chickens, among the poultry, are the main source of manure for biogas production, because more than any other poultry they are "intensively raised". Large numbers of turkeys are raised in the United States as a result of the peculiar festival known among the natives as Thanksgiving, but these turkeys are not always intensively raised—e.g., indoors where their manure can be collected.

There are however, individual situations where geese, ducks, doves, pigeons, and other birds are raised in "fowl" situations which allow some or all of their manure to be collected and used in a biogas generator. For the most part, these manures are comparable to chicken manure in their behavior in the biogas generator, and so we will discuss only chicken manure and assume that it covers the others.

Chickens drop both feces and urine in the same load, so the composition of the manure will not vary as a result of feces only or feces and urine collection, as there is no difference. The composition of the manure and its response to anaerobic digestion will vary depending on diet, degree of confinement, and whether the birds are kept in wire-bottomed cages or on litter.

One unique system of manure collection under wire-bottom cages consists of glass plates onto which water is fogged and windshield wipers which scrape the wetted manure into a collection trough (L. John Fry, 1974). Another method under wire cages is to let the droppings accumulate on concrete and collect them periodically by shovel or mechanical means. With this second method, some ammonia nitrogen, valuable in the final fertilizer byproduct of the biogas process, is lost by evaporation. One final interesting method (although not a collection method) applicable to any animal kept on litter or bedding, is to shred the bedding straw in one central location and transport it pneumatically (via air pressure in tubes) to the pen or stall. This has the advantage for biogas generation of adding carbon to balance the C/N, and of resulting in a short fiber length substrate which is ready for continuous-feed generation (if the generator is properly designed).

Chickens in intensive production situations are often fed medicated feed, which is likely to disrupt the biogas process. Certain antibiotics have a definite "life"; in other words, a certain number of days after injection or ingestion beyond which they no longer are effective. If medicated feed is being used, try an experiment using the manure after the prescribed number of days for the chemicals involved, if it will not digest when it is fresh. If the manure will digest when well-seeded, and after the medicine is supposed to be ineffective, then use it; but not otherwise. If a generator is "killed" by manure from medicated animals, it can generally only be re-started by completely replacing the contents.

When digesting poultry manure, Hart (1963), found an average of 765 cubic centimeters per gram VS destroyed, but a low VS destruction (average) of 32% brings this to 247 cubic centimeters per gram VS added. Savery and Cruzan (1972) reported a much more encouraging 560 cubic centimeters per gram VS added.

Summary

A very complete source of information on the anaerobic digestion of livestock wastes to produce methane, is a book of that name. If you are interested in setting up a continuous-fed generator operating on livestock wastes and you want a lot of information on this subject send $2.00 to J. A. Moore, Agricultural Engineering Department, University of Minnesota, St. Paul, MN 55108. The book is not an attempt to explain the mysteries of biogas production. It's an annotated bibliography, with sometimes cryptic and frustrating references, but which overall has much good information.

To predict, in a general way, the amount of biogas available from your animals, you will need to know the amount of manure produced, and the expected gas production per weight of manure. Table 16.2 gives general averages.

When you see a chart like Table 16.2, neat and tidy, and very certain looking, you should immediately be suspicious. Where did the author get his figures, and what assumptions has he made?

The chart is based on averages of values reported in many places, and a bit of extrapolation. It assumes 100% collection of manure plus urine, and a stable and adequate digestion process. Figures in the first column are not averages, but rather can be considered as theoretical maximums. That is, if 100% of the VS is decomposed, and each gram thereof produces biogas at the maximum reported amount per gram VS decomposed, then we will be able to achieve this figure; not otherwise.

Frankly, the figures are of little use, if we expect precision. For example, since the gas production figures are averages, they lie somewhere between what we might expect from a continuous-fed generator with a short HRT and what we might expect from a batch-fed generator. Because of the long HRTs involved, batch digestion generally yields more biogas per unit weight of volatile solids than does continuous digestion. Notice also the high average percentage of VS destruction for poultry manure. Such a figure could not have come from studies where ammonia toxicity, due to high percentage solids in the slurry, was a problem.

Most of us can expect to come somewhere near these figures. If your gas production is well below these aver-

Source	Mcc/gm	%	Acc/gm	Kg	m³	Cal x 10³	gm/l	l/da
Cattle, beef	900	48	458	2.7	1.2	6.8	1115	23.6
Cows, dairy	800	31	248	4.0	1.0	5.2	1020	37.8
Human	1060	58	610	0.5	0.3	1.6	1010	10.0
Pig	1100	49	538	2.7	1.6	8.2	945	28.3
Poultry	600	61	367	5.9	2.2	11.3	960	28.3

Mcc/gm: Normal maximum (not average) cubic centimeters of biogas produced per gram of VS destroyed.

%: Percentage VS destruction.

Acc/gm: Average cubic centimeters of biogas produced per gram of VS added.

kg: Kilograms of VS produced per animal unit (450 kg) per day.

m³: Cubic meters of biogas produced per animal unit per day.

Cal x 10³: Thousand Calories of net heat (methane) energy produced per animal unit per day (assumes 65% CH_4 in biogas).

gm/l: Density of manure, feces, and urine; grams per liter.

l/da: Volume of manure produced per animal unit per day, in liters.

Table 16.2 Biogas from Manure

Source	cc biogas/ gm VS added	m³ biogas/ animal unit/day	Cal x 10³ methane/ animal unit/day (assumes 65% CH_4 in biogas)
Cattle, beef	458 cc/gm	1.2 m³	6.8 Cal x 10³
Cows, dairy	248 cc/gm	1.0 m³	5.2 Cal x 10³
Human	610 cc/gm	0.3 m³	1.6 Cal x 10³
Pig	538 cc/gm	1.6 m³	8.2 Cal x 10³
Poultry	367 cc/gm	2.2 m³	11.3 Cal x 10³

Table 16.3 Biogas Production Figures

Source	Normal maximum cc biogas/ gm VS destroyed	% VS destruction	kg. VS produced/ animal unit/day
Cattle, beef	900 cc/gm	50%	2.7 kg.
Cows, dairy	800 cc/gm	30%	4.0 kg.
Human	1060 cc/gm	60%	0.5 kg.
Pig	1100 cc/gm	50%	2.7 kg.
Poultry	600 cc/gm	60%	5.9 kg.

Table 16.4 Manure and VS

Source	Manure, feces, urine density gm/l	Volume manure produced/ animal unit/day
Cattle, beef	1115 gm/l	23.6 l/da
Cows, dairy	1020 gm/l	37.8 l/da
Human	1010 gm/l	10.0 l/da
Pig	945 gm/l	28.3 l/da
Poultry	960 gm/l	28.3 l/da

Table 16.5 Manure Statistics

Source	HRT (days)	VS load rate (grams VS per liter of generator each day)	TS load rate (grams TS per liter of generator each day)	Comments
Cattle, average	15	3.8	4.5	10 day HRT
highest	15	8.0	9.6	
Human, average	25	1.8	2.5	Municipal sewage plants,
highest	10	4.7	7.9	Morgan 1954
Pig, average	13.5	3.2	3.8	
highest	15	4.0	4.8	
Poultry, average	14.6	3.0	3.6	

Table 16.6 Loading Rate and HRT

(Note that "highest" refers to the highest loading rate; this is not always accompanied with the shortest retention time. The above assumes VS = 80% TS, except for human TS loadings, where the figures are taken from Morgan, 1954.)

ages, then your process is probably not healthy.

Now we'll give you average and minimum reported successful HRTs, and average and maximum reported loading rates for different manures—because we promised. The success of a particular combination of HRT and loading rate can be measured in many ways, among them, percent VS destruction. Many studies (but not all) show a peculiar pattern: when the HRT is decreased and the loading rate increased, percent VS destruction goes up. We might expect that if the anaerobic bacteria had less time (a decreased HRT), and more food (an increased loading rate), that they would respond with a smaller VS destruction rate, rather than greater, as is often the case.

Here too, the particular factors involved are open to speculation. Obviously, as HRT is decreased and load rate increased still further, the process becomes unstable and the generator will become stuck in the acid production stage. However, each situation seems to have slightly different optimums, and it is wise to experiment. These numbers are useful primarily for establishing the ballpark, as before. You may be able to do better, or not as well, depending on your generator and situation. (We recommend figuring your loading rate on the basis of TS, rather than VS, for your own use, since VS is hard to determine without the proper equipment, and since it's only tangentially useful anyway.)

All of the studies from which these figures were taken were for continuous-fed generator of below 10% solids slurry and held at 35°C, or thereabouts. Agitation varied from some to continuous. Higher loading rates and shorter HRTs seem more successful if:

1. Agitation is regular—say 15 minutes per hour.
2. Feeding is more or less constant (e.g. hourly) as opposed to once daily or several times a week.
3. Ammonia toxicity is dealt with: by dilution, or by gradual acclimatization, by CO_2 recirculation, or by adding chemicals to combat the ammonia. See Chapter 13.

With this in mind, let's look at what has been reported in the literature in Table 16.6.

Terms

Ruminants: Four-legged, fur-bearing, cud-chewing, milk-giving, cloven-hoofed, horn-possessing beasties.

Dung: Manure, usually without urine.

Soil amendment: Fertilizer.

Questions

None

Problems

None

17: Plant Substrates

Information about the anaerobic decomposition of plant substrates is meager. This is primarily because practically no one out there thinks such research would have a significant economic impact. This is, upon analysis, a peculiar outlook. Each day, 17 times more energy is stored in plant matter by photosynthesis than the world now uses (Wilson 1976). We have no energy crisis. What we have is an overdependance on, and a shortage of, crude oil.

In the experiments done on anaerobic decomposition of plant wastes reviewed, the following materials were used:

- Artichoke Tops
- Banana Skins
- Barley Straw
- Bracken Fern
- Cabbage
- Carrots
- Cornstalks
- Dog Food
- Flax Straw Shives
- Garbage
- Grass Clippings
- Kelp
- Leaves
- Oat Straw
- Paper (all kinds)
- Papyrus
- Potatoes
- Rape Seed Cake
- Rice Straw
- Seaweed
- Water Hyacinth
- Wheat Straw
- Wood Wastes

Many of these studies used sewage sludge as the initial substrate, to get things going. Gas production figures, in these cases, are only near-guesses, since stable digestion using the plant waste alone was almost never achieved. So, unfortunately or fortunately, the figures given below, or at least most of them, are actually reflections of the amount of biogas produced by sewage sludge and the plant material, minus the gas produced by the sewage sludge when it was bubbling away by itself.

Unless stated otherwise, please assume that the following represent batch experiments conducted at mesophilic temperatures. Most of these experiments lack pH control, and some lack adequate seeding.

Notice carefully that gas production figures are given in several different ways, the most popular being VS added and VS decomposed.

Artichoke Tops

Nelson, Straka, and Levine (1939) experimented with artichoke tops ground so finely that they referred to the resulting material as a flour. (As other experiments show, grinding a substrate this finely can have a marked effect on increasing its decomposition). In these experiments, 573 cubic centimeters of gas were produced per gram VS added in 32 days. Ninety percent of that amount was available (from a seeded slurry) in 19 days. At thermophilic temperatures, 539 cubic centimeters were produced per gram VS added in 14 days, with 90% of the gas available in 6 days. Methane percentages, respectively, were 49.1 and 52.6.

Banana Skins and Stalks

Gas yield for banana skins per gram of TS added is reported (Buswell and Boruff 1933) as 360 cubic centimeters in 30 days in batch experiments. Use of the continuous-fed fibrous generator resulted in 250 cubic centimeters per gram TS added, or about 278 cubic centimeters per gram VS added. Some provision for scum breakup needs to be made.

Fowler and Joshi (1920) reported a miserable 41 cubic centimeters per gram TS added under the same conditions of temperature and time in batch experiments.

Barley Straw

Acharya (1935), experimenting with barley straw and a number of other materials which we will speak of below, showed that the addition of N helped stabilize the pH without the addition of chemical buffers, and the amount of biogas when N was added was 33% above that produced when digesting a barley straw slurry lacking the added N. With added N, gas production was 150 cubic centimeters per gram TS added, or approximately 175 cubic centimeters per gram VS added. Since only 40% of the VS was decomposed (biogas production at 440 cubic centimeters per gram VS decomposed), indications are that better yields are possible.

Acharya's studies were all of 6 months duration. Where he did not add N to his substrate, pH control was attempted with $KHCO_3$ (potassium bicarbonate) initially and NaOH (sodium hydroxide) subsequently.

Bracken Fern

Acharya also tried bracken fern, high in lignin (31% of TS). Extremely poor results were obtained. The natural lignin fraction is so high in mature bracken fern that it seems to depress decomposition.

Cabbage

Straub (1943) added various vegetable wastes to sewage sludge. He produced 792 cubic centimeters of gas per gram cabbage VS decomposed. The gas had an intolerable smell.

Carrots

Straub also tried carrots. Whereas carrots only produced 693 cubic centimeters per gram VS decomposed, the carrots decomposed more completely. Although he does not so state, it appears from his data that carrots out-produced the cabbage by giving 150% of the biogas per gram VS added that cabbage produced. In other words, much more of the carrot VS decomposed than the cabbage. CH_4 content for biogas from both vegetables was just below 60%.

Cornstalks

If any of you are considering using cornstalks in your generator, you're in luck. This seems to be the all-time favorite plant material for research. Buswell (1930) gave excellent and startling reasons for using this widely available substrate:

> The present estimate is that from 5 to 10 cubic feet of gas can be obtained per pound of cornstalks, and that the rate of production will be from 1/2 to 1 cubic foot of gas per day per cubic foot of tank volume. Taking the lower figure, a ton of cornstalks would furnish gas for 400 people for one day, allowing 25 cubic feet per capita per day. From the data given by Weber for yields from regions where 30 percent of the land is planted to corn, an area with an 8-mile radius will produce enough cornstalks to supply a city of 80,000 inhabitants continuously. In other words, the cornstalks from one acre will produce the gas for one person for a year.

Buswell refers to Weber, *Industrial Engineering and Chemistry*, 21, 270, 1927. Buswell's figure of 5 to 10 cubic feet per pound translates to 312 to 624 cubic centimeters per gram. We assume this refers to dry weight (TS) added.

Getting down to cases, we find that the amount of gas produced per gram VS added varies widely. Once again, however, Buswell and Boruff (1930) come through with the goods, one reason for the variations. When cornstalks are shredded and digested, a lot of scum is produced, which soon stops gas production. They show that by soaking cornstalks in limewater for four days (and then neutralizing the mass before digestion), a tremendous increase in gas production is experienced (1929).

Whereas our good friends Nelson, Straka, and Levine (1939) give commendably complete data, showing a yield of 360 cubic centimeters of gas from each gram VS added in 30 days at mesophillic temperatures, Buswell and Boruff completely upstage them with an indicated (approximately) 620 cubic centimeters of gas per gram VS added over a period of 50 days, or approximately 480 cubic centimeters in 30 days, for the soaked cornstalks.

The digestion of presoaked, shredded cornstalks resulted in an estimated 60% VS reduction, whereas shredding without soaking apparently gave only about 35% VS reduction in Buswell and Boruff's 1929 study.

A most fascinating generator, designed specifically for continuous digestion of cornstalks and related materials, is presented in Buswell and Boruff 1933. In their experiments, this generator and a cornstalk substrate gave 142 cubic centimeters of gas per gram VS added. These figures are not as good as they achieved in batch experiments, but the HRT which is not given in this paper, may have been short.

Dog Food

In a series of interesting but poorly reported experiments, researchers at UARL (United Aircraft Research Laboratories 1974) used Purina Dog Chow as a start-up material on small experimental digestors. They report no gas production figures; however, the idea is valid. That is, for a substrate for a small experimental or demonstration digestor, this material may serve to limit substrate variations so that other parameters can be explored. The vital statistics of P.D.C. are listed in the C/N table.

Flax Straw

Seed flax straw was tested by Nelsen, Straka, and Levine (1939) and flax straw was tested by Buswell and Boruff (1933). Buswell and Boruff reported 300 cubic centimeters biogas produced in 20 days per gram added. If we assume this is TS, and take the VS of flax straw at 90% (close enough), this would be 333 cubic centimeters per gram VS added. No figures are given on percentage VS decomposed.

Nelsen, et al, reported the same 333 cubic centimeters per gram VS added, produced in 34 days. 63%

was produced in 6 days. (See Appendix 4.) At thermophilic temperatures, 242 cubic centimeters were produced in 21 days, 63% in 3 days.

Flax shives decomposed by the Messers. B. and B. produced 210 cubic centimeters per gram TS added, in 20 days.

Garbage

In general usage, this word means organic waste, rather than trash composed of glass, metal, and the like. The numbers reported below must be considered very general, since whatever it is that you call garbage may differ markedly from what these researchers called garbage. Of course, the same admonition applies to any category or kind of substrate we may discuss, but the ball park marked out is probably the same one you'll be playing in.

In batch experiments, Keefer and Kratz (1934) found that garbage digested well in proportions of 1 to 1 with sludge effluent. They found a variation of from 600 to 700 cubic centimeters of biogas produced per gram of VS added. Digestion was considered complete in 30 to 40 days.

Trying again (1934), this time with continuous-fed small generators, they experimented with variations in the loading rate. Although it's difficult to be sure from their figures, it seems that they varied this parameter from 0.23 grams VS per liter of generator to 1.14 grams VS per liter. VS digestion varied between 45% and 42%, at the different loading rates. Biogas was produced at 962 and 614 cubic centimeters per gram VS decomposed, or 433 and 258 cubic centimeters per gram VS added. All these figures are for an undefined mixture of sewage sludge and garbage, but, at a guess, it may be that they were using a 1:1 ratio as before.

Researchers found that garbage substrate, continuous-fed generators were able to handle high loading rates better than sewage-fed generators (SERL 69-1). Garbage-fed generators had greater population diversity and a greater buffering capacity. Working with what they called "green-garbage," they found that even feedings of 100% garbage could be maintained for a long time without adverse effect. It is probable that no paper was included in the garbage.

Grass Clippings

Acharya (1935) worked with what he called "lawn mowings," and he found that with the addition of N, the grass gave 4% less biogas, but, just a shade more total CH_4. (Adding N seemed to shift the volatile acids content away from acetic towards butanoic— butyric —acid.) He found about 315 cubic centimeters of 59% CH_4 biogas produced in 6 months per gram VS added. VS percentage of grass he lists at 74.4%.

SERL in its Third Annual Report (70-2) tells us that grass clippings seemed to resist digestion when the waxy surface of the grass had not been bruised. (Drying may help break down this surface.) When that surface was broken, however, the grass was easily digested, using a slurry of 50% grass, and 50% of either chicken manure or sewage sludge. These experiments were conducted with continuous-fed generators. Grass proved 73% (of TS) digestible, producing an average of 487 cubic centimeters per gram of VS added (743 cubic centimeters per gram VS digested) for the grass-sludge mix. The grass-chicken manure mix fared less well, with the figures (respectively) at 368 and 800 cubic centimeters. Obviously, although the grass-chicken manure produced more biogas per VS gram destroyed, VS destruction was not as great—45.5% as compared with 66.2% in the grass-sludge mix. Good quality biogas was produced in both cases, of nearly 70% CH_4. The relative failure of the grass-chicken manure slurry to perform better is not explained by the authors of the SERL report. It may be that the C/N ratio was low, but insufficient data are given to be sure.

Kelp

Several, if not many studies have been done on the anaerobic decomposition of Macrocystis Pyrifera. However, only one has been obtained for perusal prior to writing this book, done by United Aircraft Research Laboratories, hereinafter referred to as UARL, (1974). The study was interesting; these researchers not only tried using kelp in a continuous-fed freshwater generator, but in a seawater generator. The substrate was pure kelp. The seawater generator was gradually acclimated to increasing concentrations of seawater, and a stable and successful ecosystem was established in one of the generators thus converted. Gas production was in the neighborhood of 530 cubic centimeters per gram TS added. The freshwater generator averaged about 580 cubic centimeters per gram TS added.

David Chenowyth of the Gas Technology Institute has done studies on this substrate, and he reported that low concentrations of N were a problem with a completely kelp-fed generator. Digestion was unstable, and gas production was low. Also, one of the main components of kelp, mannitol, a sugarlike chemical, rapidly decomposed to produce a volatile acid "burst". Thus, continuous feeding was almost a necessity with kelp. The addition of N, on the basis of Chenowyth's reports, would seem to be wise. He reported that production of Macrocystis Pyrifera was estimated to be in the area of 75 tons per acre per year.

Leaves

Few authors seem to regard leaves as suitable material for digestion studies. The author has satisfactorily generated biogas with unshredded white oak leaves, using urine as the N source, at 16% solids. No measurements of gas production were taken.

Hussey, Row, and Allison (1934) reported 243 cubic centimeters per gram TS added. Leaves which do not do well in aerobic compost, such as eucalyptus or walnut, will probably not do well in biogas production.

Oat Straw

In the decomposition of oat straw, Acharya found that N seemed to very slightly depress the production of biogas. Oat straw without N added (but with buffering) produced 172 cubic centimeters of gas per gram VS added. Volatile acid production was not matched by biogas production, a sure sign that the MF bacteria were unhappy.

Paper

Many studies are available on the anaerobic digestion of paper of various kinds. For example, SERL (69-1) tried digestion studies on mixes of sewage sludge and kraft paper pulp. Kraft paper is what paper bags are made of, and it's about 99% cellulose. High grade toilet paper or filter paper are also essentially pure cellulose. As you may remember, cellulose is 44% C, 0% N. In the studies under discussion, mixtures of 50% pulp, 60% pulp, and 68.2% pulp were tried. They had gas production, respectively, of 805, 755, and 693 cubic centimeters per gram VS decomposed (this includes VS of sewage sludge). Cellulose breakdown was over 90% in the 68% mix, but the generator got stuck and failed, probably because the C/N ratio was so high—just over 50. With pH control, this mix of 68.2% kraft paper pulp may have worked better.

Further research was done with kraft pulp and chicken manure. With chicken manure making up the remainder, the slurries were 50% and 69.3% pulp. The respective C/Ns were 30 and 60. Both mixes averaged 65% VS destruction. Data are lacking for gas production figures.

Also tried were other mixes with increasing percentages of kraft paper pulp, the highest being 74.3% pulp, 25.7% chicken manure, C/N 75:1. Their conclusion was that a mix of 70% pulp, 30% chicken manure would digest satisfactorily.

In another annual report (70-2), SERL reports studies done on newspaper. Newspaper is much more resistant to decomposition than kraft paper, and they report their newspaper as being about 75% ground wood pulp and 25% kraft type paper. It assayed at 88.5% cellulose, C/N at 800, N at 0.05%, C at 40.6%. Again using slurry mixes of paper and sewage sludge, mixes of 10%, 20%, 30%, and 50% newspaper were tried. The generator receiving 50% newspaper was not agitated, and the remaining 50% of the slurry was chicken manure. This generator achieved only 256 cubic centimeters gas production per gram VS added, but 774 cubic centimeters per gram VS digested, indicating that VS digestion was low. For the slurries with 10%, 20%, and 30% newspaper plus sludge mixes, the figures were 618, 549, and 468 cubic centimeters biogas produced per gram VS added.

It seems evident that—at least with the particular substrates used in the SERL studies—the chicken manure plus paper mixes could tolerate a much higher C/N (around 45) without adversity than could sludge and paper mixes. Newspaper is not as well digested as kraft paper.

In the digestion of paper, the addition of lime and P (phosphorus) has been shown to be helpful. B vitamins, and minor amounts of molybdenum (Mo), cobalt (Co), or magnesium (Mg) seem to stimulate the decomposition of cellulose, which is a major constituent of paper and wood. Some newspaper is printed with mercury—containing ink, which could prove toxic to the process.

Papyrus

Papyrus has gained fame, along with the pyramids and the sphinx, because of the Egyptians. It's a tropical plant, growing in great abundance in Africa, and it has many cousins which are grown—mainly as ornamentals—in the United States.

Visser (1963), who lived in Uganda, reports that almost 40,000 square kilometers there are covered by Cyperus papyrus, and he thought that it might prove a good source of biogas. Although few of you may have such an abundance of papyrus, Visser's information may be helpful in producing ideas for other situations and substrates.

He added dried papyrus in mixtures with various clays, ash, charcoal, and swamp soil to flasks of 500 milliliters (half a liter). Unfortunately, he did not seed these mixtures with active bacteria, nor did he attempt pH control. He carried the experiment on for 500 days.

His results indicate that in this substrate, the addition of some burned papyrus (which may have helped in pH control), the addition of the swamp soil, the addition of Kaolinite, Halloysite, or Illite (all clay products), or the addition of asbestos fibers, all seemed to stimulate the production of biogas. Although Visser used them, the use of asbestos fibers in any capacity is not recommended. They are a known carcinogen. In any case, the asbestos fibers may have stimulated the biogas process solely because they gave the bitty biogas buddies a place to stay. They much prefer to rest on a solid base than to be floating around.

The gas yield from fresh papyrus was disappointing, but upon analysis, it was found that the most rapid production of volatile acids occurred with fresh papyrus, and lacking adequate seeding or buffering capacity, such a slurry can be expected to fail. The fresh papyrus also had an intolerably high C/N of 65, to which no source of N was added. The addition of swamp soil had a marked effect on increasing the number of fungi and important groups of bacteria in the slurry.

Gas production was in the range of 230 to 310 cubic centimeters per gram air-dried solids added. It appears that 470 cubic centimeters biogas were produced per gram VS digested, but from the figures given by Visser, it is nearly impossible to tell.

Potatoes

Hindin and Dunstan (1963), working with settled potato processing waste, (which is different than straight potatoes) found 480 cubic centimeters gas produced per gram VS added, from a mixture of 75% settled potato processing waste, 25% sewage sludge. Biogas quality was good, with somewhat less than 60% CH_4.

Straub (1943) experimented with whole potatoes and found 430 cubic centimeters per gram VS decomposed (notice the differences). Straub's figures are, however, given minus the gas produced by the sewage sludge, whereas Hindin and Dunstan include sewage sludge gas. These are the kinds of variations that make library research such a thankless task. The mix used by Straub was about 85% potatoes to 15% sewage sludge, TS to TS weight.

Potato skins do not seem to digest as readily as the pulp. Potatotes produce a slimy effluent, which would make dry digestion difficult.

Rape Seed Cake

Rape seed is an oil seed, and rape seed cake is the residue left after the oil has been expressed from the ground-up seeds. Acharya (1935) found that rape seed cake slurry rapidly accumulated volatile acids, and thus the production of biogas was inhibited. No N addition was needed. Only a paltry 62 cubic centimeters per gram VS added was produced. Had an adequate seed been used, rape seed cake might have produced 460 cubic centimeters per gram VS added, and possibly more. This figure is based on Acharya's data and Buswell and Boruff's volatile-acid, biogas-products formula. Comparable performances can be expected with any residue from vegetable oil processing which has not been treated with toxic chemicals.

Rice Straw

Using rice straw, Acharya discovered almost 30% increase in biogas production when N was added, so that in the N-enriched slurry, 455 cubic centimeters per gram VS added was produced. Forty percent of the VS added was decomposed (approximately) and this gives a whopping 1,110 cubic centimeters per gram VS decomposed. However, it should be noted that the biogas was but 50% CH_4, when corrections were made for CO_2 dissolved in the slurry. Nevertheless, that's still a good turkey shoot.

Seaweed (Ascophyllum Nodosum)

UARL has done work with this variety of seaweed which proved intractable to digestion. This does not mean that it cannot be digested, but merely that it was not well digested. After pretreatment with an alkali (see Water Hyacinth, below), gas production hit 300 cubic centimeters per gram TS.

Water Hyacinth (Eichhornia Crassipes)

This abundant water weed was studied by apparently anonymous researchers (no names are given) from UARL (1974). They found it difficult to digest, and so they tried alkaline hydrolysis, meaning that they boiled the shredded hyacinth in a solution of 0.25 grams of sodium hydroxide (lye, NaOH) added for every 0.1 liter of water hyacinth slurry. The slurry had 4 grams TS per liter, so this represents about 0.6 grams of NaOH per TS gram of hyacinth—an abundant amount.

Faced with a similary resistant material (cornstalks), Buswell and Boruff merely soaked them for 4 days in limewater (of unreported strength). This seems a more reasonable way to initiate digestion of resistant plant materials. Another idea is to keep a pile of such materials damp and let them get moldy. The mold fungi will accomplish much the same thing naturally as limewater and lye accomplish chemically. It is a little known fact that most plants are attacked first by molds, and then by decomposing bacteria, even in your garden variety compost pile. The molds break down the cell walls of the plants, and then the bacteria can get to the goodies. In the absence of molds, plants are highly resistant to attack by bacteria.

Since molds, like biogas bacteria, may need to be found and cultured, look around for mushrooms or molds growing in likely places, or on similar resistant plant materials. Then, keep a portion of the moldy substrate handy to infect future piles of the same material.

Water hyacinth, pretreated as described with lye, still performed poorly for UARL, with the best available gas production at 163 cubic centimeters per gram TS added. The study was done at 48°C. All you water weed fans, don't lose hope though. Your experiments may turn out entirely differently. The HRT was only 10 days, and this may have a lot to do with the poor gas production.

Wheat Straw

With wheat straw, Acharya found a marked difference in gas production when N was added, so that the figure of 227 cubic centimeters per gram VS added exceeded by 120% the gas production of wheat straw in the absence of N. As was the case with oat straw, volatile

Plant	Cubic centimeters of biogas per gram VS added	Cubic centimeters of biogas per gram TS added	Comments
Algae		240	Assumes 70% combustible gas (?)
Artichoke tops	573		
Banana skins		360	
Barley straw		150	
Bracken fern	0		
Cabbage	792		Per gram VS decomposed
Carrots	693		Per gram VS decomposed
Cornstalks	620		
Flax straw	333	300	
Garbage	500		Average
Grass clippings	390		Average
Kelp		243	
Oat straw	172		
Paper	475		Average of reported newspaper values
Papyrus	470		(?)
Potatoes	480		Value limited as reported
Rape seed cake	62		
Rice straw	455		
Seaweed		300	
Water hyacinth		163	
Wheat straw	310		Average, chopped, mesophilic
Wood		500	

Table 17.1 Plant Wastes and Gas Production

acids were in excess abundance in the effluent. That means a lot of our potential biogas is going down the drain.

The purpose of Acharya's study was to find a treatment which would better measure the amount of material actually available to anaerobic decomposition. In a sense, Acharya was looking for a more accurate determination of VS. His conclusion was that a 5% H_2SO_4 in water solution would give a fair equivalent to anaerobic decomposition carried out over a period of six months, when the substrates were soaked in that solution for one hour.

Several other studies of wheat straw are available. Buswell and Boruff (1933) found 254 cubic centimeters of gas produced per gram TS added in batch experiments, and 156 cubic centimeters per gram TS added (average for wheat and flax straw) using the continuous-fed fibrous, substrate generator mentioned before.

Nelson, et al (1939), shows 342 cubic centimeters of gas per gram VS added (chopped wheat straw) and 388 cubic centimeters per gram VS added (ground wheat straw) produced. At thermophilic temperatures, these researchers found 365 (chopped) and 355 (ground) cubic centimeters produced in 21 days, per gram VS added.

Wood

Few studies are available on the digestion of this ubiquitous material. To some degree, the digestion of newspaper should be similar. In natural situations, however (e.g., in the real world), wood, as is true for plants in general, is almost never attacked by bacteria—it is attacked by fungi first, and later by bacteria. This is because some of the natural constituents of the wood, such as tannins and turpentines, are toxic to bacteria. For this reason, wetting the wood and letting it grow moldly before using it should help digestion. Wood is designed to resist breakdown because it lives for so long. Bark is particularly resistant to breakdown, and certain woods—redwood and woods with high resin contents– are more resistant than others.

Lime, nitrogen, and thermophilic temperatures seem to help the breakdown of cellulose. Our old arch-nemesis, lignin, is also present to a great degree in wood. It should not surprise us then, that many studies done have proven less than satisfactory.

Those busy folks at SERL (69-1) tried monterey pine sawdust. They found it to be inert to digestion. Monterey pine contains a very high portion of pitch and resin. Not yet content, they tried again (70-2), this time with white fir sawdust, which has a lower content of pitch and resin. Still no luck. The sawdust didn't interfere with the process, but neither was it digested.

However, Romashkevich and Karelina (1961) using sawdust of foliate trees produced 500 cubic centimeters per gram TS added, from a 5% solids slurry. This is one study reported in the book previously mentioned, *Anaerobic Digestion of Livestock Wastes.* The paper is in Russian or Rumanian, and we have no further information than this. Anyone out there speak Russian or Rumanian?

Alkaline hydrolysis, before digestion, cold or hot, may help wood decompose; or try good old mold.

Plant Wastes Gas Production

Table 17.1 must be used in conjunction with the information just presented on various plant wastes. The figures have a degree of value, and so they are presented, but without the previous information, they may be quite misleading. The figures on algae are derived from the information in the next chapter. A question mark in the Comments column indicates that the figure given is an educated guess.

Terms

None

Questions

None

Problems

None

18: On Growing Substrates

The biogas process is essentially a means of collecting secondhand solar energy. The sun shines, plants grow, we make the plant matter into biogas. This being the case, the question naturally arises: Can we make use of this process to grow energy? Can we, in other words, turn a certain area into a biological solar collector?

As with any process where energy is transferred or transformed, a certain amount of energy is lost during each step. In the sun-plant-biogas process, it might look like this:

1. Sun to plant conversion efficiency: 2% maximum
2. Plant to biogas conversion efficiency: 65% average

Let's look at each step in this energy accounting.

Sun to plant conversion efficiency depends on a number of things. Incoming radiant solar energy is only about 43% visible light. The rest is heat (infrared radiation) and other kinds of radiation. Plants use only the visible portion of the spectrum for photosynthesis. Here, then is the first limitation. Plants—at 100% conversion efficiency of the visible light—could only make 43% of the sun's radiation into the stored energy of plant matter.

However, plants do not make 100% of the sunlight into plant matter. Some of the energy is used to support the life processes of the plant, some is not even gathered. The upshot is that only a very small portion (1% to 5% under ordinary conditions) of the visible light is converted into plant matter. All told, this means that only (0.43 x 0.05) 2.15% of the incoming solar radiation maximum is converted into plant matter under ordinary conditions. (Plant matter is also known as biomass.)

From this, we must subtract further energy. The energy required to make the necessary farming equipment (which may not even resemble a tractor); the energy used to manufacture, transport and apply the herbicides, pesticides and fertilizers; the energy required for irrigation; the energy we must invest in planting, cultivation, harvesting, storing and—perhaps most important—transporting our substrate plants to the generator; as well as the energy consumed in building, maintaining and heating the biogas generator; must all be subtracted from the only energy we care about: the energy stored in the bonds of the CH_4 molecules in the biogas.

Many authors seem to concentrate on the growth and utilization of algae as the plant for conversion into biogas. Indeed, algae has shown the advantages, in small-scale studies, of having excellent conversion efficiencies (from 2% to 18%) of the incoming light into the potential energy of plant mass, and of having high yields of 100 to 160 pounds per acre (160 pounds per acre is about 12.5 grams per square meter) per day.

Algae, however, have not been grown on a very large scale. The facilities for large-scale algae growth will be very expensive, and the energy involved in harvest and processing will be high. Algae do not digest well at ordinary mesophilic temperatures, primarily because the conditions in a mesophilic generator do not kill the cells—which therefore remain intact and unavailable for digestion. Thermophilic digestion is required for intact algae, and this requires higher generator temperatures, and therefore more total energy and/or a higher cost in insulation. Killing the algae cells before use, by heat, ultrasonics, microwaves, fungi or drying would make mesophilic digestion feasible, but these processes must be energy efficient to be competitive with thermophilic digestion.

Because of the high protein content of algae, it produces a very colloidal jello-like, slimy effluent which is hard to dry. This need not be a drawback where all the effluent is recycled into ponds used for growing algae, but the higher water content of such an effluent makes any use that requires dewatering or transport more expensive. A possible use for an algae effluent is dilution and direct irrigation, but this has not been investigated.

Algae cultures must also be cooled in any situation where intense sunlight is available. Photosynthetic green plants do not make use of the sun's heat, yet where sunlight is strong, heat is usually also intense. The algae pond, filled with water, has a high capacity for

Species or Plant	Location	Tons per acre per year	Yield in metric tonnes per hectare per year
Jerusalem artichoke	Russia	13.5	30.3
Exotic forage sorghum	Puerto Rico	30.6	68.6
Forage sorghum (irrigated)	Kansas	12	26.9
Kenaf	Florida	20	44.8
Water hyacinth	Florida	16	35.9
Sugarcane	Mississippi	20	44.8
Sugarcane (state average)	Florida	17.5	39.2
Sugarcane (best case)	Texas (south)	50	112
Sudan grass	California	15–16	33.6–35.9
Bamboo (4 yr. stand)	Alabama	7	15.7
Algae (fresh water pond)	California	8–39	17.9–87.4
Tropical rainforest (average)		18.3	41
World's oceans (primary productivity)		6	13.5
Sugar beet (best growth)	England	24	53.8
Potatoes (experimental hydroponic)	U.S.	60	134.5

Table 18.1 Biomass Yield

heat, and avidly collects it. As temperatures climb above 27°C (80°F), the growth of algae suffers. Either this water must be cooled, or the heat must be extracted before it arrives at the pond surface. Both options require further equipment and therefore more money and energy. The use of a "filter" made of a solution of water and alum (potassium aluminum sulfate: $KAl(SO_4)_2$ x 12 H_2O) will allow visible light to pass, yet stop the infrared (heat) radiation, but a simple, inexpensive and foolproof method of using this information has not been devised. So growing algae is expensive and complex.

For these reasons, a low-technology approach to growing biomass for biogas will for the time being probably have to be based on the higher plants. The disadvantages of algae do not put it out of the running, but they tend to legislate against its use in small-scale situations.

In Table 18.1, we give some figures on plant yield per acre per year. However, in agriculture, climate is all important. Some crops suitable in one climate will not grow in another. Notice that the tropical climate crops outperform the temperate crops. This is at least partly because yields fluctuate in temperate climates from summer to winter, but yields in tropical crops are more or less constant.

Remember as well that these figures represent a broad spectrum of farming practices and that each plant has a different suitability to the biogas process as well as different final biogas heat value yield per unit weight. Much more experimentation is needed in this area, for biogas is such a lovely fuel, so well suited to many different uses, and yet the production of the substrate can be very "low technology" if necessary. This makes biogas production suitable to many more primitive situations where energy is otherwise scarce, and the local technology is undeveloped.

The hydrophonic growth of potatoes is of course an extremely energy inefficient process; it is included here in Table 18.1 to indicate that the yield of the higher plants can equal or exceed anything reported for algae, under similar highly controlled laboratory conditions.

For those interested in any case in experimenting with algae, some points of information will prove helpful.

Whenever effluent is used for the nutrient base in the pond, the algae will grow best in association (symbiosis) with certain kinds of bacteria. These bacteria break down more complex nutrients into the simple molecules which provide food for the fastest algal growth rate. Once the algal pond is established, only a small percentage of the total nutrients need to be imported in the form of manure, sewage or other materials. The rest can come from the effluent of the algae fed generator. As more of the algae goes to other purposes—such as animal feed—more nutrient imports will need to be made.

All the available byproducts from associated processes should, where possible, be returned to the algal pond. If the algae is used for feedstuff, the animal manures should be returned. If the algae is used for fertilizer, the crop wastes should be returned. As well, the gases "scrubbed" from the biogas, and whenever possible, the combustion byproducts of the burning of biogas should be returned to the algal pond—particularly CO_2.

CO_2 is important to the algae because the growth rate depends—among other factors—on what is known as the "limiting nutrient". Photosynthetic plants, as you may know, require a great many nutrients, the main ones being carbon (C), oxygen (O), hydrogen (H), nitrogen (N), phosphorus (P) and potassium (K). Any plant, other conditions being favorable, will grow as well and as fast as it can as long as the necessary nutrients are available. Ideally, growth continues until one of them is used up. This nutrient is the limiting nutrient because the amount of this one (relative to the plant's needs) is less than any other. Plants need CO_2, much as we need oxygen. However, the air contains only 0.03% CO_2. In an algal pond, CO_2 is often the limiting nutrient. Returning this CO_2 to the algal pond may stimulate the growth of algae.

CO_2 concentrations of 0.1% to 5% (by volume) in air, when bubbled through the culture water, will markedly increase the growth rate. At the higher concentration, 6.25 milliliters of the gas mixture per liter of culture per minute has, in one experiment (Geoghezon, 1953), proven sufficient.

When using organic substrates such as the recycled algae effluent or sewage, CO_2 is produced when aerobic bacteria break down these substrates, supplying to some degree the need for CO_2 in the culture water. Aerobic compost also produces abundant CO_2—but capturing it may be difficult.

As has been mentioned, the temperature of the pond is important. Temperatures in excess of 27°C (80°F) are inadvisable. Cooling can be accomplished by evaporation in a cooling tower, where the pond water is pumped up to a height and allowed to drain down surfaces while exposed to air. Water lost to evaporation needs to be replaced. It seems a shame, however, to waste all that lovely heat.

Another solution to the problem of high pond temperatures is to grow temperature tolerant species, but this is a high technology approach since the cultures may need to be kept pure (only one species of algae) by the use of chemicals or sterile equipment.

Algae grows best if the temperature is varied from day to night. In fact, high temperatures which might otherwise depress the growth of algae (30°C, 86°F) can prove a stimulant to growth if the temperature is also lowered during the night (20°C, 68°F). Sometimes temperatures as high as 45°C (113°F) are used to control pest microorganisms such as rotifers, but these temperatures will also damage algae, and even if they are not killed, growth will be hampered for days. The temperature limits and optimums depend on the species involved.

The full intensity of sunlight is more than algae can efficiently handle. All the studies showing very high efficiencies were the result of experiments using low intensity light or intermittent light. A young culture can even be killed by full sunlight, and so it is good practice to give an algae culture partial shade when it is becoming established.

Turbulence in the culture water will, at sufficient concentrations of algae cells, cause the individual cells to experience varied light intensities and the distribution of light energy input will be more even among all the cells in the culture.

In cultures which are nourished by organic substrates, continued stirring or turbulence is not beneficial, since the particles of substrate and the aerobic bacteria will disperse throughout the culture water and absorb too much light. When stirring ceases, these particles settle and most of the algae continues to float. According to experiments done by Oswald and Golueke (1960) using an algae pond such as is described below, mixing during the day lowers the pH, which later rises as the algae use up nutrients in the pond water. This is probably a result of the changing concentrations of CO_2 that such mixing might cause. A lower pH would result from an increased concentration of CO_2. Mixing during the night allows the settled bacteria and substrate to become reoxygenated and does not interfere with photosynthesis. Unless mixing occurs at least once in 12 hours, the settled bacteria will use up the oxygen available in the pond bottom, which will gradually become anaerobic. For these reasons, mixing for a half hour around 1:00 PM, and for 2 to 4 hours starting at midnight was recommended by and proven satisfactory for these researchers.

For a great many chemical and biochemical reasons, a pH range of 6.0 to 6.5 is best. When the nutrients in the culture come from organic sources, the pH generally remains more stable, but it may need to be manipulated to bring it into this range, either by mixing or by chemical means.

Young cells have a higher protein content and more vitamin B_1, and older cells (14 to 28 days) tend to have more fats. If the nitrogen content of the culture water is restricted, the fat content of the algae cells increases, but total biomass production is lower. Whenever cell division ceases, while cell growth continues, fats increase. These facts may hold promise for biogas production, as fats produce a better quality and greater quantity of biogas than protein. As was mentioned before, the high protein content of algae produces a very colloidal effluent. A higher fat content algae may produce an effluent which can be more easily dewatered.

Maximum biomass yields per unit area per unit time occur in 2 to 5 days, but if the fat content of the cells markedly affects biogas production, the maximum biogas or methane production per unit area per unit time may take longer to develop.

It appears that the simplest low technology method of growing algae is a pond, with curbs or dikes in it so that the water, pumped from one end causes a flow

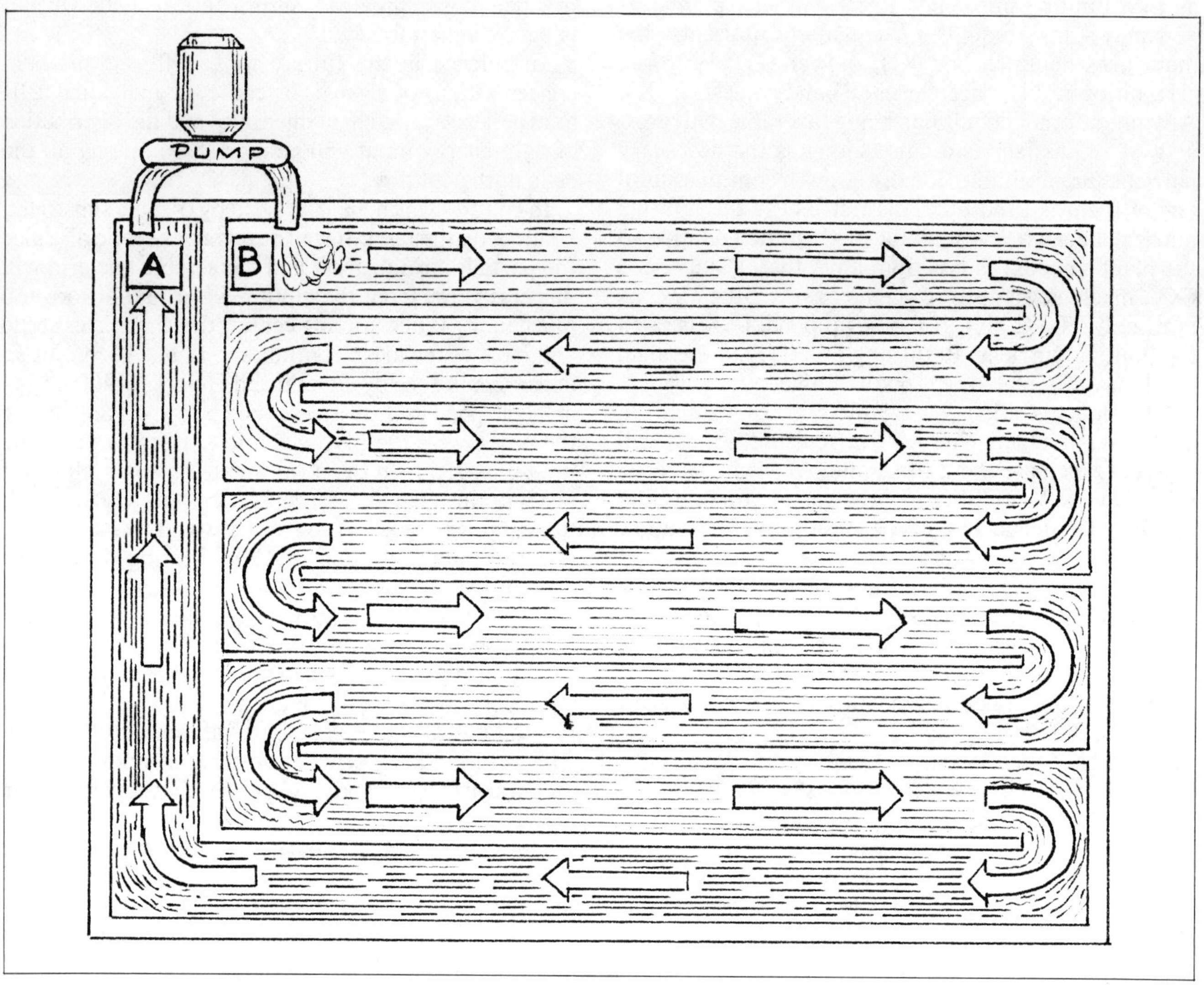

Fig. 18.1 Algae Pond Design

along the whole length of the pond. See Figure 18.1 for a schematic drawing of such a pond viewed from directly overhead. If water is pumped from A, the outlet, B would be the inlet end.

This pond design is essentially the same as that proposed and used by Oswald and Golueke (1960). The reason for using such a design is primarily that it is easier to mix the pond culture for aeration in this design versus a pond without channels and dikes. A flow velocity of 30 centimeters (one foot) per second, or more, will mix the substrate and the pond water sufficiently to produce aeration. (This velocity is the minimum required to stir up the organic matter, causing it to mix with the algae and become reoxygenated; lower velocities may lengthen the maximum time required between mixings by causing some oxygen to infiltrate the organic material on the pond bottom, without stirring it up so much that it blocks the light.)

The pond should be 20 to 30 centimeters (8 to 12 inches) in depth, and have a channel length, width, and bottom smoothness such that the energy needed to drive the pumps to cause that flow will be at a minimum. Specific mathematical information is available from Oswald and Golueke (1960), and in publications on hydraulic flow.

Harvesting the algae provides another problem. Filtering, according to Oswald and Golueke, is too difficult, and centrifuging is too costly. Chemical coagulation is feasible with the use of lime, adding that until the pH rises to 11.3, rapidly mixing the culture for a brief period, then 3 to 5 minutes of gentle stirring to encourage the formation of coagulated groups of algae cells, known as floc particles. The process, of course, is then flocculation.

Draining off the top liquor before dawn will help, as the algae tend to settle at night. But a natural flocculation process was observed in the Richmond, California algae pond. During the afternoon of sunny days, when the temperature of the pond had increased several degrees above the morning level and the pH had increased to

10 or 11 (as a consequence of the changing CO_2 concentration, as noted earlier), the algae clumped together and settled to the bottom. It was recommended that special ponds 7.6 to 15 centimeters (3 to 6 inches) deep be constructed to take advantage of this natural process. The culture water should be pumped into the flocculation pond early in the morning, and after the algae have settled, the supernatent liquid should be returned to the culture pond, and the concentrated algae transferred to another pond for further concentration, or to a sand drying bed. Natural flocculation does not occur in ponds of greater than 60 cm (2 feet) depth.

As mentioned, algal slurries should be digested thermophilically unless the algae are first killed. Hydraulic retention times of 11 to 20 days can be used. Experiments have shown that from 150 to 225 cubic centimeters of combustible gas (CH_4 + H_2) is produced from each TS gram of algae introduced into the generator (2.4 to 3.5 cubic feet per pound). Loading rates of 1.2 to 2.3 grams VS per liter (0.09 to 0.18 pounds VS per cubic foot) have been tried and do not overload the generator. On a dry weight basis, algae is 85% to 90% VS when it is grown chemically. Effluent-grown algae has a lower VS content of 80% to 85%. Generator sizes of 9.1 liters per square meter of pond (0.27 cubic feet per square yard, 1,300 cubic feet per acre) should prove adequate. Further information can be found in the Bibliography.

Terms

Flocculation: The process of natural coagulation.

Questions

None

Problems

None

SECTION IV

Uses

Biogas (as you no doubt are aware) is a combination of several different kinds of gases—CO_2, H_2S, H_2, CH_4, (sometimes even N_2) and other gases. The main important component of biogas is CH_4, methane. In this section the focus is on combustion, and on methane, for four reasons: (1) Methane, because of its quantity and quality, is the major source of the energy available in biogas. (2) The combustion of methane is directly comparable to the combustion of H_2, the only other reasonably useful source of energy in biogas, and so general information about the combustion of CH_4 is general information about the combustion of H_2. (3) The ordinary combustion or burning of methane is the only low-tech method by which to make use of the energy locked up in the methane molecule. (4) We will discuss combustion not only so that we can produce and control it when we want it, but also so that we can avoid it whenever that is appropriate. Biogas can kill you, and it's best not to forget that.

To start with, then, we'll talk about scrubbing biogas. The term "scrubbing" refers to any process that removes any gas component from the biogas, leaving us with more nearly pure CH_4. Then, we'll move on to specific uses—illumination, heating, etc.—and discuss these subjects in a fair amount of detail.

Where possible, we want to give you enough information to enable you to make the device described. As well, such detail will enable you to understand how to use the devices efficiently, so that you don't waste your precious biogas. Last, such information will enable you to evaluate the amount of biogas needed to power certain devices and meet certain needs, so that you will be able to find out if you can really do what you want to, with biogas.

Woven into each discussion is general information about physics and chemistry, some of which relates to the understanding of the next subject. If you have a solid background in these subjects, feel free to skip around. If not, it is suggested that you read through the whole section, sequentially, at least once.

19: Scrubbing

Why Scrub?

Biogas consists primarily, in most cases, of CH_4 and CO_2. A long retention time will increase H_2S, hydrogen sulfide. High-carbohydrate substrates and certain manipulations of the biogas process increase the H_2 (hydrogen gas). We also find N_2 (nitrogen gas) in the biogas if the slurry has been contaminated by air. Water vapor will also be present.

We may, for various reasons, wish to remove some of these gases from the biogas so that what we have left is more nearly pure methane, or "pure" methane and hydrogen. CO_2 and N_2 are inert; they do not, as far as we are concerned, enter into combustion reactions. (Some N_2, in an internal combustion engine, does react to form various oxides of nitrogen, symbolized NO_x.) These inert gases are fillers, meaning they add little or nothing to biogas as an energy source. When we use biogas as a source of concentrated heat, these inert gases get in the way because they steal heat from the reaction. We have to heat them up, as well as whatever else we are trying to heat. Further, they dilute the energy in the biogas. A liter of 66% methane biogas has about 5.3 calories of heat energy available. A liter of pure CH_4 has about 8 calories of available heat energy, respectively, 590 and 900 Btu per cubic foot. Figures like this are often given with reference to STP–Standard Temperature and Pressure, which is O°C and sea-level atmospheric pressure.

In attempting to store biogas, we should remember that if we can remove the inert ingredients, our storage facility can be smaller, and therefore less expensive. The money we save in storage, however, must be greater than the expense of a scrubber. It is good to remove H_2S from our biogas if we are going to use it in an engine. About one tenth of one percent by volume is acceptable. Above this amount, it will acidify the oil and corrode the internal parts. Therefore, to deal with all of the above problems we can scrub the biogas.

But wait. Let's not assume that scrubbing biogas is inevitable. If we know something about what causes the variations in the composition of the biogas in the first place (*e.g.*, in the generator) then we may be able to manipulate the process a little to come up with a biogas composition which is more acceptable to us, without resorting to scrubbing at all, or without resorting to as much of it as we would need otherwise.

Gas Composition

In brief, Table 19.1 gives us the reasons for changes in composition.

CO_2 is produced mainly in the acid-forming stage of anaerobic digestion, whether or not the second (methane-forming) stage is operating. With a high C/N, it is difficult for the ecosystem of the generator to establish an adequate buffer system, and the response it makes is to try to stabilize the pH by throwing off excess CO_2, thus lowering the C/N and raising the pH. (See Chapter 7 for more on this subject.)

The addition of a buffer chemical, or increased digestion time, or the addition of old sludge or effluent to new slurry, all help create a more balanced pH and a better buffer system, and this causes the CO_2 to be retained in the slurry, which in turn means that less shows up in the biogas. Oxygen contamination enhances acid digestion and harms the methane-forming bacteria, thus increasing CO_2. The amount of CO_2 in the biogas, and pH factors in the slurry are intimately related.

CH_4 increases and decreases as shown. It should be pointed out that there is a difference between a relative increase, and an absolute increase in the amount of CH_4. A relative increase (percentage increase) in CH_4 could be due to a decrease in CO_2. One cause of this would be that more CO_2 is, for some reason, being dissolved in the slurry. If this were the only cause of the CH_4 increase, the total amount of biogas would decrease but the amount of CH_4 would remain the same. That's a relative increase.

On the other hand, an absolute increase in CH_4 simply means greater volume of CH_4 produced during a given period of time. What we want is both a relative increase (an increase in the percentage CH_4) and an absolute increase (an increase in the volume of CH_4). The more the merrier.

CO_2	increases with a high C/N. increases with a lower pH. increases with O_2 contamination of the slurry. increases with almost any upset of the digestion process. increases during (first stage) acid digestion. decreases with a longer digestion time. decreases with a higher pH. decreases with the addition of a buffer chemical (e.g. lime).
CH_4	increases with longer retention times. increases whenever the digestion process is biologically stable. increases whenever certain substrates (e.g. grease and fats) are decomposed. decreases whenever the digestion process is upset. decreases when the slurry is contaminated by air.
H_2	increases with a low pH. increases during (first stage) acid digestion. increases when cellulistic substrates such as foliate tree sawdust or cotton are used. decreases during well established (second stage) methane digestion.
H_20 vapor	increases directly with the temperature of the slurry. decreases (condenses) as the temperature of the biogas drops.
H_2S	increases with a low pH. increases with a rise in digestion temperature. increases with time. increases with any upset of the digestion process. increases with increased seeding with old sludge. increases with greater amounts of protein in the substrate. increases with more "natural" sulfates in the dilution water. increases when chemical sulfates are added to the slurry.
N_2	increases with contamination of the slurry by air. increases with a low C/N (possibly).
NH_3	increases with low C/N. increases with thermophilic temperatures.

Table 19.1 Gas Composition

An increase in the amount of H_2 in biogas is usually associated with a decrease in the amount of CH_4. What little is known about the subject is covered in detail in Appendix 2.

Water vapor, as mentioned in Table 19.1, will be present in the biogas according to the temperature of the water or slurry over which it lies. Certain scrubbing processes require water or solutions containing water, and these may add to the water vapor carried by the biogas.

H_2S increases in thermophilic digestion and at the higher mesophilic temperatures. It is unfortunate since these are the very conditions that we want to produce when an increase in the rate of biogas evolution and a corresponding decrease in generator size is sought. What is, is. Of the proteins, albuminous proteins (sulfur-containing—like egg yolk) seem to be most responsible for H_2S increases.

Henry (1961) claims that at mesophilic temperatures and pH 5.0, 98% of all of the sulfides dissolved in the slurry show up as H_2S, whereas at pH 7.2, this is reduced to 24% with an approximate 50% reduction for each 0.2 pH unit increase thereafter (*e.g.*, 12% at 7.4, 6% at 7.6, 3% at 7.8, 1.5% at 8.0). Where scrubbing H_2S is desirable but unrealistic for some reason, pH adjustment may be the answer—but take care not to overadjust. A further idea is the addition of the ferric ion of iron; Henry used ferric chloride at the rate of about 2 grams per liter. This tied up the sulfide ion and no H_2S was produced in a digestor with this high dosage of ferric chloride.

As regards N_2 and NH_3, it has been stated by no lesser authorities than Buswell and Boruff (1932) that it is impossible that the free gaseous nitrogen (N_2) found in some biogas can come from any source other than contamination by air (70% N_2). Nevertheless, N_2 in biogas might come from a low C/N—as other authorities believe—and NH_3 (ammonia) certainly does. Since scrubbing N_2 is not easily done, and since N_2 and NH_3 are very rarely present in biogas evolved from a slurry with a balanced C/N and no atmospheric contamination, scrubbing these gases will not be discussed.

CO_2 Scrubbing

CO_2 is the second most abundant gas in biogas. CO_2 and CH_4 ordinarily make up 95% to 99% of the volume of biogas. Thus, in a great many listed analyses of biogas, only the methane content and the CO_2 content appear.

The heat available from a unit volume of biogas can be increased rather dramatically if the CO_2 is removed. In a particular sample of biogas which is 55% CH_4, 45% CO_2, completely removing the CO_2 component will increase the heat available from a unit volume of that biogas by a factor of 1.8—more than one-and-a-half times the original heat value.

Scrubbing the CO_2 is also helpful if you are planning on large-scale storage. Where biogas is generated for a period of days and used intensively at the end of that period, or where pressurized storage is used, scrubbing the CO_2 will reduce the capital cost of storage equipment markedly.

As you will see in the discussion on internal combustion engines, CO_2 dilution will also decrease the suitability of biogas for use in such equipment. For example, a biogas of 50% CH_4, 50% CO_2 has as much total energy in two volumes as does one volume of 100% CH_4, but the energy in those two volumes of diluted methane is used (on the average) with 54% less effect Neyeloff and Gunkel, 1974). In other words, it will require 3.7 volumes of the 50% CH_4 biogas to equal the performance of 1 volume of 100% CH_4.

However, for many uses and in most small-scale situations, CO_2 scrubbing may not be necessary. In any case, other approaches to increasing the CH_4 content of the biogas should be tried first (*e.g.*, liming the generator).

CO_2 can be removed from the biogas by bubbling the gas through water containing any alkaline chemical, such as calcium hydroxide—$Ca(OH)_2$.

This is because CO_2 and water form carbonic acid:

$$CO_2 + H_2O \longrightarrow H_2CO_3$$

Then, alkaline chemicals such as calcium hydroxide combine with and remove the carbonic acid. In this case, these compounds form calcium bicarbonate:

$$2H_2CO_3 + Ca(OH)_2 \longrightarrow 2H_2O + Ca(HCO_3)_2$$

When the solution is exposed to the air, calcium carbonate is formed:

$$O_2 + 2Ca(HCO_3)_2 \longrightarrow 2CaCO_3 + 2CO_2 + 2H_2O$$

This, to some degree, regenerates the solution so that it can be used again. Is your head reeling? The short form is that CO_2 can be scrubbed from biogas by any alkaline chemical. When calcium hydroxide is used to do this, the solution can be partially regenerated by exposure to the ordinary and free air. Got it?

H_2S

Hydrogen sulfide is a gas with a distinct smell of rotten eggs. Its presence in biogas can be dangerous, because H_2S is poisonous, and this should never be forgotten. H_2S is unsafe above concentrations (in air) of only 13 parts per million. (It can be detected via nostril power at about 1 part in 10 million.) It is not immediately fatal, nor is it highly toxic. Poison, however, is poison. H_2S is also poisonous to vegetation, and if a generator in a greenhouse were very leaky and its biogas had a significant percentage of H_2S, this might repress the growth of the plants.

H_2S poisoning affects the nervous system. Its powerful smell can often be a warning, but its continued presence can paralyze the olfactory nerve, and when this happens it can no longer be smelled. It can still be detected by noticing the early symptoms of H_2S poisoning—slight headache, burning eyes, and clouded vision. If, at any time, when working with or near biogas, these symptoms are noticed, go immediately to wherever there is fresh air. Another person who has not been exposed to the biogas will be able to smell the H_2S if it is present. Remember these symptoms: slight headache, burning eyes, and clouded vision.

H_2S is flammable or explosive between 4.5% and 45% in air. Since biogas rarely attains to more than a 1% concentration of H_2S, it is impossible for these concentrations to develop unless the H_2S is separated from the biogas and then released. Even this situation is unlikely, but if it should happen, the chief danger would be explosion, since the ignition temperature of the H_2S/air mixture is very low—around 250°C (480°F). In turn, the flame temperature of the resulting flash would be above the ignition temperature of biogas, and could therefore cause a further explosion (of an air/biogas mixture). Such a situation is at the end of a long chain of ifs, but an awareness of such circumstances—as unlikely as they are—should make them even more unlikely. If you are wise, it will make them impossible.

Up to this point, we have not discussed the dangers

of the biogas process. But let's face it, anything that can cook your meals, run an engine, or heat your house can also kill you. The point here is—stay alive. Don't be careless with something precious—you. Refer to Chapter 43 for further ideas.

When H_2S burns, in a correctly proportioned air/biogas flame, most of it is converted directly to sulfur:

$$2H_2S + O_2 \longrightarrow 2H_2O + 2S$$

As air is added in greater and greater excess, some or all of the H_2S is converted to sulfur dioxide. Even when the air/biogas mixture is perfect, some of the H_2S is converted to SO_2:

$$2H_2S + 3O_2 \longrightarrow 2H_2O + 2SO_2$$

This in itself is not too bad. Sulfur dioxide *as* sulfur dioxide is mildly toxic, but the real problem arises when SO_2 meets up with ubiquitous H_2O:

$$SO_2 + H_2O \longrightarrow H_2SO_3$$

H_2SO_3 is sulfurous acid, which is bad enough, but still worse, when more oxygen gets into the act, the ante suddenly doubles:

$$2H_2SO_3 + O_2 \rightarrow 2H_2SO_4$$

This villian is the infamous sulfuric acid.

These reactions, and others equally undesirable, form a group of other reasons for removing H_2S from biogas. However, for most applications, the removal of H_2S is not critical. Concentrations of H_2S are generally low—lower than 0.1%. The most critical use of biogas as far as its H_2S content is concerned is fueling internal combustion engines. This will be covered in greater detail in the chapter on that subject. H_2S removal can also be desirable where the by-products of biogas combustion come into contact with metal, as in venting through metal pipes. In any case, as with any other addition to the biogas process, the addition of an H_2S scrubber must return in benefits more than it costs in time, energy, and dollars.

Rust Removal of H_2S

The most common method of removing H_2S is by the use of a quantity of rust, enclosed in a container through which the biogas flows. This is called the dry box method. Fe, shown below, is the chemist's shorthand for elementary iron:

$$FeO_3 + 3H_2S \rightarrow 3H_2O + FeS_3$$

$$Fe_2O_3 + 3H_2S \rightarrow 2FeS + 3H_2O + S$$

One of the molecules found in rust is Fe_2O_3, ferric oxide. When this reacts with H_2S, it produces ferric sulfide (Fe_2S_3). Less often, the reaction produces ferrous sulfide (FeS) and sulfur (S).

Both ferric sulfide and ferrous sulfide are very unstable, and in the presence of oxygen, they experience total nervous breakdown:

$$2Fe_2S_3 + 3O_2 \longrightarrow 2Fe_2O_3 + 6S + \text{heat}$$

$$4FeS + 3O_2 \longrightarrow 2Fe_2O_3 + 4S + \text{heat}$$

These reactions are accompanied, as the above equations indicate, by substantial heat. Therefore, great care should be taken when exposing these sulfide by-products to the air. Small amounts of air, released gradually into our scrubber will do the trick, but to expose the Fe_2S_3 and FeS to air immediately and totally would be highly unwise. Also, the reactions depend on what is called hydrated forms of ferric or iron oxide. This simply means that they have water molecules associated with them. If heat evaporates the water, the rust no longer works for scrubbing H_2S.

The rust used for this process can be produced by mixing cast iron or mild steel powder or shavings (found, one might hope, at a nearby machine shop) with moist sawdust or ground corncobs, or any moist, porous material. Use 20 to 25 pounds iron to one cubic foot of base material—about 1.5 to 2 kilograms per liter. Try to get cast iron, or mild steel, as these will produce rust of a more suitable type than a more exotic alloy. If allowed to remain moist for 3 to 4 weeks, this mixture will then be very suitable for H_2S scrubbing. The addition of salt to the water will cause the rusting to occur faster, but it will also produce more heat, and in this case the material should be carefully watched. If it overheats, it will harm that batch for H_2S scrubbing. Spreading such a mixture thin and keeping it moist will help avoid these problems when salt is used.

Three qualities are called for in rust used for H_2S scrubbing: (1) activity: a measure of *how rapidly* the rust combines with H_2S); (2) capacity: a measure of *how much* H_2S can be captured by the rust, either at first or the total amount upon repeated use); and (3) regenerative ability: a measure of *how many times* and to what degree the first two parameters can be regained when the used material is exposed to the air.

Activity and capacity seem to trade off, so where one rust sample will be very active, it will lack capacity, and where another will show poor activity, it will have great capacity. Much depends then on how much H_2S we have, how many scrubbers we have, and what the conditions of scrubbing are, in considering what kind of rust will be best suited to our needs.

The phrase "kind of rust" may puzzle you if you think that rust is rust is rust. The chemistry of H_2S/Fe_2O_3 scrubbing is a bit peculiar, but it seems that there are different kinds of iron oxides because they combine with different amounts of water and have different impurities. The particular factors which affect rust in its ability to scrub H_2S are not well understood, but the differences in activity and capacity, while not completely predictable, are at least partly recognizable by the various colors of the rusts (due to varying water content and origin).

American	Materials	Metric
1 bushel (16 gallons)	Shavings (sawdust)	16 liters
25 pounds	$FeSO_4$ (ferrous sulfate)	25 kilograms
5 to 6 pounds	Lime	5 to 6 kilograms
1 pound	Rusted iron bits, powder or flakes	1 kilogram

Table 19.2 H_2S Filter Sponge Recipe

The yellow and yellow-brown rusts tend to have low activity and fairly high capacity. The reds, reddish brown, and darker rusts tend to have greater activity and less capacity. (For examples of the two different kinds of rust, examine rusty metal. Yellow rust is often found floating on radiator water. Red and darker rust is formed on old metal which has been underground).

For excellent information on H_2S/Fe_2O_3 scrubbing, refer to the Illinois Engineering Experiment Station Bulletin number 119, *Some Conditions Affecting the Usefulness of Iron Oxide for City Gas Preparation*, by William. A. Dunkley (February, 1921). Dunkley made tests on which much of the above information is based. One oxide which showed excellent activity, capacity, and regenerative ability was manufactured by using the recipe in Table 19.2. The metric and pound amounts are not equal, but the ratios, in each case, are equal.

Dissolve the ferrous sulfate (known as copperas) in hot water. Mix lime with sufficient water to be able to sprinkle the resulting mixture. Spread shavings on the ground about 18 inches (45 centimeters) thick. "Dust" with all the rust, sprinkle with half of copperas solution and half the lime solution. Turn with a shovel to mix. The next day, sprinkle the remainder of the two solutions and turn again. Mix daily thereafter until the mass cools.

Since the ability of these rusts to capture H_2S varies so much, little can be said about how much to use for a given amount of H_2S. Theoretically, one unit weight of rust will snatch up 0.639 units weight of H_2S. (Note that in Dunkley's mixture, not all of it is rust, and thus even theoretically one unit weight of Dunkley's mix will not snatch 0.6 plus units of that bad gas H_2S.) This theoretical amount, however, will never be achieved, if for no other reason than the fact that the rust will not be available molecule by molecule. The rust is in lumps, which are really only active on their surfaces. You can expect that finely divided, well prepared oxide will generally be able to hold (or remove) only 20% to 30% of its theoretical H_2S capacity. Activity varies so much that it can only be tested. As far as regenerative ability, four times is a general maximum according to Dunkley. Each time the rust is reused, its capacity falls off.

The size of the rust particles or lumps is of some consequence, since, while very fine particles have a greater surface area and thus remove more rust per unit of weight, very fine particles also resist the flow of gas. When the resistance grows too great, then we have to add a pump to the scrubber, increasing its cost and complexity. So, if we use very fine particles, then we must also use either several or many thin layers, or we must add a filler material to allow the biogas to pass by the particles and yet still be filtered. The Dunkley mixture adds a filler: sawdust.

Sizing of the particles is done the same way Sally sifted seashells by the seashore—with screens. The U.S. National Bureau of Standards has set up standard sizes, and the numbers associated with screen size refer to these standards. Fine particles are those which will pass through a number 40 screen (less than 417 microns in size) and which will be retained by a number 100 screen (more than 147 microns in size). That's pretty small—dustlike, even. Mixtures made of particles of this size are fine mixes, mixtures made of particles above these dimensions are granular mixes (Seil).

Seil also tells us the moisture content is very important. Less than 17% or more than 55% of moisture in the rust/shavings mixture will retard scrubbing. Best is 30% to 50%. The pH should remain alkaline or neutral. Round shapes for scrubbers are better than angular (for example, square) shapes, since it is easier then to pack the scrubber. Corners are likely places for the unscrubbed biogas to sneak through, resulting in lower efficiency. If a new scrubber, containing freshly packed material, is placed on line, the oxygen in it may cause overheating when it first operates. Further, the volume of oxygen could be sufficient to cause explosive mixtures in the storage system. Take care when using a scrubber.

The Fe_2O_3/H_2S process—the dry box method—is common in small-scale situations because the material used (rust) is easy to find or make and this filter material is easily recycled for further use.

There are many other ways of scrubbing H_2S, some of which are suitable to small-scale situations, some of which reach into the upper echelons of high technology. These methods have been developed because H_2S is a common industrial contaminant, and a lot of money has been available to solve the problem of its presence. Alas, that every question involving the lovely biogas process could not have been favored by such a rich uncle.

Na_2CO_3

One method of H_2S removal which is not as selective as the dry box method—since it affects CO_2 as well as H_2S—is to bubble the H_2S containing gas through a solution of sodium carbonate.

$$Na_2CO_3 + H_2S \longrightarrow NaHCO_3 + NaHS$$

Here, the sodium carbonate (Na_2CO_3) is turned into sodium bicarbonate ($NaHCO_3$). The other byproduct is sodium hydrosulfide (NaHS). This is a one-way process unless the byproduct solution is washed with steam, which will release H_2S and set matters straight again. The CO_2 component of the biogas will also enter into the above reactions, and thus the CO_2 content of the biogas will also be reduced. This CO_2 reaction, however, will remove some of the sodium carbonate, making the solution less able to remove H_2S:

$$Na_2CO_3 + CO_2 + H_2O \longrightarrow 2NaHCO_3$$

Thus, if sodium carbonate is used to remove H_2S, this should occur after removal of CO_2, perhaps by running the biogas first through a more concentrated solution of sodium carbonate followed by a more dilute solution.

NaOH

H_2S is also removed via a solution of sodium hydroxide:

$$H_2S + 2NaOH \longrightarrow Na_2S + 2H_2O$$

If the process continues, the sodium sulfide (Na_2S) is further reacted and produces sodium hydrosulfide:

$$Na_2S + H_2S \longrightarrow 2NaHS$$

The sodium hydroxide (NaOH) also reacts with CO_2 forming our old friend Arm and Hammer, sodium bicarbonate:

$$NaOH + CO_2 \longrightarrow NaHCO_3$$

Again, either remove CO_2 before using NaOH, or double up. In either of the above (Na_2CO_3 or NaOH) scrubbing processes, the sodium bicarbonate formed might be used as a buffer solution in the generator. However, the addition of the sulfur-containing byproducts will probably cause the concentration of H_2S in the resulting biogas to increase, so don't use the solution unless the H_2S content of the biogas is unimportant.

Plain Good H_2O

The final method for H_2S scrubbing mentioned here uses water. This method is based on the fact that H_2S is soluble in water to the tune of 2.6 volumes of gas in one volume of water at 20°C. Countercurrent flow (gas and water flowing in opposite directions) and violent agitation will help this process achieve its right and true end—the removal of H_2S. CO_2, once again, is also dissolved in water. How much? See Appendix 7.

Water and Chlorine

Henry (1961) states that the addition of chlorine gas to the biogas, prior to scrubbing with water, reduces the need for large volumes of water. He states:

> With chlorine addition, scrubbers can be made smaller, simpler, and lighter. They may be made almost entirely of fiberglass and plastic.
>
> The trials were made without recirculation of scrubbing liquid. In this operation, at relatively high pH, the chlorine demand was about 5.1 lb/hr for 100 cfm of gas input, containing 100 grains/100 cu ft of hydrogen sulfide. The chlorine demand depends on the amount of hydrogen sulfide. This amounts to about a 1:2.5 molal ratio, midway between complete oxidation to sulfate and partial oxydation to free sulfur. At low pH's, the reaction goes to free sulfur only (Black and Goodson, 1952). This last reaction can be achieved by recirculating the scrubbing liquid to maintain a low pH and would result in lower chlorine costs as well as possible production of byproduct sulfur and hydrochloric acid.

Other Possibilities

More selective chemicals for the removal of H_2S are potassium permanganate solution, or a buffered solution of sodium dichromate and zinc sulfate. The high-tech possibilities are numerous but generally inappropriate for small-scale situations. If you like to dream and experiment, you might think about the fact that there are certain forms of bacteria which utilize sulfur. Although biological removal of sulfur could not be used where the gases must flow rapidly, it might be tried where a slow flow over a long distance could be made to move through a substrate that could mechanically support sulfur-loving bacteria.

Water Vapor Removal

Water vapor can be removed from biogas in three general ways: (1) When the biogas is pressurized, some or much of the water vapor condenses. (2) When biogas cools, water condenses. (3) The water vapor can be removed from biogas by using a chemical or substance which is deliquescent, meaning that it grabs any water vapor that happens to be handy, and hangs on to it.

For most purposes, water vapor need not be removed. It is annoying to have condensed water in gas lines, but in any case, gas lines should be laid on a slope, with valves to remove the water from low spots. Drying the biogas will not change this necessity, since there is no such thing as biogas, produced in a low-tech situation that is 100% free of water vapor. Some people claim that greater heat is available from dry biogas than from water-saturated biogas, and this is true—but the difference generally amounts to less than 2% of the total heat value of the biogas. Water vapor in biogas may condense and help cause corrosion in conjunction with H_2S and CO_2, but this also is nearly inevitable since H_2O is a byproduct of the burning of methane (as is CO_2), and very often biogas is stored in contact with water. As long as provisions are made for condensation, and against corrosion, there is no compelling reason to remove water vapor from biogas except where the biogas is stored under pressure.

Should you wish to remove the water vapor, either solid sodium hydroxide or calcium cloride ($NaOH$ or $CaCl_2$) are deliquescent. They can be regenerated for use by heating them to drive off the water.

Scrubber Design

For scrubbing methods which use solid materials as gas filters (for example, the dry box method of H_2S scrubbing), a large (4 to 6 inch) pipe, 12 or more times as long as wide, packed with active scrubber material, can be used. The active scrubber material should be packed in thin (3 or 4 inch) layers and tamped each time before another layer is put in. This slow and tedious packing method, if carefully done, will result in better use of the active scrubbing material (filtrate). What happens in the real world when the biogas is forced through the filtrate is that it seeks the path of least resistance. Thus, some of the material is exposed to all of the biogas, and some to none, since the biogas bypasses it entirely or almost entirely. This fact and others conspire to make it impossible for a given amount of filtrate to react completely with a given amount of biogas. Thus, even if the chemical formulas lead us to believe that two parts of this will react with and remove one part of some contaminant, it may in reality take three, four, or more times that amount of filtrate to really do the job.

Another way to overcome this difficulty is to use two-stage scrubbing. In this scheme, two pipes are used. Biogas flows through one and then the other. When tests indicate that the filtrate is no longer doing what it should, the second pipe is placed in the number one position, and a new pipe is put in the number two place. Your tests may indicate that used filtrate (even of a kind that cannot be regenerated like the rust mixtures above) may still have some scrubbing capability left. This can be the case since, when it is removed from the pipe, carefully mixed, and carefully repacked, new material, previously bypassed by the biogas, may become available.

Scrubbers making use of countercurrent flow and water, or water-based solutions, can be made out of larger pipe. Ordinary garden hose spray nozzles can be used to provide a spray in order to mix the biogas and the scrubbing solution together.

Summary

Each of the undesirable gases in biogas can be removed, should you wish, but the reason for removal should be good enough to make the investment worthwhile. Water vapor, if it is removed, should be removed last, because many of the other scrubbing processes will tend to saturate the biogas with water vapor. CO_2, if it is removed, should usually be removed first, because then the volume of the biogas will be greatly reduced, allowing everything thereafter to be designed for just that much less gas volume or throughput rate. Smaller gas volume will also increase the relative concentration of the other contaminants, and thus it will make their removal more efficient.

Terms

Scrubbing: Removing other gases from biogas so that it ends up being more nearly pure methane.

STP: Standard Temperature and Pressure.

Questions

None

Problems

None

20: Combustion

Combustion is chemical oxidation, or in mildly inaccurate terms, the combining of free oxygen with a molecule, resulting in one or more different byproduct molecules and the liberation of heat.

As your experience may confirm, when simple or complex substances burn, they produce different colored flames, different amounts of heat, and burn at different rates. If the rate of combustion is rapid enough, there will be an explosion. If it is slow, the substance may smolder or glow without flames being produced. If the combustion—chemical oxidation—is very slow, then other results (rust, aerobic composting, wood rot) may appear.

Down where the molecules dance and sing, regardless of the process, certain things occur, which we can visualize as follows:

1. A force, usually heat, causes the original molecule to break apart. Methane, subjected to a temperature of around 610°C, will break apart into C and H. Energy stored in the molecular bonds is liberated. More complex substances, like wood, will break down into gaseous molecules before such high temperatures are reached and then these gaseous molecules will further be heated until they break down—disassociate—like methane. In this sense, wood doesn't burn. What happens is the wood is heated until it gives off a gas, and the gas burns.
2. If free oxygen is available in sufficient quantities, the oxygen will combine with the carbon and hydrogen atoms (from "shattered" methane molecules) to produce CO_2 and H_2O. This recombination into still more stable molecules requires some energy to accomplish, but not as much as the original methane molecule had stored. The excess energy is given off as heat.

   If free oxygen is not available in sufficient quantities, the hydrogen will steal most of the oxygen to make water, and the free carbon will be heated to incandescence. This reaction is what causes the yellow part of the candle to glow. Often, carbon monoxide (CO), that deadly and odorless gas, is produced.
3. The energy liberated from the bonds of the methane molecule which is not required in forming the new bonds in the water, carbon monoxide, or carbon dioxide molecules, serves to heat other methane molecules to the ignition temperature of around 610°C, and the reaction continues as long as there is sufficient fuel (methane), sufficient oxygen, and sufficient heat. In the case of some other molecules, the heat required to stimulate combustion is minimal. Coal, for example, oxidizes at almost any temperature when oxygen is present. Some kinds of coal must be either kept away from air or stored so that there is not a heat buildup, otherwise, combustion may accelerate to the flash point, and fire will break out.

The Most Heat

The amount of energy liberated from methane is fixed by the nature of the bonds on the methane molecule, which is supposed to look like this:

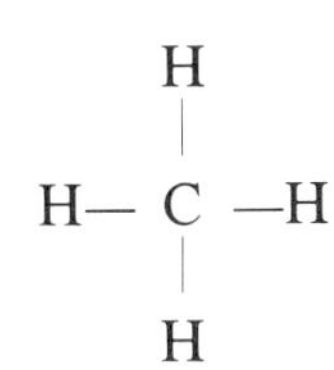

However, the amount of energy that ends up being useful is variable, depending on several things. Theoretically, it takes 2 cubic feet of pure O_2 per cubic foot of pure CH_4, for complete combustion to take place. In practice, more oxygen is required because in a flame (where gas is moving into an area of heat and then the byproducts are rapidly ventilated away) there is not always a spare oxygen molecule for each and every waiting methane molecule. So, more than 2 cubic feet of oxygen for each cubic foot of methane is required to fill in wherever the gases are less than perfectly mixed.

As well, we do not have either pure methane or pure oxygen. Two units volume of oxygen is found in 9.53 units volume of ordinary air. A unit volume of methane is found in 1.67 units volume of biogas (for a 60% methane biogas). The result is that to burn 3 units volume of reactants ($CH_4 + 2O_2$), we need to heat over 12 units volume of our air-and-biogas mixture to the kindling temperature of the methane (at 610°C). Of course, much of this energy comes back to us as heat, but the point is that it is more difficult to extract a given amount of heat from a large volume of gas than it is to extract that heat from a smaller volume of gas. As a result, some of the heat is lost to us. In this case, more is less.

A higher temperature (and therefore easier heat extraction) is easier to obtain with smaller volumes of gas, than it is with larger volumes of gas, if both have the same amount of heat. If this application confuses you, go back and review this in the chapter on energy.

Speed of Reaction

In the real world, everything takes time. An object not only has three dimensions in space, it also has a dimension in time. For example, my snowman lasted two hours before it melted.

Events (even though more complex than objects) also have a dimension in time. "All of a sudden," is still not instantaneously. It is difficult for us to imagine a billion years, but it is also difficult for us to imagine a billionth of a second. Nevertheless, there are events which occur in both of those time frames.

The combustion of methane is a chemical reaction which has certain limits in time. Methane has a slow reaction velocity as compared with the combustion of hydrogen—or almost any other inflammable gas.

Suppose we have a one-inch diameter glass tube, filled with 9.8% pure methane, 90.2% air. (Note that whenever we refer to a percentage of gases in this book, we are speaking of percentage by volume. Percentages can be measured as a ratio of weights, which is how we refer to percentages of solids and liquids. The reason is that solids and liquids are easy to weigh, but often hard to measure volumetrically. Gases are easy to measure volumetrically, but hard to weigh.) If we light that mixture, its flame will travel 0.67 meters per second. In the same glass tube, however, a mixture of 37% H_2, 63% air will travel 4.88 meters per second. (Both of these experiments occur at standard temperature and pressure. Enlarging the tube diameter or increasing the initial temperature of the gas mixtures will cause these speeds to increase.)

Both of the mixtures described are the optimum mixtures for combustion. The conclusion is that methane burns seven-and-one-quarter times more slowly than hydrogen. The slow flame velocity of methane means that flames such as pilot lights will not be as faithful in a biogas system as they would be in some sort of hydrogen-powered device. They might go out because the flame itself could not travel down the rising column of gas and air as fast as it need to, in order to keep burning.

Limits of Inflammability

Methane is inflammable or explosive, between 5.3% and 14% in still air, or between 5% to 15% in turbulent air. In pure oxygen, methane is combustible from 5% to 59%. The theoretical optimum combustion mixture is 9.8% methane in air, as we mentioned.

Combustion Reactions

Now, all you chemistry and physics fans, it's once again time to make the atoms sport and play. Complete combustion of methane occurs as follows:

$$CH_4 + 2O_2 \longrightarrow CO_2 + 2H_2O$$

There results from this 11.90 net Cal per gram CH_4.

Incomplete combustion of methane gives us carbon and carbon monoxide (CO):

$$CH_4 + O_2 \longrightarrow C + 2H_2O$$

(and 4.05 net Cal per gram CH_4)

$$2CH_4 + 3O_2 \longrightarrow 2CO + 4H_2O$$

(and 9.48 net Cal per gram CH_4)

Both C and CO, above the proper temperature and with enough oxygen, will further burn:

$$C + O_2 \longrightarrow CO_2$$

(and 7.86 net Cal per gram C)

$$2CO + O_2 \longrightarrow 2CO_2$$

(and 2.43 net Cal per gram CO)

Methane, composed of only carbon and hydrogen, will give off characteristic colors when it burns with various proportions of air. The particular chemical reaction which is taking place is indicated by the flame color. Of course, when these different reaactions are happening simultaneously in one flame, they are difficult to distinguish.

$$C + O_2 \longrightarrow CO_2$$

(nearly colorless blue)

$$2H_2 + O_2 \rightarrow 2H_2O$$

(nearly colorless yellow or white)

$$2C + O_2 \rightarrow 2\,CO$$

(bright blue)

$$2CO + O_2 \longrightarrow 2CO_2$$

(pink or rose)

$$CH_4 + 3O_2 \longrightarrow CO_2 + 2H_2O$$

(essentially colorless or white)

When the air and gas mixture is correct (slight excess of air over theoretically perfect mixture of CH_4 = 1 volume, air = 9.5 volumes) and when the flame is small relative to the volume and pressure of the gas, then the flame will burn at its hottest. The maximum theoretical flame temperature of methane is 2,000°C (3,640°F), but this is never achieved in practice. This planet, after all, is not the realm of perfection.

Terms

Combustion: Burning, loosely speaking.

Questions

None

Problems

None

21: Illumination

Burnable gases can be used in two general ways to produce illumination. They can be burned to give light directly, as the wax for a candle is burned; or they can be burned to give heat so that another material, brought to incandescence, will glow. In the same way, electricity heats a filament which then incandesces, and fuel in a Coleman lantern or Aladdin lamp is used to heat a mantle to incandescence.

Luminous Flame

The amount of illumination given off directly by the flame of a hydrocarbon fuel such as methane is due to the fact that hydrogen, in burning, gives off a great deal of heat which causes the remaining carbon to achieve incandescence. This phenomena can be seen in the flame of a candle. If you have a candle available, get it and light it so that you will be able to follow this discussion and compare it with the reality of the flame in front of you.

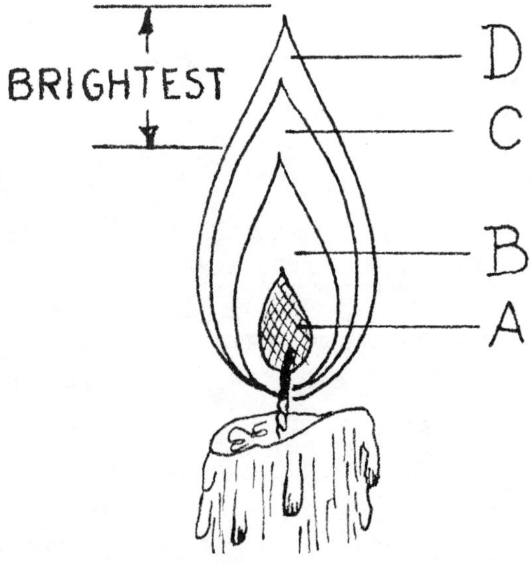

Fig. 21.1 Parts of a Flame

The flame of a candle is very similar to the flame produced when methane burns straight out of an ordinary pipe—in other words, when no air or oxygen is mixed with the methane before burning. This flame has four parts or cones. While they are not strictly isolated from each other—as Fig. 21.1 suggests—they can be identified as areas of different activities. Cone A is bright blue but it emits very little light. In this area of the flame, the gases are not burning. Rather, they are being heated up to kindling temperature. No free or available oxygen is present in Cone A.

Cone B is a darkish yellow transparent area. Here, the molecules of the gases are further heated, reach kindling temperature, and begin to break down into atoms. However, no oxygen is yet present to cause chemical oxidation.

Surrounding this darker cone is a cone of brilliant yellow (Cone C). Here, the reaction $CH_4 + O_2 \rightarrow C + 2H_2O$ takes place. Insufficient oxygen is available to cause complete combustion, and so the hydrogen—more greedy, if you will, for the oxygen—is oxidized to water, and this reaction heats the remaining free carbon to incandescence. The brightest part of the flame is indicated. Here the unburned carbon is hottest. The shell of this cone can be seen in the candle flame extending down around Cone B.

Last, and very difficult to see in the candle flame, is the area where the reaction $C + O_2 \longrightarrow CO_2$ takes place. Here, in Cone D, more heat is generated but little visible light is given off.

Free oxygen is supplied to the flame from the surrounding air. This oxygen has to penetrate the cones of the flame, and most of it is intercepted by Cone D (the CO_2 layer) before it can come to Cone C (the H_2O light-giving layer). Unless methane can burn in stages, no direct light will be given off, for the complete combustion reaction ($CH_4 + 2O_2 \longrightarrow CO_2 + 2H_2O$) yields little or no usable light.

Increasing a Flame's Light

A biogas flame, burning without premixed oxygen, is an "open flame." Such flames can be produced by simply allowing the raw biogas to burn from a hole, or out of the open end of a pipe. An open gas flame will be more luminous as it becomes hotter. The larger the surface area of the flame (given that the amount of gas present for fuel is equal), the more heat it will lose to the surrounding air and thus the cooler it will burn. For this reason, larger flames are generally cooler and darker flames. As well, if the pressure of the gas is too high for the size of the hole from which the flame is burning, it will either blow off, or air will mix with it so completely that combustion goes immediately to completion. When this happens, free carbon does not exist in the flame long enough to glow and emit light—therefore the flame is of low luminosity. For these reasons, the flame must not be made too large in relation to the size of the burner hole, and in relation to the amount of fuel consumed.

If the flame is too small, sufficient oxygen will be taken up into the flame and the free carbon produced will pass out of the flame as smoke before it can be heated enough to burn. This condition can be dangerous because one of the byproducts of incomplete combustion is CO, carbon monoxide, a poisonous gas. This smoky condition can be eliminated by moderately increasing the supply of gas to the flame in order to increase its surface area.

The ability of an open flame to produce light directly will be enhanced by a glass chimney, of the kind commonly used on kerosene lanterns. This chimney increases light because it helps the flame stabilize, since drafts cause the flame to flutter and smoke. The glass shields the flame from drafts. The light is also increased because the glass reflects radiant energy back into the flame, maintaining a higher temperature, and the chimney will help create and maintain a flow of air upward around the flame so that oxygen is mixed with it properly.

Open Flame Burner Designs

The simplest open flame burner is merely a hole in a gas pipe, or a small tube, from which gas issues and is directly burned. On a comparative scale with other burners, in which 100 is the maximum expected light output for a given amount of gas input, the hole or tube flame burner rates at 4.

A more efficient burner, and not difficult for the average individual to manufacture, is the bat's wing burner. The bat's wing burner is simply a cap on a tube in which slots or slices have been made instead of a hole. On the 0-100 efficiency scale, this burner rates a 10.

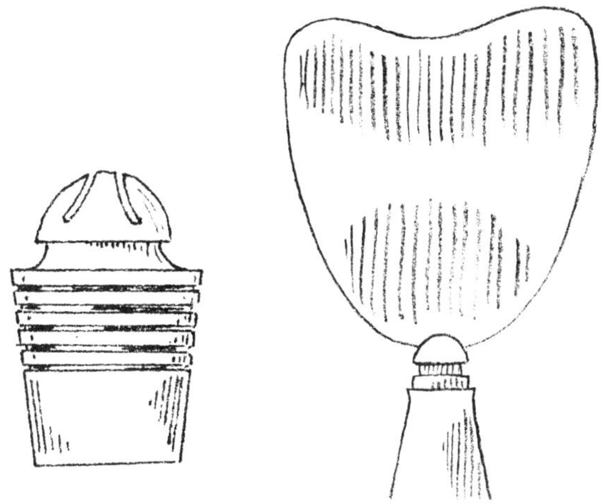

Fig. 21.2 Bat's Wing Burner (2 views)

The best open flame burner is known as the Argand burner. Shaped similarly to the top part of a light bulb socket, the Argand burner has little holes placed in a ring. The gas coming out of these holes burns together into a circle or cylinder of flame. Air comes up the center of the ring and also flows around the outside. Oxygen is available to both the inner and outer surfaces of the cylinder of flame. The whole is surrounded by a glass chimney.

On some Argand burners, the gas is preheated before burning. This is accomplished by coiling the gas input pipe above the flame. Be very careful if you try this, since if you have oxygen in the biogas, it could begin to burn inside the supply pipe, travel back to the gasholder, and blow up. These modified burners are referred to as regenerative burners.

Unmodified plain Argand burners achieve a rating of 12 or 13. Regenerative burners achieve a rating of 28 to 40.

Incandescent Illumination

In incandescent burners, the heat of the burning gas is used to bring a woven net of the "ash" of certain kinds of compounds to incandescence. While the idea in open flame illumination is to increase the light of the flame, the idea behind incandescent illumination is to increase the heat of the flame, so that the material which is made to glow will be as hot as possible.

This is accomplished primarily by adding air to the gas before it begins to burn. The two main types of premixed burners which are used are the bunsen burner and the fisher burner. These will be discussed in greater detail in the next chapter.

Incandescent burners, then, are essentially bunsen, or better, fisher burners, over which a cone made of material which will incandesce when heated, is hung.

These mantles, as they are commonly called, are made of fibers soaked in solutions of compounds like thorium oxide and cerium oxide. Once the mantle is exposed to flame, it becomes quite brittle and can be easily broken. For this reason, these mantles come in two varieties. One is the type commonly used in Coleman lanterns. It is flexible, and is sold in a small plastic bag. To prepare it, the mantle is tied around the burner outlet and burned either with the gas or with a match. This burns away the soft material in the mantle and leaves the skeleton of the incandescent ashlike material.

The second type is exemplified by the Aladdin lantern mantle. This mantle is preburned, and not flexible. After the mantle fabric is manufactured, it is put on a wire frame and burned. Then the mantle is dipped in a solution which dries and holds the mantle together—makes it shockproof—during shipment. To use the mantle it is placed on the lantern and the dried solution is burned away, exposing the skeleton. These mantles are generally more brittle than the first type.

An incandescent biogas-powered lamp can be constructed out of one of the two burners (bunsen or fisher) and a mantle. It is also possible to buy lights that operate on butane or propane. Check Appendix 18 for other sources. Lights which operate on propane will be designed for an air/fuel ratio of about 25. Those designed for butane will have an air/fuel ratio of about 33. In order to use them for methane, less air should be mixed into the gas, so that the air/fuel ratio will be below 9, depending on the design of the burner. Any lamp designed for natural gas—which is principally methane—should operate quite well on scrubbed biogas. Many decorative lamps which operate on natural gas have been available, but these are usually cumbersome and inefficient, because they are designed to burn continuously and at a lower temperature both to conserve gas and extend the life of the mantle. However, your local natural gas dealer (in the nearest city) might be a good source of information on where to find such lamps (the guts might be usable, even if the whole lamp is not). In some places, natural gas lights are not easily available because of the shortage of natural gas.

Any lamp can be made more efficient by regenerative heating. These incandescent lamps can rate from 70 to 100 on our efficiency scale.

Type	Rating
Simple tube or hole	4
Bat's wing	10
Argand	12—12
Regenerative Argand	28—40
Incandescent	70—95
Regenerative incandescent (high pressure)	90—100

Table 21.1 Lamp Ratings

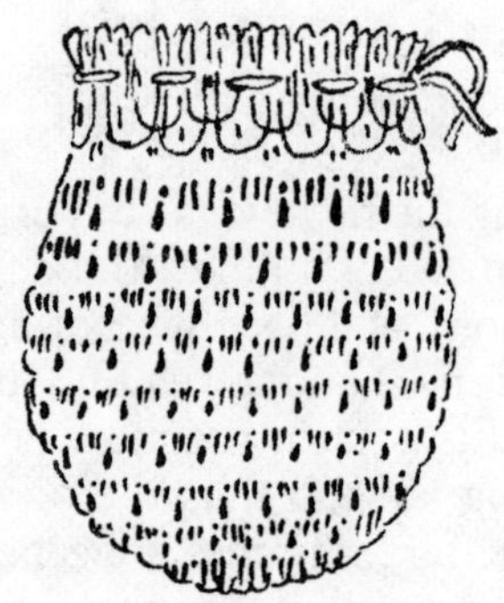

Fig. 21.3 Coleman Mantle

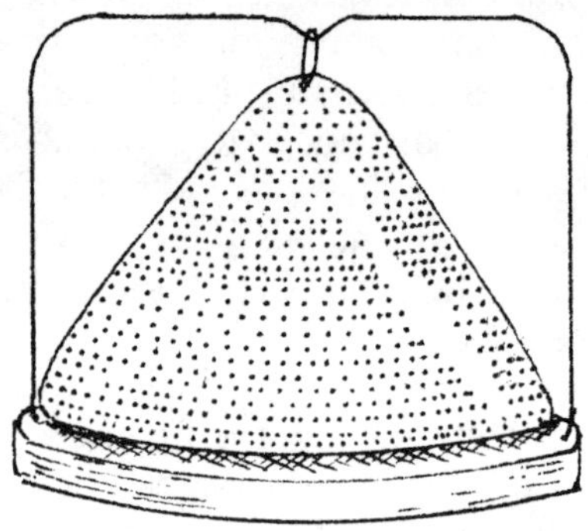

Fig. 21.4 Aladdin Mantle

These ratings can be related to the light output of a lamp per heat energy unit per hour. Gas lamps of good design (inverted mantle, high pressure), can achieve 2.3 lumens per Calorie (of energy input as biogas) per hour (0.58 lumens per Btu per hour).

Compared to electric illumination, this is definitely not big potatoes. A 100-watt bulb will give 14.2 lumens per Calorie (of energy input as electricity) per hour. But even this is outclassed by fluorescent lamps, which can give up to 73 lumens per Calorie per hour. At 100% efficiency of conversion, heat into light, we would expect 721.5 lumens per Calorie per hour, so nobody can really brag.

At any rate, the light equivalent of a 100-watt bulb (using the above figures) will cost us 355 Cal (1,410 Btu) of heat energy per hour. At 5.8 Calories per liter of biogas (650 Btu per cubic foot), 100 watts of light would be obtained from 60 liters of biogas (2.2 cubic feet) burned each hour in a very good incandescent mantle lamp. This can, however, vary a great deal. One propane light listed in the Sources appendix only achieves a 50-watt light output at a cost of 452 Cal per hour (1,800 Btu).

This is approximately a "best case" and "worst case," and most inverted mantle manufactured natural gas or biogas lamps give light and burn gas in this range. The efficiency of burning can be increased by scrubbing the biogas.

In the absence of scrubbing, the potential for the

biogas to give good light can be enhanced by passing it through a container which has gasoline in it. It is not necessary to bubble the biogas through the gasoline, according to Ram Bux Singh. Even so, it is dangerous to mix biogas and gasoline vapor and any device which is constructed to accomplish this should be viewed with great suspicion. It is a potential bomb. Whenever we open it up to put in more gasoline, we also let air in the biogas lamp feed line, possibly producing explosive mixtures.

One larger lamp, using the same amount of gas as several smaller lamps, tends to produce more light. Inverted incandescent lamps (the mantle is below the main body of the lamp) with good reflectors, deliver more of the light they produce than upright lamps of similar design. Heat accumulation can be a problem with these lamps, and if someone wishes to design and manufacture his own, heat-resistant materials should be used.

Terms

Incandescent lamp: A gaslight with a mantle.

Mantle: Cone of material which flows to emit light, when heated.

Open flame burner: A gaslight which gives light directly from an open flame.

Questions

None

Problems

None

22: *Burner Design*

In the last chapter, we discussed open flame burner designs and mentioned premixed burner designs—the bunsen and fisher burners.

Bunsen was a German chemist who became interested in combustion, when—as the story has it—a pot he was heating over an open gas flame came too close to it and soot (unburned carbon from Cone C—CH_4 + O_2 ——> C + $2H_2O$ was deposited on the bottom. Upon realizing that the black stuff was unburned carbon, Bunsen deduced the cause and decided to make his burner more efficient by premixing air with it. (Any burner which does this is now called an atmospheric burner.) See Figure 22.1 to get an idea of Bunsen's design.

The lettered parts are A, the gas injector; B, the mixing tube; C, the air inlets; and D, a collar which can be moved around to open or close the air inlets.

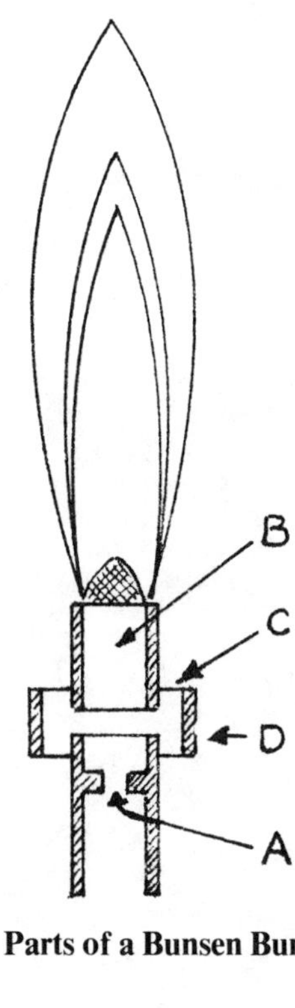

Fig. 22.1 Parts of a Bunsen Burner

As gas is forced out of the injector orifice into the mixing tube, it will pull air into the air inlets and the gas and air will mix. If the air/gas mixture is not moving too fast, nor too slow, and if there is not too much air in the mixture, the flame on this burner will stay at the top of the mixing tube.

The flame of the bunsen burner derives only part of its oxygen from the premixed air and the rest from the surrounding air. The reason we cannot feed the gas all the oxygen it needs from the premixed air is that if the gas/air mixture coming up the mixing tube were more nearly perfect, the flame would flash back, traveling down the tube to the injector, where it would rest, open flame style. For this reason, the bunsen burner is still not the highest and best type of gas burner in terms of its ability to deliver the utmost in concentrated heat. That is, to some degree, it still acts like a candle, and the various cones of its flame compete for oxygen.

However, varieties of the bunsen design are nearly universally used for stoves, ovens, griddles, and the like, because of their simplicity of design and ease of cleaning.

Flashback can happen in any of these burners when the gas pressure falls, or as the gas valve is turned down, because the supply of air remains great while the supply of gas decreases. Thus the proportion of air in the mixture increases until flashback is possible. Three solutions to this problem are: (1) Use a valve which regulates both air and gas. (2) Use each size of burner within a restricted range of gas pressure and volume and therefore of heat output. (3) Use a burner design which mixes air into the gas in proportion to its velocity and abundance.

The velocity/mix burner design modification is primarily a modification of part A in Fig. 22.1. As shown, part A (the injector) is not very efficient at *entraining,* mixing the air into the gas stream. Fig. 22.2 shows the modified design.

The injector is threaded inside to accept O, the orifice disk which is a flat plate with a hole in it. The orifice disk can be screwed into or out of the body of the

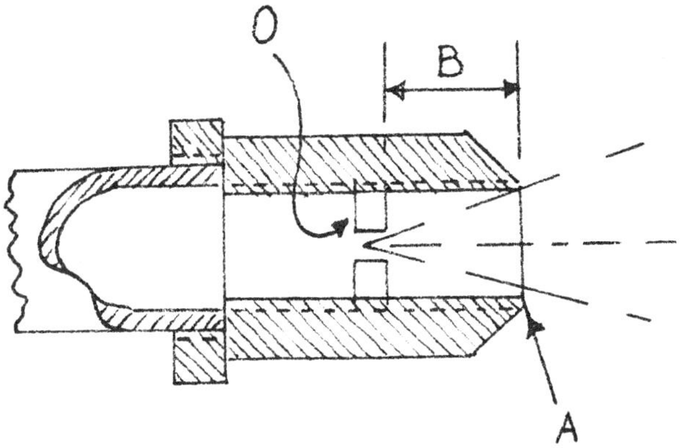

Fig. 22.2 Injector

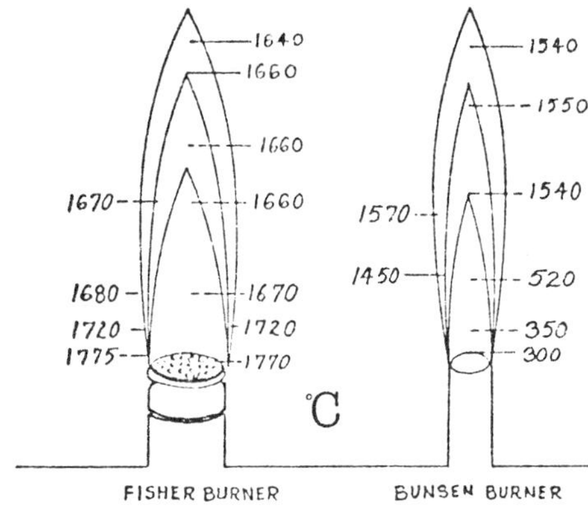

Fig. 22.3 Flame Temperatures Compared for Fisher and Bunsen Burners

injector, so that B, the distance between the injector lip and orifice disk, can be changed. This injector will entrain a certain amount of air depending on the velocity of the gas stream and the distance B. Using this injector, the total area of the air inlets is not as critical as with the previous unmodified injector.

The amount of gas any given injector orifice will release is a function of several factors: the area of the injector orifice, the pressure of the gas, and the specific gravity of the gas, a measure of its density (weight per unit volume) relative to air. A formula relating these variables, and tables for changing orifices used with other gaseous fuels into orifices suitable for use with biogas, are found in Appendix 7.

Beyond changing the injector, the basic bunsen design can be greatly upgraded by one simple change—putting a screen of finely woven wire on top of the mixing tube. The wire should be made of brass or copper, woven with 11 or more strands to the centimeter (28 or more to the inch) and having no less than 122 crossings of the wire per square centimeter (800 per square inch).

If you have a piece of wire gauze lying around that you suspect fits these criteria, don't ruin your sanity and eyesight trying to count itsy, bitsy wires. Get a linen tester, which is an inexpensive device specifically made for such jobs, basically a magnifying glass mounted over a square hole, the edges of which are marked off in millimeters or small divisions of the inch. Many places have them, but if your resourcefulness quotient is low, check Appendix 16.

This wire gauze conducts heat away from the bottom of the flame (which doesn't quite touch it). It is cooled largely by the gases flowing up the mixing tube, and it in turn shields those gases from the energy of the flame. This means that the gas/air mixture in the mixing tube will not reach ignition temperature, and thus a burner with this wire gauze modification can mix air in the burner tube, in the proportions necessary for complete combustion, without danger of flashback.

A bunsen burner with one or more gauze fire checks is a fisher burner. The fisher burner can develop higher heat in a smaller area because of its better air/gas mixture. More information on burners is found in the next chapter.

Unscrubbed, 60% CH_4 biogas	Scrubbed, 95% CH_4 biogas	Comments
6.3	10.5	Too lean
5.8	9.1	Best
4.8	8	Too rich

Table 22.1 Air/Fuel Ratios for a Fisher Burner

Terms

Bunsen burner: Simply designed, universally used burner.

Entraining: Mixing.

Fisher burner: Bunsen burner modified with a fine-meshed screen to stop flashback.

Injector: Orifice.

Orifice: Injector.

Questions

None

Problems

None

23: Cooking

Some simple, but fascinating, experiments on cooking were done by Dr. Rudolph Hauschka and reported in his book *Nutrition.* Dr. Hauschka boiled distilled water in pots made of different materials (clay, aluminum, gold, etc.) and used different sources of heat (electricity, coal, wood, etc.). Then, cooling the waters thus treated, he sprouted seeds with them. He found that both the material of the pot and the source of the heat affected the amount of growth the seeds made. The worst results were for aluminum pots and electric heat, the best results were for gold pots and the heat from burning straw. Wood heat and earthenware pots also gave very good results.

Mention is made of this to point out that the benefits of biogas—a very natural energy source—may go beyond those postulated by the modern science of thermodynamics and its theories. If this seems too farfetched to you, why not try the experiments yourself?

Cooking with scrubbed biogas is simple, since its burning characteristics are so similar to those of natural gas. Therefore, a natural gas stove will generally work without modification with scrubbed biogas. The burner on a stove works the same way a bunsen burner does, as the air/fuel mixture will be slightly less than optimum to insure that the flame does not flash back into the mixing tube.

Cited by almost everyone writing on biogas are the experiments conducted by the General Laboratory of the Gas Council, Watson House Centre, in Great Britain. We have not been able to track down the original report of this research, so the figures reported below are from Ram Bux Singh (1975). (From whence, it seems, everyone else is obtaining the figures.) Working, apparently, with biogas of 68% CH_4, and a heat content of 6.1 Calories per liter (678 Btu per cubic foot), researchers found that stable flames could be obtained using a burner with 36 ports or holes, each 3 millimeters—Singh says 0.114 inches—in diameter.

Under this burner, two injector hole sizes were successful—1 and 1.04 millimeters in diameter (Singh: "0.038 and 0.041 inches")—and the gas was delivered at pressures varying from about 3 to 20 centimeters on a water manometer (1 to 8 inches).

The Watson House Centre found that a ratio of total burner hole area to injector hole area should be 300 to 1. Using Singh's report, we can calculate that the ratios varied from 324:1 to 278:1. Further, Singh reports: "The heat input under these conditions ranged from 3,360 to 11,000 (Btu per hour per square inch) of flame port area." At 678 Btu per cubic foot, gas heat value, and with the reported flame port area totaling about 0.37 square inches, this means that, at the lowest heat input the burner was consuming (3,360 ÷ 678 x 0.37) about 1.8 cubic feet of biogas per hour. At the 11,000 Btu input rate, this amounts to almost 6 cubic feet per hour. No correlations of pressure of gas delivery, and injector to flame port area ratio are given.

For those who wish to run similar experiments on their own homegrown biogas, a table giving a translation of drill sizes into areas is included in Appendix 7. It should prove relatively simple to find out how your biogas responds to variations in the ratio under question. The simplest method of testing might be to get a natural gas burner and fiddle with the injector orifice size—which will have to be drilled larger or even taken off when using unscrubbed biogas. (In a low-tech situation, the difficulty will come in determining efficiency, in terms of heat value input (in biogas) versus heat output. Stiff upper lip and all that.)

If anyone has access to the Gas Council report, please send us a copy. Refer to Appendix 7 for more information.

Terms

None

Questions

None

Problems

None

24: Water Heating

Heat is transferred in three ways: (1) conduction; (2) convection; and (3) radiation. When your hand gets burned as a result of touching a hot stove, that's conduction—heat is transferred directly between two surfaces in contact. Whenever a fluid medium (gas or liquid) is heated, it expands, becoming less dense than the surrounding fluid. This causes heated fluids to rise. One well known example of convection is when exhaust gases escape up a chimney. Sunlight is probably the most universal example of radiation. But radiated energy need not be visible. Radiated heat, of the kind that warms us near the fireplace for example, is not visible. Conduction, convection, radiation.

All three of these kinds of heat transfer occur in gas-fired water heaters. We want to encourage each of them to transfer heat into the water, and discourage them from transferring heat out of it. Gas-fired water heating has a long history and there are many designs for such water heaters because this is a subject which has been near to the human intellect and dear to the human heart for some time—hot water.

Among these designs, several seem worthy of note. Among these we have:

Quick Recovery
Slow Recovery
Direct Contact
Instantaneous

Most libraries have plumbing books which describe these heaters, so we won't duplicate the effort here. Each of these is suited to a particular task, and of the above, the first three are storage types, while the last generally has no storage facility. All storage heaters take advantage of the fact that hot water rises (convection), by tapping the hot water from the top of the tank. The water, in layers, becomes cooler as we move from the top to the bottom, and this is called stratification.

One problem with water heating is sedimentation. Sediment occurs in the water heaters because of the characteristics of the water. Briefly, sediment occurs when water containing certain salts of calcium or magnesium—called hard water— is heated above 65°C (150°F). Under these conditions, the salts are deposited on any surface which is hot, such as a pipe, or the inside of a water heating tank. When this happens, our hot water heater and its pipes develop arteriosclerosis. That is to say, they begin to clog up and close down.

Thermosiphon

One very important proccess, which makes use of convection, is known as thermosiphon. Thermosiphon, as Fig. 24.1 shows, is basically a convection process. Hot water (in picture 2) rises out of the right side of the pipe into the tank, and cooler water is drawn in to take its place. The cooler water is heated, it rises, and the cycle continues. Thermosiphon is easily slowed or stopped by numerous bends, or small diameter pipe,

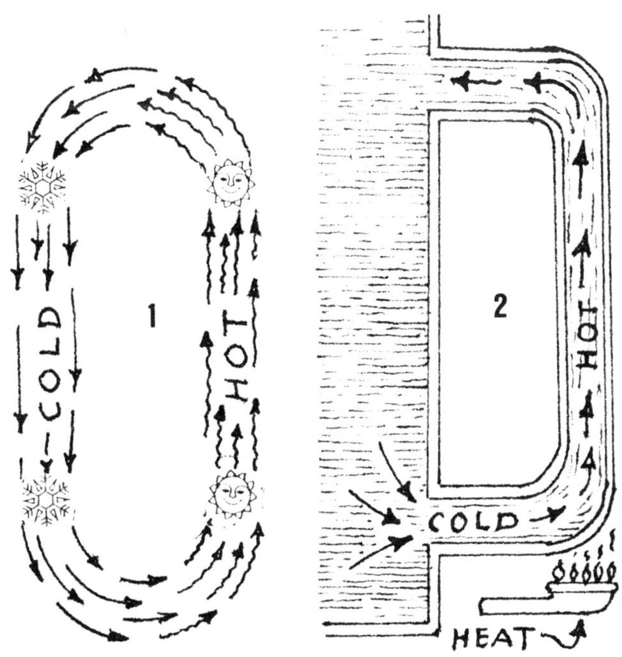

Fig. 24.1 Thermosiphon

and so, if you plan to use it in any way inside of pipes, use it where the bends are few and the pipes relatively large. Note that thermosiphon does not, in any popular sense, pump water. It merely moves hot water up and away, and causes cooler water to take its place. If the water level were below the top outlet of the pipe, then there would be no thermosiphoning action. Remember too, that thermosiphoning works both ways. That is, if the flame goes out and the pipe cools down, then warm water is drawn into the top of the pipe, where it is cooled, and the circuit reverses itself.

A Few Notes About Safety

If you build or buy a water heater, make sure it has a pressure relief valve and that the thermostat works and continues to work. It's cheap insurance. Otherwise, your water heater may be a real blast. (They have been known to explode, you know.)

Because hot water is dangerous above 70°C (160°F)—it can give severe burns—it is not wise to heat it above this temperature for storage, since a child's mistake in not mixing hot with cold at the faucet could be a tragic one. If hot water in greater volume is required, the best option is to heat larger quantities to lower temperatures, as it is generally easier to efficiently heat and store water at a lower, than a higher temperature. High temperatures also accelerate precipitation of scale and corrosion of pipe.

It has been reported (*Sewage Works Journal*, volume 16, page 628) that the deposits formed on the inside of biogas-fueled boilers are deliquescent. That is, they snatch moisture from the air and become sticky, or in some cases, drip down the sides of the combustion chamber. This may or may not happen to you, but be aware.

Direct Contact

While we said that we wouldn't spend time explaining water heaters to you, at least one should be mentioned. For water not intended for consumption—as for example, the water in a space heating system—a marvelously efficient method of heating is to mix the water with the hot exhaust gases. Heaters of this design are known as direct contact water heaters.

In September 1931, the Purdue University Engineering Experiment Station published Research Series Paper number 38, *Development of a Direct Contact Water Heater* by L.A. Scipio. In this paper, Scipio describes a series of experiments directed toward developing a direct contact heater and his success along the way. The research is marred by the fact that Scipio seems to have had access to a complete foundry and machine shop, and so the design he finally settled on is of Rube Goldbergian proportions.

There can be little doubt that it did the trick, however, since Scipio reports from 76% to 96% efficiency of conversion. This, dear friends, is a beautiful anomaly, as most water heaters are nowhere near this efficient.

The basic design idea involved mixing the exhaust gases in a downward flowing stream of water, to create an artificial draft. The essential elements of the design were the burner, the burning tube, the water nozzle, and the water/exhaust gases mixing tube. The water heater operated (necessarily) at ambient air pressure, so a pump was required to operate the entrainer nozzle and push the hot water around to wherever it was needed.

A fisher burner, with its air inlets exposed and its mixing tube surrounded by a water-jacketed combustion chamber (lined, if possible, with some refractory material to enhance combustion by preventing rapid cool off of the exhaust gases) could operate as a burner plus burning tube, should someone desire to duplicate the basic idea of Scipio's heater in a simpler form.

The nozzle might be an ordinary shower head, all metal, adjusted for a medium fine spray and a 60 degree angle of spray. The water/exhaust gases mixing tube should be 10 to 15 cm (4 to 6 inches) in diameter, and filled with smaller tubes to increase the surfaces of contact. Its bottom should not rest in the stored hot water. Note that Fig. 24.2 is a schematic picture; if you were to build such a heater exactly as shown here, the burner tube would rapidly fill with water. Either lower the water nozzle or turn the burner tube so that it is upright.

No increase in the hot water's acidity due to the mixing of combustion gas was noticed, but the gas burned apparently had little or no H_2S.

Another interesting design for a water heater is the instantaneous variety. Basically, these heaters have no storage capacity and they are designed to heat water as it is used. They are generally not efficient when heating the water, since they operate so rapidly. However, in places where a little hot water is required occasionally, which are a good distance removed from any other point of hot water use, an instantaneous type heater may be useful, and even more efficient, since they will not need to hold the water at some (warm or hot) temperature, meanwhile losing heat.

Instantaneous heaters are basically constructed of coiled (generally copper) tubing on which a flame burns whenever the faucet is fiddled. (Nero would have loved them.) Design details are left to your imagination. Let it roam. (Copper tubing can be bent in spirals, but it is better to fill it with sand before attempting to bend it, as when it is packed with sand it will not as easily collapse. Remember the principle of countercurrent flow.)

Venting

In venting the exhaust products from hot water heaters or furnaces, if the exhaust is cooled below about 60°C (140°F), the water vapor in the exhaust will condense.

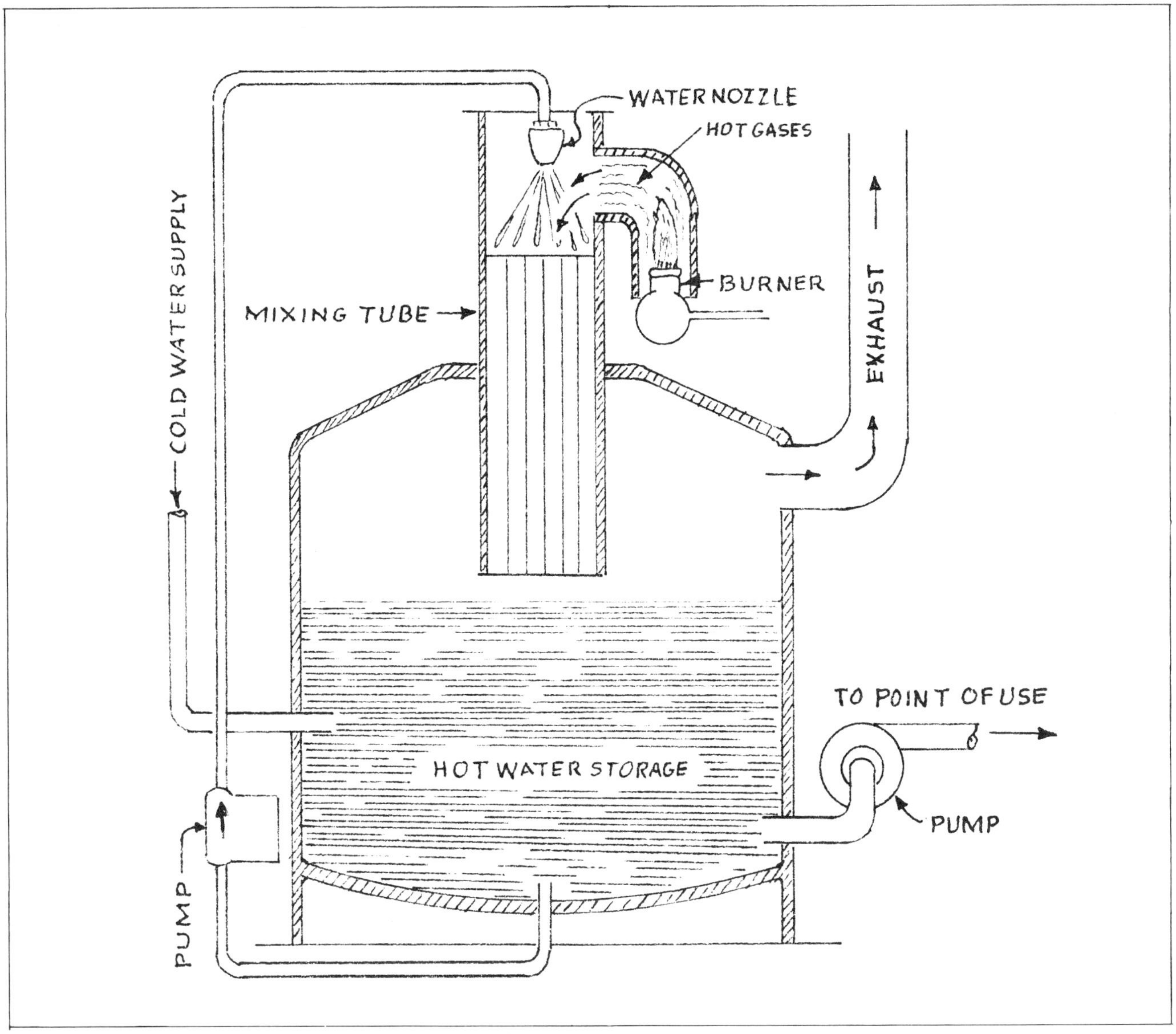

Fig. 24.2 Direct Contact Water Heater

(The exact temperature at which this condensation will take place—called the dew point or condensation temperature—depends on the amount of water vapor present, and the composition of the exhaust gases.)

The condensed water will corrode nearly any metal it comes into contact with, since it will contain varying amounts of carbonic and sulfurous acids.

The three solutions to this problem are: (1) Use no metal in the venting chimney. (2) Keep the temperature of the exhaust products above 60°C (more or less). (3) Dilute the exhaust with warm but dry air (which will decrease the percentage moisture in the exhaust and air mixture and thus cause the condensation temperature to drop). This last is done by using a back draft diverter, several examples of which can be seen at your local furnace dealer.

Heat Requirements

The amount of heat required to push the temperature of a colder quantity of water to some higher temperature is related to the specific heat. It requires one Calorie to heat one kilogram of water through one degree Celsius (1,000 calories). This is the specific heat of water. In American, 1 Btu is required to heat one pound of water through one degree Fahrenheit. The reason these numbers are so round is that these are very nearly the definitions of Calorie, and Btu.

When biogas is burned to accomplish the heating of water, only a portion of the energy tied up in the molecules of the methane shows up as the heat in water. This portion determines the conversion efficiency. To put this all in the sweet language of mathematics, the SI units:

$$Vb = \frac{S\ V\ w\ \Delta t\ c}{Hv\ E}$$

where:

Vb = volume of biogas in liters
Hv = heat value of biogas in Calories per liter
E = overall conversion efficiency as a decimal number
S = specific heat of water in Calories per degree Centigrade per kilogram
Vw = volume of water (not Volkswagen) in liters
Δt = change in temperature of water (either desired or experienced) in °C
c = conversion factor, kilograms of water to liters (1 kg = 1 liter); no units

The funny little triangle is the Greek letter, *delta.* It refers to the change in whatever it is next to; in this case, temperature. "Delta t", then, means the change in temperature. In a formula like the one above, this is simpler than saying "subtract the final (higher) temperature from the original (lower) temperature". The conversion factor is required since we need to change the units of the final answer from kilograms to liters.

For example, if Hv = 5.8 Calories per liter of biogas, and we want to heat 30 kilograms of water (or 30 liters, since one kilogram equals one liter) from 10°C to 50°C (Δt = 40°C), and if E = 60%, then:

$$Vb = \frac{(1\ Cal\ °C^{-1} kg^{-1})(30\ kg)(40°C)(1)}{(5.8\ Cal\ l^{-1})(0.60)}$$

$$Vb = \frac{1{,}200\ Cal}{3.48\ Cal\ l^{-1}}$$

Vb = 345 liters (biogas required)

In American, the equation is:

$$Vb = \frac{S\ Vw\ \Delta t\ c}{Hv\ E}$$

where:

Vb = volume of biogas in cubic feet
Hv = heat value of biogas in Btu per cubic foot
E = overall conversion efficiency (no units)
S = specific heat of water in Btu per °F per pound
Vw = volume of water in gallons
Δt = change in temperature of water in ° F
c = conversion factor, pounds of water to gallons of water (8.3 pounds per gallon); no units

The inclusion of specific heat and conversion factors makes these formulas more complex than they need to be, but they are included to show that—as always—the units will cancel to leave us an answer in the desired units. In a simplified form, using the meanings previously assigned to each symbol, the equations are:

SI	American
$Vb = \frac{Vw\ \Delta t}{Hv\ E}$	$Vb = \frac{8.3 Vw\ \Delta t}{Hv\ E}$

One method of estimating hot water requirements is by the use of Table 24.1. Better estimates can be obtained for your own situation by timing how long it takes a faucet to fill a container of known volume. This will allow you to calculate the flow rate for that faucet. Armed with this knowledge, you can measure or estimate the number of minutes (average) you and your family have that faucet on each day, and then multiply the rate of flow per minute by minutes used per day, to find volume of use per day, from that faucet. Hot water appliances such as dishwashers generally have some indication of their water consumption rate written on them, or this figure can be gotten from the person who services such machines. Your own use patterns can then be used in calculations to determine the hot water consumption per day, week, or month. In fact, this is recommended, since as you may realize, the Table 24.1 is based on wasteful habits, and it will not give you estimates as accurate as those you can make with a small bit of work.

The total heat required for hot water in a certain situation will be determined by the factors already discussed (Δt, Hv, etc.) plus the heat required to maintain the water in the water heater at the thermostat temperature (for storage-type water heaters).

The energy required for temperature maintenance is equal to the heat energy lost through the water heater's insulation. We will be discussing methods for estimating heat losses in Chapter 28. Refer to that chapter for more accurate and specific details.

But, for a quick and dirty estimate for a hot water heater with average insulation, the approximate amount of energy required each day to maintain the temperature of the water, in SI units, can be found by:

$$Vb = \frac{2.7\ Vw\ \Delta t}{Hv}$$

where:

Vb = volume of biogas in liters
Hv = heat value of biogas in Calories per liter
Vw = volume of heater, in liters
Δt = difference between the air temperature surrounding the water heater, and the water temperature, in degrees Celsius

(Notice that, once again, we have left out S and c factors for simplicity.) In American:

$$Vb = 22 Vw\ \Delta t / Hv$$

where:

Vb = volume of biogas, in cubic feet
Hv = heat value of biogas in Btu per cubic foot
Vw = volume of heater, in gallons
Δt = difference between the air temperature surrounding the water heater, and the water temperature, in degrees Fahrenheit

Average Hot Water Use

	Liters	Gallons	Comments
Bath, adult	25–40	7–10	
Bath, baby	10	3	
Shower	95	25	5 minutes
Laundry	75	20	
Dishes (per person per meal)	2	0.5	Washed by hand
Personal toilet	20	5	Per day
Other cleaning	20	5	Per day
Shower head	20	5	Per minute
Bath faucet	23	6	Per minute
Sink faucet	17	4.5	Per minute

Table 24.1 Average Hot Water Use

	Liters	Gallons	Comments
Shower	95	25	
Bath	26	7	
Laundry	10.7	2.9	Per day
Dishes	12	3.2	
Bath	5	1.5	Per day
Total	149	40	(rounded off)

Table 24.2 Daily Hot Water Use for a Family

The estimate gained from the use of the formula for heating the water, plus the estimate gained from the formula for maintaining that heat, will give the final estimated heat requirement or biogas volume requirement.

Try an example:

For a family in which Dad takes a shower each day, Mom takes a bath every day, laundry is done once a week, and baby gets a bath every second day (and if other hot water use is kept to a minimum), we have the daily consumption rate of Table 24.2.

If the incoming water is 5°C (40°F), and if the water heater is set at 60°C (140°F), where biogas of 5.8 Calories per liter (650 Btu per cubic foot) heat value is available, at an overall conversion efficiency of 60%, then:

$$Vb = \frac{(149)(60 - 5)}{(5.8)(0.60)}$$

$$Vb = 2{,}350 \text{ liters}$$

If the water heater is of 150 liters (40 gallon) capacity, and kept in the house where the surrounding air temperature is 20°C (68°F), then the heat required to keep it warm is estimated at:

$$Vb = \frac{(2.7)(150)(60 - 20)}{5.8}$$

$$Vb = 2{,}790 \text{ liters}$$

Total Vb is (2,350 + 2,790) 5140 liters per day.

That's a lot, in some people's terms, as it requires a generator of (very roughly) that same volume in liters, or about 180 cubic feet—twenty-three 55 gallon drums.

Biogas might be an excellent backup for a solar hot water heater. Another method of estimating heat requirements for hot water is still simpler and even less accurate and is found in Chapter 29.

Terms

Conduction: Transfer of heat via contact.

Convection: Transfer of heat via currents of air or water.

Radiation: Transfer of heat via radiant energy.

Questions

None

Problems

None

25: Steam

The production of steam from water heated by biogas can be useful in many applications from steam engines to steam heat. In the last chapter, we talked about raising the temperature of water—but not boiling it.

Vaporizing the water, or boiling it, introduces a new factor. The heat required to change the water from a liquid to a vapor is considerable, even after the water is at its boiling point. Consider an example.

Suppose we have a liter of water at 99°C. To raise the temperature further, we must add heat. For a liter of water at 99°C, we must add about one Calorie to raise the temperature to 100°C. (Assuming, of course, that we're not losing any heat meanwhile.) As we continue to add heat, the temperature does not change, but more and more of the water is converted to steam—water vapor. To change the entire liter of water at 100°C into steam at 100°C, we must pump in 539 Calories! This is more than five times the heat required to bring that liter of water from a temperature of 0° to a temperature of 100°C. This means that a certain quantity of water, as steam, will be able to give off many times more energy than that same quantity as water. It is because of the tremendous energy released during the sight change of temperature that occurs whenever materials change phase (go from solid to liquid or from liquid to vapor and vice versa), that change of phase materials are so exciting to solar enthusiasts. While the temperature doesn't change a lot, nevertheless a great deal of heat stored and released during these critical transformations.

The heat stored during changes of phase is called latent, because it does not show up in the temperature of the material. When the material vaporizes (as opposed to the change of phase that occurs when it moves from solid to liquid), this hidden heat is called the latent heat of vaporization. When it moves back down the scale of phases, from vapor to liquid, or from liquid to solid, this latent heat is released.

It is difficult at best to calculate the exact amount of energy available in a quantity of steam. Factors that we haven't discussed come into play, and so it is impossible to get accurate answers without explaining many new concepts only dimly related to biogas. Let's say then, that the heat available from steam (or required to produce steam) can be found by:

$$Ht = S\,W\,\Delta t + W\,L$$

where:

Ht = total heat available
S = specific heat of water.
W = weight of water which has been converted to steam.
Δt = temperature difference between the boiling point of water and the temperature of whatever is receiving the heat from the steam.
L = latent heat of vaporization.

We're assuming here (among other assumptions) that the steam is being converted back to water—all of it. If we want to find the heat available from 3 kilograms of steam (or 3 kilograms of water that has been made into steam), when it condenses in a radiator to heat a room which is at 10°C, then:

$$Ht = (1\ \text{Cal}\ ^\circ C^{-1} kg^{-1})(3kg)(100^\circ C - 10^\circ C) + (3kg)(539\ \text{Cal}\ kg^{-1})$$
$$Ht = (3\ \text{Cal}^\circ\ C^{-1})(90^\circ C) + 1{,}617\ \text{Cal}$$
$$Ht = 270\ \text{Cal} + 1{,}617\ \text{Cal}$$
$$Ht = 1{,}887\ \text{Cal}$$

Notice that by far the largest share of the heat (86%) has come from the condensation of the steam, not from that portion of the heat used to heat the water to 100°C.

The volume of biogas required to convert a certain amount of water entirely into steam is:

$$Vb = \frac{S\,W\,\Delta t + W\,L}{Hv\,E}$$

where:
Vb = volume of biogas
Hv = heat value of biogas

E = overall conversion efficiency
S = specific heat of water
W = weight of water (one liter = one kg, one gallon = 8.34 pounds)
Δt = temperature difference between boiling point of water and temperature of what is being heated
L = latent heat of vaporization of water

Terms

Change of phase: Changing from a solid to a liquid, or from a liquid to a vapor.

Latent heat: That heat stored during a change of phase.

Latent heat of vaporization: That heat stored specifically during the change of phase from a liquid to a vapor.

Vapor: The gaseous state.

Questions

None

Problems

How much heat is required to boil away 5 liters of water, starting from an initial temperature of 20°C and assuming a boiling point, and a final temperature, of 100°C?

26: Refrigeration

Oddly enough, biogas, as a source of heat, can be used (in the absence of electricity) to refrigerate and even freeze things. This is done in what is called an *absorption cycle refrigerator* (ACR).

Knowledge of some basic physics is required to understand just how an absorption cycle refrigerator powered solely by the heat of a flame, and using no mechanical (pump) energy, can cool things. Since it is unlikely that you will have the resources required to design and manufacture such a refrigerator, it is probably best not to describe the details of such a beautiful device. For those interested in further pursuing this subject, many articles and books are available.

The elegance of the ACR is threefold: (1) It is simple. (2) It requires no mechanical energy input (but only a heat source—one can be run, for example, on solar energy). (3) Wonder of wonders, it is quiet—silent!

Any one of these three qualities, but especially the last, are sufficiently rare in any modern appliance that they should compel us to further study this refrigerator. Anyone who has been kept awake at night while his or her up-to-date, thin-wall, frostfree refrigerator/freezer maniacally continued to churn out crescent-moon ice cubes will deeply appreciate this fact.

The general use of these lovely machines has faded, chiefly because they are not as efficient as well designed refrigerators with mechanical pumps. If you are without electricity, however, the greater efficiency of a unit requiring electricity will not help you. Even if you have electricity, if you have to sleep near the kitchen, you may welcome an ACR.

If you are interested in finding an ACR, you have two options. The best idea is to find a Servel Electrolux, Ice-o-lator, or another brand of old ACR. The gas-fired variety will, of course, be easier to convert than those which obtain their heat from electricity. Large cities often have an abundant supply of these oldies-but-goodies lying around in nooks and crannies; however, you aren't the only one without electricity. Many vacation cabins lack this new convenience too, and the owners will pay handsomely for old Servels. This, unfortunately, drives up the price for all ASE fanciers as well. The older design (water-cooled) ACRs require about 35 Calories per liter of interior space per day to run, plus 60 liters of cooling water per liter per day.

Some of the newer, propane-powered ACRs could be converted to biogas. Their efficiency in such conversions is open to question. However, one manufacturer (Norcold) lists data for its propane or electric-powered ACRs, meanwhile urging you to buy only the electric compressor models. From this data, we can see that the largest refrigerators are the most efficient per liter of total (based on outside dimensions) space, at 13.7 Cal per liter of space per day. This compares with 300% greater efficiency of energy use in the electric compressor model of the same type, for full-time operation; further, the compressor model only operates 50% of the time under adverse conditions, and need not be level to do its thing.

But, we have what we have. The least efficient Norcold model (surprisingly, not the smallest, but the next-to-smallest) operates at about 21 Calories per liter of total volume per day.

Terms

ACR: Absorption cycle refrigerator.

Condensation: Gas becoming a liquid.

Equilibrium: In this context, that point reached when condensation equals evaporation.

Evaporation: Liquid becoming a gas.

Thermal oscillation: Molecular shufflin' and jiving.

Van der Waals forces: Weak forces of attraction between atoms or molecules.

Vapor: Gas molecules.

Vapor pressure: The pressure exerted by a vapor. A measure of vapor's ability to evaporate at a certain temperature.

Questions

Does this chapter leave you cold?

Problems

None

27: Engines

Methane (or well-scrubbed biogas) makes an excellent fuel for ICEs (*I*nternal *C*ombustion *E*ngines) for the following reasons:

1. It has a very high octane rating
2. It leaves little or no carbon deposit in the cylinder or on the piston
3. It greatly reduces the amount of sludge buildup in the oil, and thus means longer distances between oil changes
4. It does not dilute the oil on the cylinder walls during engine startup like liquid fuels do, and (with numbers 2 and 3) promotes longer engine life
5. It has no tetraethyl lead to foul spark plugs and pollute the air
6. It mixes better with air than (liquid) gasoline, resulting in a better detonation in the cylinder;
7. It results in less valve burning
8. It burns clean and without as many harmful pollutants as other fuels
9. It is more fun to drive a car operating on historical, versus prehistorical sunlight

But with all this said, the other side of the story should be presented as well.

Ever since Harold Bates broke into the headlines with his "chicken powered" car, people have had a lively interest in powering internal combustion engines on biogas. Mr. Bates, however, left a lot out of the newspaper articles. Indeed, the fact that he is selling a regulator (available from Earth Move) makes the omissions somewhat suspect. The biogas generator he displayed was a smallish affair of (being generous) 680 liters (24 cubic feet). Apparently it was heated by biogas (which would, of course, consume some of the output). But ignoring this, an average output for a high-rate generator is one volume biogas produced per volume of generator, each day (or 680 liters of biogas in this case). A figure possible of attainment with high technology and fudge factors is 1.2 volumes of biogas per day per generator volume. All told then, it seems the maximum amount of biogas he could have attained is 815 liters of biogas each 24 hours.

Now, assume that Mr. Bates' Hillman auto has a 35 horsepower engine. (It does not. The engine is larger and would require more biogas than a 35 HP engine.) Further assume that the biogas he generates has 75% methane. (That figure would be exceptional in a high-rate generator giving 1.2 volumes of biogas per day per generator volume.) An engine of 35 horsepower would require (at 25% efficiency) 14,440 liters (510 cubic feet) of biogas per hour.

In other words, even with every benefit of the doubt in excess of reality, Mr. Bates would have to generate biogas for 25 days to power his car for one hour. It appears that he is a better publicist than inventor—perhaps his car runs on newspaper rather than biogas. (He may, however, buy his methane ready-made. But this is not what he has implied.)

This example also serves to introduce the subject of biogas fuel for internal combustion engines by pointing out the vast gap between the hopes of many and the reality of the situation. ICEs are big, noisy, and inefficient. ICEs run on biogas are still big, noisy, and inefficient. You will not change a misshapen monster by feeding it a vegetarian diet.

Nevertheless, there is hope. Stationary ICEs can be powered on biogas at fairly good efficiency. As well, there is something to be said for running limited range vehicles—like tractors—on biogas. Let's look at it.

There is a direct relationship between pressure and temperature. When the pressure goes up, so does the temperature, and when the temperature goes up, so does the pressure. This is exactly what happens in the cylinder of an ICE. In the gasoline ICE, a fuel/air mixture is let into the engine cylinder, and the piston compresses it. The spark plug fires, there is a detonation, and the hot gases formed by the burning fuel expand and push the piston down.

At the very bottom of the piston's travel, the cylinder space has its greatest volume. At the very top of the piston's travel, the cylinder space is as small as it can be. The ratio of the largest volume to the smallest volume is called the compression ratio. If the compression ratio is 4 to 1, the fuel/air mixture will be compressed by a factor of four. Or, to look at it another way, the detonated gases will expand by a factor of four. Simple and straightforward.

As the process of compression and firing repeats and continues, the cylinder walls heat up, and this increases the temperature of the incoming fuel/air mixture as well. As this mixture is compressed by the piston, it heats further (because of the increase in pressure) and it may reach its ignition temperature before the piston has fully compressed it. Boom, it predetonates. This is called knock. It steals power from the engine because the piston must continue upward against the forces of the explosion.

Obviously, the more we compress the fuel/air mixture, the greater will be its tendency to predetonate, since greater compression will mean higher pressures and greater temperatures. So it would seem that what we want in an engine is a low compression ratio, right?... Wrong! As was pointed out earlier the compression ratio is also the *expansion ratio*, and the more the detonated hot gases are allowed to expand, the more they will *fall* in temperature. In essence, this means that the *greater* the expansion of these hot gases in the cylinder, the more efficient the engine will be because it will convert more of that heat into the motion of the piston. Table 27.1 represents an ideal engine, that is, not taking into account friction and other such party poopers.

Compression Ratio (Calculated for a diesel engine)	Theoretical % Thermal Efficiency
3	34.3
4	41.0
6	48.9
8	53.5
10	56.9
12	59.4
14	61.2
16	62.8
18	63.9

Table 27.1 Compression Ratios and Thermal Efficiency for an "Ideal" Engine

So the trade-off is between predetonation or knock, and thermal efficiency. Note that while the efficiency increases rapidly between three and four, the increase between sixteen and eighteen is not as astonishing.

Spark engine fuels are rated by their octane number. Everybody has heard of octane but few know what it is. Essentially, the octane rating of a fuel is a measure of how well it avoids predetonation. Methane has an octane number of 120 or more. This means that it can easily be used in high compression engines, because it very rarely predetonates.

Biogas, however—methane mixed with CO_2, etc.—has a lower octane rating than methane (although it is still 100+). As well, the CO_2 in biogas acts to dampen methane's ability to detonate when it is ignited, so not as much power is available from the methane in unscrubbed biogas as from pure methane, given equal volumes of methane. Even worse, the mere fact that anything except oxygen is mixed with the methane will dilute it, so that not as much can get into the cylinder, and clearly this will further reduce the power available from each power stroke in the cylinder. So removing CO_2 will increase power available.

H_2S should also be removed from the biogas if it is present in amounts (by volume) greater than 0.1%. (For means of doing this, refer to the section on H_2S in Chapter 19.) The U.S. Bureau of Standards has said (Walraven, 1932):

> From our tests it is evident that a sulphur content of 0.040% does no harm, that 0.151% sulphur does considerable harm and that 0.458% does very great harm. Just where to draw the line is hard to say. It would seem, however, that 0.100% would be a fair dead line. No doubt, a lower sulphur content is desirable, but it is probable that a sulphur content of 0.100% would cause no appreciable injury.

It seems that the Bureau was referring to weight of sulphur. As we have already pointed out, in this book, unless otherwise noted, we refer to volumes when we talk about the percentages of various gases. If this is true, then 0.10% total sulphur by weight is 0.07% H_2S by volume. Many people report that 0.10% H_2S is still a safe quantity. The content of sulphur is sometimes also reported as grains per 100 cubic feet. A grain is 0.064799 part of a gram, or (1/7000) 0.000142857 parts of an ordinary pound. See Appendix 14 for more information. Converted to percentage volume, 60 grains per 100 cubic feet is 0.10% H_2S.

Nevertheless, good quality biogas (60% or over CH_4) can be used in an engine without scrubbing. If the percentage H_2S is too high, it will not run well for long, but it will run.

H_2S troubles can be partly overcome by replacing the valves with heat-resistant valves, and changing the thermostat in the cooling system so that the water circulates at 65°C (150°F) rather than 50°C (120°F). Sometimes, however, even these precautions are not taken and the engine will run anyway.

By way of illustration, Hazeltine (1933) reports running a model A engine on unscrubbed biogas of varying percentage CH_4, under adverse conditions (for example, the motor was left outside, virtually unprotected from the weather), with widely varying pressures (1½ to 5 inches of water), and without any reported modifications of the engine. Starting was difficult, but the engine did not stop once it was started, and under loaded and unloaded conditions, chalked up 592 hours of service with an average of 152 cubic feet of biogas consumed per running hour. H_2S percentage averaged very low in this biogas and no engine wear was noticed in the complete teardown and overhaul which followed its service.

Many of the big boys also operate engines on unscrubbed biogas. For example, the Municipality of Metropolitan Seattle runs at least four, 450 HP Waukesha engines (used for pumping sewage) on raw biogas *(Diesel and Gas Turbine Progress,* 1974). For the small-scale user, however, scrubbing the CO_2 and H_2S will rather dramatically improve performance.

Tests run by Neyeloff and Gunkel (1976) showed that using the best fuel/air ratios and average outputs for each, 100% CH_4 outperformed a 50% CH_4, 50% CO_2 mixture by approximately 86% in the same engine, all other conditions equal. (Looked at another way, the diluted methane had to provide 1.86 times the energy input to provide the same energy output.)

They were working with an engine designed for research—a variable compression ratio (1:4 to 1:16), spark-ignited engine. For this particular engine, they found that output peaked at a compression ratio of 15:1, a fuel/air ratio of 0.1 (10% CH_4 to 90% air by volume), and with the timing set so that the engine fired 30 degrees before top dead center. (Very possibly because of the slow flame speed of methane or biogas, several researchers have mentioned advancing the timing for efficiency. See, for example, McGee, 1955.)

The fact that these figures relate to one particular engine is emphasized because different engines seem to behave differently in terms of what is optimum. The main factors which change are:

1. Percentage of CH_4 in fuel
2. Timing
3. Type of engine
4. Compression ratio
5. Fuel/air ratio
6. Altitude
7. Efficiency (in terms of energy in and useful work produced)

We have discussed how changes in biogas composition affect the efficiency of the engine, and the fact that the timing must be advanced, but the other factors require further explanation.

Engine Type

The Neyeloff and Gunkel engine, and the Hazeltine Model A, were both 4-cycle (spark-fired gasoline-type engines). This is your garden variety, ordinary engine —Chevy, Chrysler, Edsel. These engines can easily be converted to run on biogas, but they tend not to have the high compression ratios which can be used with biogas.

Many very small engines—motorcycle or lawnmower engines—are spark-fired, but if they require a fuel mix of gasoline and oil, they are 2-cycle spark-fired engines. These engines are not very suitable for biogas, but they can be used. Lubrication may be a problem, since these engines get some of their piston lubrication from the oil in the fuel—of which biogas has none.

Another common engine type (but less common than the above two) is the diesel. Diesel engines do not have spark plugs. What happens in a diesel engine is that air alone is compressed and when the piston reaches the right place in the cylinder, the diesel fuel is squirted (injected) into the cylinder and the heat which was developed by compressing the air, ignites the whole mixture. Diesel fuels do not have octane ratings, they have *cetane* ratings. The word is different for diesel because the qualities needed for diesel fuel are very different than the qualities needed for gasoline type fuels. In the spark engine, we want the fuel to wait to burn until we torch it off with the spark. In the diesel engine, we want the injected fuel to burn as soon as it enters the cylinders. Ergo, cetane numbers are all about how easily the fuel spontaneously ignites in the cylinder.

Diesel engines are usually noisier than spark engines because the fuel burns with a faster and bigger bang. The compression ratios they use are higher than those with spark engines. Because of the greater stresses they encounter, diesel engines are often better built than spark engines (at least large diesels are), and they generally last longer. They get better fuel mileage, but they are also more expensive—sometimes quite a bit more expensive.

So, biogas is a great fuel for a diesel engine. It has a poor cetane number, so we can stick it in the cylinder with the air, compress it right up there, and inject a little (6-13% of the usual amount) diesel fuel into the cylinder to fire the whole mixture off. A diesel engine run in this manner is generally referred to as a dual fuel engine, and the injected fuel is "pilot oil." (If our biogas has an unusually high percentage (over 5%) of H_2, such biogas will no longer have a high octane/ low cetane rating. Rather, the situation will begin to reverse —a lower octane, and a higher cetane number. Hydrogen-laden biogas is liable to preignition.)

There is also a third type of engine, only very rarely seen in a small (under 100 horsepower) configuration —the gas-diesel. This engine is the same as the dual-fuel engine, with the exception that both the biogas and diesel fuel are injected into the cylinder. As in a diesel run solely on fuel oil, air only is compressed in the cylinder. Larger engines of all types are generally more efficient (in general) than smaller engines.

Compression Ratio, Fuel/Air Ratio

As mentioned before, the high octane rating of methane allows for a high-compression ratio. However, for perfect methane/air mixtures (1 part of CH_4 to 9 or 10 parts air), the compression ratio will have to be lower than for leaner mixtures (1 part gas to 15-30 parts air).

Crooks (1949) gives data which indicates that with these leaner mixtures, compression ratios of 18:1 are possible and desirable. Our old friends Neyeloff and Gunkel got along fine with a 15:1 compression ratio, but a compression ratio increase to 18:1 would have given the engine an approximate 20% more power, ignoring other factors.

For dual-fuel engines as well, this leaner mix may prove better. The reason is that with a diesel engine, the heat and force of the exhaust can still be made to do work by driving a turbocharger—essentially an air pump which rams the fuel/air mix into the cylinder. With a turbocharger, we start with more fuel (plus air) in the cylinder, and thus end up with more bounce to the ounce. A diesel engine with a turbocharger gives increased power, at a better overall efficiency. Crooks (1949) says that turbocharging increases output power as much as 75%.

Altitude

At higher altitudes, the efficiency of the biogas as a fuel will drop, because ambient air pressure will be less, and so less biogas + air will be able to get into the cylinder. This gives us a problem similar to the one we have when the CO_2 content of the biogas increases. For example, at an altitude of 1,520 meters (5,000 feet) the volumetric efficiency will drop by 16% as compared with the volumetric efficiency at sea level, all else being equal. (Volumetric efficiency is a term that refers to how efficient a unit volume of the cylinder is. With more CO_2 or a lower ambient air pressure, the efficiency of a given volume of the cylinder drops.) Warmer biogas will be less dense than cooler biogas, causing a similar change in volumetric efficiency to occur.

Review

Unfortunately, we have no hard figures to give you on the efficiency of biogas-powered, turbocharged, dual-fuel engines. However, there are some other figures available for discussion.

Let's briefly review what we know about efficiency so far before we go further:

1. Pure methane outperforms "diluted methane" (unscrubbed biogas) in the same engine, because with diluted methane, not as much space in the cylinder can be devoted to the methane/air mix (some must be given to CO_2, etc). Further, CO_2 and other gases steal heat from the combustion reaction, and slow it down.
2. Higher compression ratios increase the power available from biogas-powered engines.
3. Timing is often advanced in biogas-powered spark engines.
4. Larger engines, as a general rule, are more efficient than smaller engines. However, the larger engines referred to are from 300 to 700 horsepower. The amount of biogas required to run these engines removes them from our consideration. A 400 horsepower engine, for example, may chew its way through 142,000 liters (fifty thousand cubic feet) of biogas in an 8-hour day.

Load

Another factor which affects the efficiency of the engine is its load. A free-running engine may not appear to use as much biogas as an engine which is driving a water pump, but per horsepower hour delivered, the unloaded engine will actually be using more biogas. As a general rule, when the engine is giving us its "rated load," that is, the power output at a specific number of RPMs that it was designed to deliver, it will be more efficient.

Walraven (1932) estimated:

Percentage of rated load	Btu per BHP hour	Calories per BHP hour
100	10000	2520
80	10800	2720
70	11300	2850
60	12000	3020
50	13000	3275

Table 27.2 Engine Load and Efficiency *(From Walraven, 1932)*

Translating, this means that when an engine is run at its rated load (100), we can estimate that it will consume 10,000 Btu or 2,520 Calories for every brake horsepower (BHP) the engine develops, for every hour it runs. (BHP is simply measured, versus the rated, horsepower. A model A engine is rated at 22 horsepower. The measured horsepower—BHP—when the engine is run on modern gasoline, is 40.)

A liter of pure methane has about 8 Calories of heat energy available (a cubic foot has 912 Btu). If the above estimates are accurate, we can calculate the volume of methane required to run a 50 horsepower engine at 80% of full load (extracting, in other words, 40 horsepower from it), first by calculating the required heat for those conditions (10,800 x 40 = 432,000) and then finding the volume required (432,000 -: 912) —474 cubic feet (13,400 liters) of pure methane per hour.

But, of course, these figures are estimates, and they are related—in Walraven's article—to 1,000 Btu per cubic foot gas or, by implication, natural gas, a fossil

Source	hp	CR	T	cu.l.	S	E	Cal	%	Comments
Lawrie 1941	437	high		326(?)	N	D	1890	33.9	
McGee 1955	24.5	7.5	17	482	N	S	2800(?)	23.0(?)	
Hecking 1940	460			352(?)	N	D	2040	31.4	
Hecking 1940	460			396(?)	N	D	2295	28.0	at 78% load
Hecking 1940	460			404(?)	N	D	2345	27.4	at 74% load
Walraven 1932	175			422	N	G	2525	25.4	
Hazeltine 1933	40(?)	6(?)		596(?)	N	S	3350(?)	19.2(?)	5 BHP output (?)
Neyeloff and		15	30	456(?)	Y	S	2645	24.3	100% CH_4, best
Gunkel 1976		15	30	464(?)	Y	S	2690	23.9	100% CH_4, average
"		15	30	499(?)		S	2895	22.3	83% CH_4, average
"		15	30	522(?)	N	S	3025	21.2	71% CH_4, average
"		15	30	652(?)	N	S	3780	17.0	63% CH_4, average
"		15	30	758(?)	N	S	4395	14.6	56% CH_4, average
"		15	30	579(?)	N	S	3360	19.1	50% CH_4, average
"		15	30	860(?)	N	S	4990	12.9	50% CH_4, average
"		7.5	30	628(?)	Y	S	3640	17.6	100% CH_4, best
"		7.5	30	759(?)	N	S	4400	14.6	50% CH_4, best

Source: the bibliographic reference

hp: the horsepower of the engine

CR: the compression ratio of the engine

T: timing, in degrees, engine before top dead center (spark fired only).

cu.l: cubic liters of biogas per brake horsepower per hour. Divide by 28.3 (or multiply by 0.0353) for cubic feet.

S: scrubbed. "Y" indicates "Yes"; "N" means "No." A "Yes" means pure or nearly pure methane.

E: engine type. "D" for dual fuel; "S" for spark; "G" for gas-diesel.

Cal: calories per brake horsepower (multiply by 3.968 or 4 for Btu/BHP).

(%): percentage efficiency in terms of energy input versus useful work output.

If the figure in any column is followed by a question mark (?), it is an estimate, based on the information in the Source reference. Any estimate in the cu.l column is based on 5.8 Cal per liter (about 650 Btu per cubic foot) biogas. A blank space in a column indicates that the information is unknown or inapplicable.

Table 27.3 Biogas Use in Various Engines

fuel. If we extend the findings of Neyloff and Gunkel to cover these figures (e.g., diluted methane is less efficient), then using biogas of 50% CH_4, 50% CO_2, Walraven's figures for Btu or Calorie per BHP hour need to be nearly doubled.

So, how much 50/50 biogas do we need? Strictly on a mathematical basis, given what we know, our estimate would be twice what Walraven tells us: that is, 950 cubic feet (26,800 liters) of biogas per hour for a small 50 horsepower engine. Just as a question of pessimistic philosophy, it's a good idea to add 20%, or 30%, to cover contingencies, but it becomes hard to make the kind of decisions economics often requires us to make, based on such nebulous figures. Frankly, it's going to be hard to get any better. The factors change so much in different situations that it is impossible without specific tests to be any more accurate. Still, let's look at some of the performances reported in the literature. Table 27.3 summarizes the pertinent data.

(It should be pointed out that Neyloff and Gunkel had an engine that seems poorly suited for use with biogas. However, don't expect to do better.)

The variations reported are startling. It may not seem that the difference between 33.9 and 12.9 percent efficiency is very great: yet that is close to 240%. So for every horsepower hour developed by the least efficient engine situation reported by Neyloff and Gunkel (50% CH_4, 50% CO_2, average output, at a compression ratio of 15:1 the engine reported by Lawrie will develop 2.6 horsepower hours.

And where does all that heat (the excess or wasted Calories) go? If we put 2,520 Calories in as heat, and we only extract 832 Calories as mechanical energy, where is the rest of that energy? Well, for an engine which is 33% efficient, it goes something like Table 27.4.

So, 66% of our heat input becomes waste heat—some of which can be extracted and used. However, in the worst case engine figures from our chart, 87% of

Heat output as	Btu	%	Calories
Work	3,333	33	831.6
Friction	900	9	226.8
Hot water (in the cooling system)	2,700	27	680.4
Exhaust heat	2,800	28	705.6
Radiation losses	300	3	75.6
	10,000	100	2,520

Table 27.4 Work and Heat Loss *(From Audel's Engine Guide)*

our Btu input comes out as waste heat. What this means in simple terms is that when we are running on biogas at poor efficiency, our engine is going to get very hot. This is another good reason to strive for efficiency in any biogas powered ICE. Otherwise, we may burn out the engine.

Here also is the reason stationary engines operate at a better efficiency than mobile engines—we can extract some of the waste heat and thus turn more of the output of the engine into useful tasks. We might, for example, run an engine to generate electricity, and use the waste heat to bring the biogas generator up to its optimum temperature.

Putting all this together and adding a bit of information, we can come up with a list of ideas about running ICEs on biogas.

1. Go for a high compression ratio. Some common spark engines can be converted to a higher compression ratio using a special head. Remember that, among other differences, a higher compression ratio will create greater stresses in the engine, and may therefore, shorten its life. Talk to your local auto parts dealer for more information. (Diesel engines, of course, already have a high compression ratio, but their effective compression can be increased by using a turbocharger. Small diesel engines are common in marine uses, so if you live around boats, you might be able to find a used one. Many small tractors now use diesel engines. See Chapter 29 for information on storage for mobile uses.)
2. In higher compression situations, the spark plug gap may need to be set smaller. Jerry Friedburg, who is an expert in propane fuel conversions, recommends making the spark plug gap smaller for the use of this fuel. L. John Fry ran a 13 BHP, diesel-converted-to-spark engine with a spark plug gap of 0.20 millimeter—0.008 inches—which had to be regapped every two weeks.
3. Scrub the biogas of CO_2, H_2S. If your biogas has a high but not extreme percentage of H_2S in it (between 0.05 and 0.10%), change the oil often, or check it for an acid pH often. Change the oil if the pH turns acid, and if it turns acid soon and often, consider trying H_2S scrubbing. Also, change the oil whenever it thickens by 30%. You can measure this by counting the number of drops coming out of a pipette or a small hole in a can in a certain amount of time. Measure the drip rate of the oil when it is new, and measure the drip rate of the oil when used, at the same temperature and for the same depth of oil above the drip hole. When the drops come out only 70% as fast using a sample of the oil in the engine at that time, as compared to the faster drip rate of the oil when it was new, change it. Modify the crankcase plug on the engine (put in a valve instead) if you plan to use this test often.
4. Try a leaner mix in dual fuel—particularly turbocharged dual-fuel engines. Neyloff and Gunkel, working with their unique research engine, found that the CH_4/air ratio for optimum power increased slightly with increasing CO_2 dilution, from 20% CH_4-80% air with 100% CH_4, to 11% CH_4-89% air with 50% CH_4, 50% CO_2. It should be noted, however, that because the CH_4 was diluted with CO_2, not as much CH_4 got into the cylinder, and so the overall effect was still that as dilution increased, less air was used than before. For example, at 10% CH_4 to air (with 100% CH_4 fuel) there was 10% "biogas", 90% air. At 11% CH_4 to air (with 50% CH_4 and 50% CO_2 fuel) there was 18% "biogas" and 82% air.
5. If you are serious about using biogas in an engine, then the best way to find out what the optimal factors are is to use your own biogas and an engine. Run that engine on biogas and put it on a dynamometer. Details on this procedure can be had at your neighborhood garage or tune-up center.

Terms

Brake horsepower: Measured horsepower.

Cetane number: A measure of a fuel's ability to rapidly easily detonate.

Compression ratio: The ratio between the greatest and the least piston chamber volume.

Diesel: An engine type widely used in trucks and boats. Air alone is compressed and the fuel injected into the cylinder. Diesel engines have no spark plugs.

Dual fuel: An engine that requires both diesel and biogas in order to run.

Gas-diesel: An engine designed for both gas and liquid fuels. Air is compressed in the cylinder, and both gas and liquid (diesel) fuels are injected into the cylinder.

ICE: Internal combustion engine.

Octane number: A measure of a fuel's ability to avoid predetonation.

Questions

1. Do you think a home biogas system is suitable for running the average (100 horsepower-plus) car?
2. If yes, why? If no, why not?

Problems

1. If your car has a compression ratio of 15:1, runs at 70% of its rated load, and runs on biogas which has 3 parts of CO_2 for every 7 parts of CH_4, what, according to the authorities cited, would we expect its consumption of biogas to be, each hour?
2. What would its consumption of methane be, under the above conditions?
3. What would its consumption of methane be if we scrubbed the biogas to the point that it was more or less pure methane?

28: *Space Heating*

The amount of heat required to heat a building under the worst general conditions which prevail in a particular area of the country—20° below in a stiff wind, if that sort of thing happens where you are—is called the "design heat load." The "heat load," on the other hand, is the amount of energy required to keep the place warm on any given day, and either one can be approximately calculated, if you know these factors:

Δt = the expected temperature difference between inside and outside.

A = the surface area of the structure; and

K = the rate at which heat is lost through the materials which make up the structure.

For greatest accuracy, additional factors can be considered, such as the outside wind speed and the amount of air infiltration. In some situations these factors will be important enough to warrant further consideration, as for example in a garage, where a large door is opened regularly. For our purposes, the simple approximation will suffice, since what we want is a ball park figure that will indicate to us whether or not biogas will be able to heat our space. Two further points deserve mention before we launch into the body of the discussion: (1) If we want, we can fulfill our heating needs from several sources—biogas need not "fail" if it can only supply a portion of our heat needs. (2) The skills that you can learn in this chapter will be helpful in considering the math of heating generators, so even those of you who are happy with your wood stove will wish to read on.

As mentioned above, K is a symbol for the rate of heat loss. Materials which allow heat to pass through them easily are called "conductors," while those which resist the flow of heat are "insulators." It's a relative thing, of course. One man's conductor is another man's insulator.

The K value of a material is measured according to the amount of heat transferred through a unit thickness of material, per unit area in a certain time, for each degree of temperature difference. In SI units, K can be conveniently measured in large Calories (heat), centimeters (thickness), per square meter (area), hour (time), and degree Celsius difference (temperature). In American, K is measured in Btu, inches, per square foot, hour, and degree Fahrenfeit.

Considering any slab of material, we can begin to understand how to deal with K. As a material is made thicker, it seems obvious that the amount of heat able to travel through it would diminish. On the other hand, as area, time and/or the temperature differential increase, the amount of heat transferred will also increase. K is measured in Cal cm m^{-2} hr^{-1} $°C^{-1}$, (that's Calories and centimeters per meter squared per hour per degree centigrade). Since we want to end up with Cal, it seems clear that we must multiply by an expression in terms of cm^{-1} m^{2} hr°C.

If we have a K value for a material of a fixed thickness—such as quarter-inch glass—then that is referred to as a "C value."

So, the K and C values of a few common building materials, in SI and American units are given in Tables 28.1 and 28.2.

Material	SI	American
Brick, masonry	101.	8.20
Insulation, polyurethane	1.86	0.15
Wallboard	2.48	2.00

Table 28.1 K Values

Material	SI	1/C	American	1/C
Air space (3/4 in., 2 cm)	5.37	.19	1.10	.91
Film, inside	8.00	.12	1.64	.61
Film, outside	29.28	.03	6.00	.17
Glass (1/4 in., .64 cm)	5.51	.18	1.13	.88

Table 28.2 C Values

The values given are only for purposes of example, and for problems found at the end of this chapter. A more extensive list is found in Appendix 9.

Here are the formulas for approximating heat loss.

$$\frac{1}{Ct} = \frac{1}{Ct} + \frac{1}{C_2} + \frac{1}{C_3} + \ldots$$

where:

Ct = total C value of wall or window or roof, etc.

C_1, C_2, C_3, etc. = C values of the materials of the wall

As mentioned above:

$$C = \frac{K}{Th}$$

where:

C = C value of material of a specified thickness
K = K value of that material
Th = thickness of that material

Finally:

$$Ht = (Ct_1 A_1 + Ct_2 A_2 + Ct_3 A_3)\ \Delta t\ T$$

where:

Ht = total heat loss, or total heat required
Ct_1 = total C value of wall
A_1 = area of wall
Ct_2 = total C value of windows
A_2 = area of windows
Ct_3 = total C value of roof (or floors, etc.)
A_3 = area of roof (or floors, etc.)
Δt = degrees of difference between inside and outside
T = time in hours

Sharpen your pencil. We have a room 5 by 5 meters, 2.5 meters high. There are two windows in the room, each 2 square meters. The wall is 10 centimeters brick, 2.5 centimeters polyurethane, and 1.25 centimeters of wallboard. The windows are 0.64 centimeter thick glass. The desired temperature is 25°C. inside, the temperature outside is -5°C. Neglecting heat losses due to air infiltration, or through the roof and floor, what is the heat loss through the walls and the windows in ten hours?

First, find the needed C values:

101 ÷ 10 = 10.1 (C of 10 cm brick)
1.86 ÷ 2.5 - 0.74 (C of 2.5 cm polyurethane)
24.8 ÷ 1.25 - 19.6 (C of 1.25 cm wallboard)

Once you have found these values, you plug them into the Ct formula. Notice, however, that you will also use the 1 ÷ C values for film, inside and outside. These values give information about how easily heat is transferred from the inside air to the inside surface of the wall (inside film) and from the outside surface of the wall to the outside air (outside film). When it's very windy, the insulating value of the outside film drops to near nothing.

$$\frac{1}{Ct} = 0.12 + \frac{1}{10.1} + \frac{1}{0.74} + \frac{1}{19.6} + 0.03$$

$$\frac{1}{Ct} = 0.12 + 0.10 + 1.35 + 0.05 + 0.03$$

$$\frac{1}{Ct} = \frac{1}{1.62} = 0.62$$

So the Ct of the walls is 0.62. In another form, with the proper units added, that's 0.62 $Cal^{1}hr^{1}m^{-2}$.

For the windows:

$$\frac{1}{Ct} = 0.12 + 0.18 + 0.03$$

$$Ct = 3.03$$

The area of the window is:

$2+2 = 4\ m^2$

The area of the walls is:

$((5)(2.5))4 - 4 = 46\ m^2$

(Remember, the window area must be subtracted from the total area of the walls.)

In this example, we'll take the Ct values times the area involved, before we multiply by the Δt and T values, just to keep the math more visible:

$$Ht = ((3.03)(4) + (0.62)(46))\ \Delta t\ T$$
$$Ht = (12.12 + 28.52)\ \Delta t\ T$$
$$Ht = 40.64\ \Delta t\ T$$

If you've followed all this carefully, you know that the value 40.64 is in terms of $Cal^{1}hr^{1}$. If all of this is still a bit foreign, go back and prove to yourself that this is indeed the case. We'll wait.

Plugging in the two remaining values:

$$Ht = (40.64\ Cal°C^{-1}hr^{1})(30°C)(10\ hr)$$
$$Ht = 12{,}192\ Cal$$

Even in such a relatively well-insulated room, where window space is at a minimum (and where we are not calculating heat losses through the roof and floor) the heat load is considerable. Let's increase the window area to a total of 12 square meters, and assume the roof and the floor have the same construction as the walls. For fun, let's also assume that the floor is losing heat, not to the outside air, but to the ground, at 7°C. Also, drop the air temperature to -20°C. Much more briefly this time, we have:

Heat loss wall:

$(0.62)(46m^2)(45°C)(10\ hr) = 12{,}834\ Cal$

Heat loss roof:

$(0.62)(25\ m^2)(45°C)(10\ hr) = 6{,}975\ Cal$

Heat loss floor:

$$(0.62)(25\ m^2)(18°C)(10\ hr) = 2{,}790$$

Heat loss windows:

$$(3.03)(12\ m^2)(45°C)(10\ hr) = 16{,}362\ Cal$$

Total heat loss:

$$12{,}834 + 6{,}975 + 2{,}790 + 16{,}362 = 38{,}961\ Cal$$

(If we can cut out heat loss through the windows, we can save a considerable amount of heat.)

Terms

Design heat load: Heat load under worst expected conditions.

Film: The surface of heat transfer between a solid and a gas or liquid.

Heat load: Heat lost under specified environmental conditions.

Questions

None

Problems

1. Your room is a cube, 5 meters on a side. One window is all you have, of 2 square meters. The walls are made of brick, with a 2 centimeter air space and 5 centimeters of polyurethane foam, and you want an interior temperature of 20°C. When the outside air is at 10°C, what will be the heat loss through the walls, in 10 hours? The brick is 10 centimeters thick. (Assume the door—if you built one—has the same Ct as the walls.)

29: Gas Requirements

The requirements for gas-consuming appliances can only be estimated, since different appliances will have different actual rates of consumption, and use patterns will deeply affect the total consumption. For example, people with more relaxed lifestyles will use less energy. If your coffee water must be heated quickly, more heat will be required for that speed than if you move more slowly. Once water is boiling, a small flame is sufficient to maintain the boil, but many people don't come back to the stove and turn it down. And so on.

Table 29.1 is given in terms of heat (Calories or Btu) and a discussion follows on how to change this into biogas volume.

This is all straightforward enough. If you've gotten this far then Table 29.1 should be crystal clear to you. Notice that the values are full bore values. In other words, gas output is at a maximum. If you use the device for less time, or at less than full output, the final amount of heat required is less. For example, if you use a regular stove burner for 12 minutes at half of its full output, this will be $(60 \div 12)(0.5)(9{,}000)$ Btu, or 900 Btu.

Obviously what we need now is some data on the heat value of biogas, per unit volume, so that we can find biogas volume needs. Quite simply, the heat value of a particular batch of biogas can be related to its percentage of combustible gases. (Unfortunately, there is no simple low technology way of determining this percentage. Nevertheless, see Appendix 1.)

Neglecting any H_2 that may be in the biogas, the gross heat value of dry biogas is:

$$Gh\,(\text{Btu ft}^{3}) = 10.12\ Pm \qquad \text{gross dry value}$$

$$Gh\,(\text{Cal l}^{1}) = 0.09\ Pm \qquad \text{gross dry value}$$

where:

Btu^{3}ft = British Thermal Units per cubic foot of biogas
Gh = gross heat value per unit volume
Pm = percentage of methane
Cal 1^1 = Calories per liter of biogas

Calculated at 15°C and 760 mm Hg (60°F and 14.7 psi). This is not STP.

Gross heat value is the amount of heat available from the burning of biogas if the water formed when it burns is completely condensed and its heat entirely extracted. You will remember that liquid water converted to steam is mighty greedy for heat, and likewise, when steam is condensed, a lot of heat is made available. If the biogas is used in an engine or a steam boiler, we will most likely never see that heat which is tied up in the water vapor, and so we will only have the net heat value available—that is, the heat available from the biogas when the steam is not condensed to water. For dry biogas:

$$Nh\,(\text{Btu ft}^{-3}) = 9.12\ Pm \qquad \text{net dry value}$$

$$Nh\,(\text{Cal 1}^{-1}) = 0.081\ Pm \qquad \text{net dry value}$$

where:

Nh= net heat value

If the biogas is not dry, but wet (saturated with water vapor), then this added "vagrant vapor" will steal further heat (although not much), so that, for saturated biogas:

$$Gh\,(\text{Btu ft}^{-3}) = 9.95\ Pm \qquad \text{gross saturated value}$$

$$Gh\,(\text{Cal 1}^{-1}) = 0.089 Pm \qquad \text{gross saturated value}$$

$$Nh\,(\text{Btu ft}^{-3}) = 8.96\ Pm \qquad \text{net saturated value}$$

$$Nh\,(\text{Cal 1}^{-1}) = 0.080\ Pm \qquad \text{net saturated value}$$

A confusion of terms leads many people to report various numbers for the heat value of one volume of pure methane. The main reason for this variation is the difference between the gross and the net values.

Common percentages of methane found in biogas are from 50% to 80%. (Consult Appendix 1 for a method of determining the percentage of methane in your biogas.) Since, more often than not, you can't expect to be able to dry the biogas, and since, more often than not, you won't be able to condense the water in the byproducts and extract the heat, the net heat value for saturated biogas will give reasonable heat values. Many of the examples in the foregoing pages are based on the net heat value of saturated biogas of about 73% CH_4.

Appliance	Intermittent Use 10^2 Cal hr^{-1}	10^3 Btu hr^{-1}	Comments
Burner, oven	20	8	per ft^3 oven space
Burner, oven	713	283	per m^3 oven space
Burner, regular top	23	9	
Burner, large simmer	30	12	
Dryer, clothes	55	22	
Engine	19	7.5	per bhp, best
Engine	27	10.7	per bhp, average
Engine	50	19.8	per bhp, worst
Heater, room, open flame	3.2	1.25	per flame tip
Heater, room, radiant	5	2	per "glower"
Heater, water, side arm	378	150	4 gallons or 15 liters per minute
Heater, water, instantaneous	567	225	6 gallons or 23 liters per minute
Heater, water, instantaneous	755	300	8 gallons or 30 liters per minute
Light, Argand	10.6	4.2	per 100 watt light equivalent
Light, inverted mantle	3.5	1.4	per 100 watt light equivalent
Light, inverted mantle	3.8	1.5	per mantle
Stove, commercial	157	62.5	four-burner
Stove & oven, commercial	270	107.5	six-burner
	Daily Use		
Hot water for 2 people	131	52	or as calculated
Hot water for 3 people	164	65	
Hot water for 4 people	209	83	
Hot water for 5 people	227	90	
Refrigerator (ACR)	0.2	0.080	per liter space
Refrigerator (ACR)	5.7	2.25	per cubic foot space (median)
Space heating			as calculated
Other needs			as calculated

Table 29.1 Gas Consumption by Appliances

(73)(0.08) = 5.8 Calories per liter

(73)(8.96) = 654 Btu per cubic foot

Notice the difference when we calculate for the gross dry heat value:

(73)(0.08) = 6.5 Calories per liter

(73)(9.95) = 726 Btu per cubic foot

About 12% more heat is released. This may not seem like much, but it's like getting a larger generator for the same cost. Nevertheless, we can generally only expect to obtain the net heat value. Elimination of saturated or added water vapor—whether from the net or gross values—will only increase the energy available by about 2% (when the gas is saturated with water vapor at 15°C).

To determine the volume of biogas needed to power a device, when the heat value of the biogas and the heat (or energy) requirements of the device are known, use this formula.

$$Vb = \frac{Hr\ T}{Hv\ E}$$

where:
Vb = volume of biogas
Hr = heat requirements of the device or need in question, per unit time
T = amount of time the device is used
Hv = heat value of the biogas, per unit volume
E = Efficiency of conversion

The E factor can only be used where we have a figure for the final useful energy required, and the efficiency of the device.

For example, the last chapter spoke of heating a room, but didn't discuss the efficiency with which the actual heater converts the potential energy in biogas into kinetic and useful warmth in the room. Suppose that conversion efficiency is 75%. That is, of 100 units of potential energy in the biogas, 75 units will be present in the room as heat. Further, suppose that we need 5,000 Calories of heat per hour to maintain the desired internal temperature. This is a consequence of leakage. Our biogas has a net heat value of 5.5 Calories per liter. Then:

$$Vb = \frac{(5{,}00)(1)}{(5.5)(0.75)}$$

Vb = 1,212 liters of biogas (each hour)

If, on the other hand, we know what the useful energy output is for some certain input, then we can ignore E. In Table 29.1, an inverted mantle light requires 380 Calories per hour. To run it for 3 hours on biogas with An Hv of 5.3 Calories per liter requires 215 liters of that biogas.

$$Vb = \frac{(380)(3)}{5.3}$$

Vb = 215

To calculate the total amount of heat available in any quantity of biogas where the percentage of methane is known, use the following formula:

$$Th = Nh\ Pm\ Vb$$

where:

Th = total heat available
Nh = net heat value
Pm = percentage of methane
Vb = volume of biogas

Terms

None

Questions

None

Problems

1. What is the net heat value in Calories per liter of saturated biogas of 72% CH_4?

   For problems 2,3, and 4, assume you are trying to run a 60 liter ACR, and an inverted mantle light (100 watt equivalent). The refrigerator runs continuously; the light is used for 3 hours per day.
2. What would your daily Calorie needs be?
3. If your biogas assays at 65% CH_4, what volume would you need?
4. At 78% CH_4, what volume?

Section V

Troubleshooting

Before we begin the Design Section, a review of generator startup, finding and development of cultures, scum, foaming, sediment, and other common problems will help us gain insight into the whys and wherefores of generator design.

Troubleshooting chapters will not include such information as has already appeared elsewhere. The control of pH, for example, is not discussed.

30: Cultures

The world we live in is aerobic. Wherever we go, the air has been there before us. Oxygen is so universal that we may never have thought about places where it is absent.

However, when we want to try anaerobic decomposition of organic matter as a means of indirectly using solar energy—e.g., biogas production—we need to consider where in the world we can find a source of anaerobic bacteria to start the whole process going.

Anaerobic bacteria are found in many places: in the digestive systems of insects and animals, in the bottom mud of still waters, in sewage lagoons, in wet rotting garbage, in garden soil, and nearly anywhere something once alive has been kept away from light, and saturated with water for awhile.

So, finding a culture—a population of anaerobic bacteria—is not hard. If we fill a bucket with water and kitchen wastes, a culture will develop in it... along with some characteristic odors. But, it has been repeatedly demonstrated that one major law of ecology applies to the biogas process. This is the law of diversity.

In any ecosystem, as environmental conditions change, some species of plants or animals will be able to better adapt than will others. The more diverse the array of species available, the more harmoniously and fully the ecosystem in a generator will be able to thrive on the different kinds of wastes it is fed.

In establishing a culture for our generator, gather anaerobic populations from several sources to ensure the most diverse population possible.

Notes on Collection

Anaerobic bacteria in the soil give it a characteristic blue-green, almost irridescent tinge. Lack of this tinge of color is not an indication that the bacteria are missing, but it does indicate that they are not present in great numbers.

Human excrement (and indeed the feces of most creatures) is a good source of thermophilic bacteria. Dried manure is not a suitable culture material, and the fresher the feces are, the more likely they are to provide good cultures.

For example, goats are notoriously nonchalant about where and when they deposit the remains of their last meal. Warm water (preferably water that has been boiled) can be put in a jar along with some of your intended substrate material, and direct and immediate collection of some goat manure should prove easy—if the goats are tame—using this jar. Be discreet, however. Such activities may arouse unusual speculations about you to circulate in your neighborhood. The sight of you chasing goats around with a jar of warm water held ready for the moment when chance will allow you to catch the falling and final product of a goat's digestive process, will be either amusing or frightening to the uninformed.

All cultures should be kept moist and warm and they are usually best collected the same way. Don't put the lid tightly on your culture jar, because it, too, is a biogas generator. (The reason a jar is good to use is that then you will be able to see the bubbles form.)

Once you have a small amount of your substrate well cultured, carefully transfer the contents of the jar to a larger batch of substrate, or directly to your generator, if it is not very large. This material , the culture plus substrate, is called "seed" or "seed material."

Unless a culture is established, generator startup will be prolonged. With plant substrates, no biogas production may occur for months. (Adding seed to such a generator may result in foaming, because the eager MF bacteria will make sudden headway on the volatile acid buildup, with violent gas production).

Experiments with Culture

Although it could not be called likely, it may be that there are natural anaerobic cultures which could produce an increase in biogas production (either in general or only for certain substrates), over and above what is now usual. Somewhat more likely is the possibility that combinations of natural sources of vitamins, minerals, or enzymes, and different cultures will react favorably together. If you wish to pursue this worthwhile line of investigation, remember that it may require time and reculturing for a combination to establish itself. A thorough investigation of the literature will give you familiarity with the methods of investigation, and hopefully will give you the ability to distinguish between results that are valuable and applicable and those that "begin in mere words, and end in words."

Who knows? Maybe hummingbirds, (or more likely, termites), carry around in their gut a group of MF bacteria yearning to get to work in the world's biogas generators.

One likely source of bacteria is trees. In certain trees, the heartwood—the central, often darker colored wood at the core of the trunk—for some reason becomes saturated with water. It is then called "wetwood." Sometimes in this wetwood, populations of methane-producing anaerobic bacteria—methanogens—are found. James C. Ward (1978) has told us:

> . . . of the various western trees, the cottonwoods are the most consistent sources of methanogens. These species of cottonwood are: Black Cottonwood (Populus trichocarpa), Desert Cottonwood (P. fremontii). Eastern Cottonwood (P. deltoides), Plains Cottonwood (P. deltoides var. occidentalis), Lance-leaf Cottonwood (P. acuminato) and Narrow-leaf Cottonwood (P. augustifolia) . . . Thus cottonwood trees can be considered a ready source of methanogens for innoculum cultures in wood waste digesters since the infected wood contains wood-degrading and other anaerobes needed to support growth of methanogens in such an environment. I believe that this mixture of microorganisms enables these methanogens to produce large amounts of methane in tree stems under relatively cool temperatures.

These cottonwood species were found "on poorly drained soils, particularly on low lying areas surrounding lakes and rivers." Zeikus and Ward (1974) report that the fluid from the wetwood had an odor very similar to the odor of a cow's stomach (rumen), and that, at times, the fluid in the wet wood was ejected by the gas pressure. Samples were taken with small boring tools, similar to drills with hollow cores. It should be possible to use a small drill to collect a sample of wetwood chips or preferably some of the wetwood fluid, for use as a starter culture. Obviously we want to avoid serious damage to living trees.

Terms

None

Questions

None

Problems

None

31: Startup

Batch generators are considerably easier to start than continuous-fed generators. Put the substrate in, add a suitable culture, pay attention to the parameters, and let it happen.

High-rate generators are continuous-fed generators with fairly low hydraulic retention times (HRT), high load rates, continuous agitation, and careful control of temperature and other parameters. These are the hardest kind to start, and an examination of the startup of these generators will provide ideas for starting generators of any design type.

Seeding

Starting high-rate generators is not hard when heavy seeding (e.g., 50:50, seed to slurry ratio) is used, detention times lengthened (say, by a factor of 1.5), and load rates reduced comparably. However, this is not always possible, especially with large high-rate generators, since enough seed material may not have been developed.

Compound Interest

In cases where the volume of seed material is not large enough to adequately culture the whole volume of the generator, startup is often done by very gradually increasing the volume of slurry in the generator.

For example, if each day we add to the generator 5% of the previous day's total slurry volume, the volume will increase as follows (if no withdrawals are made until the generator is full).

The initial volume will depend on the amount of seed originally available.

A slurry: seed ratio of greater than 8 or 10 to one is not recommended for any substrate which tends to have rapid volatile acid buildup (ammonia-rich manures, plant wastes high in carbohydrates and low in proteins, etc.). These should be seeded and left without additions for a week or so (with heating) to give the MF bacteria a chance to gear up. Then additions can begin, possibly with continued efforts at culturing, and the addition of buffers.

Day	Volume Units	Day	Volume Units
1	1	60	19
10	1.6	70	30
20	2.7	80	50
30	4.3	90	81
40	7.0	100	133
50	11	110	216

Table 31.1 Percentage Additions

Sewage Starts

Obviously, the larger the initial generator loading, the more rapidly the generator will be full and active. It may be that a good load of either actively digesting or ripe (older) sludge, can be obtained from a local sewage plant for use as seed material.

A careful watch of the generator pH or (even better) the volatile acid buildup will provide guidance about whether the additions being made are too large.

Buffer

The addition of a buffer will help stabilize the generator during the startup period. In studies done by Cassell and Sawyer (1959), limewater—$Ca(OH)_2$—gave the best control of volatile acids, and proved the best stimulant of normal digestion in high-rate generators. Limewater was added in an amount sufficient to regulate the pH to 7.

Load and Hope

Cassel and Sawyer also tried startup without seed, by filling the generator with raw slurry (they were working, of course, with sewage) and varying the HRT, with and without pH control. The study was inconclusive, but indicated that using this substrate, high-rate generators would establish normal gas generation if the HRTs were "in excess of 30 days." With pH regulation, "there were some indications that a detention time of 20 days was about the minimum which would allow starting a digester..." However, no evidence of this is offered.

Percentage of Solids

A final startup technique involves a gradual increase in the percentage of solids in the slurry fed to the generator. Cassell and Sawyer did not try this, but they did study slurries of 2%, 4%, and 6% with the load and hope method, buffered with limewater. Normal conditions were established in 41, 55, and 73 days respectively. HRT in all cases was 20 days, and in all their experiments, generator temperatures of 30° to 35°C were maintained. Substantial grease digestion did not occur in any of these "percent solids" test generators until normal digestion had been established. This indicates that substrates which are oily or high in fats will be harder to start.

Beginning with very dilute buffered slurries, then, and keeping other parameters at the values for which the generator was designed, will help establish proper biogas generation. When this is accomplished, increased solids loading may take place.

Terms

High-rate generators: Continuous-fed generators, kept at either the mesophilic or thermophilic optimum temperature, and agitated more or less continuously.

Ripe sludge: Effluent sludge.

Questions

Describe four ways to start a continuous-fed generator.

Problems

None

32: Scum

Composition

Scum is a general name for the layer of floating material which has a tendency to form on top of any slurry of below 7%-10% solids. Slurries of higher percentages of solids—dry slurries—may not form a scum layer since they don't as easily separate into layers.

Scum composition differs according to the substrates being digested, and to a lesser degree according to the parameters of digestion. In general, however, the materials which compose scum are of several kinds:

1. Fibers; including hair and certain plant wastes.
2. Soaps; discussed in the Substrates Section, these are the alkali salts of fatty acids.
3. Greases and oils; of animal or vegetable origin, also previously discussed.
4. Wood wastes; some are virtually indigestible (pine or fir sawdust) and others are only very slowly digestible.
5. Some animal wastes; mostly fibrous tissues (such as the peritoneum) which are found in slaughterhouse wastes.
6. Gas-bouyed paticles; these become trapped in the scum.
7. Indigestible materials; plastic, petroleum grease, and other miracles of modern science.

Most scum found in municipal sewage digesters could be technically classified as a "colloidal gel." This means that it is composed largely of extremely small particles made up of alkali soaps, greases, hair, and the like. Some of the large digesters reported in the literature have had scum buildup to a depth of 10 feet (3 meters) and the scum blankets have been so dense that they could easily be walked on.

Scum which forms above fibrous plant slurries are likely to be different in composition from sewage scums. Without the addition of kitchen wastes to the generator, scums which form above these slurries will lack any significant amount of grease, although they may contain soaps, in hard water areas. The main component, not surprisingly, will be fibrous plant material.

Even so, all scum, to some degree, will be formed partly of the alkali soaps of fatty acids. In fact, the major controlling factor here is the existence of calcium and magnesium ions in the water. Hard water causes more scum, because it causes more soap to form.

The bottom line, as Schlenz (1947) has pointed out, is that: "We are mainly interested in the grouping of scum forming matter into two general classifications, namely: (1) nondigestible accumulations, and (2) digestible accumulations.

Of the materials listed first above, all are digestible excepting some fibers and wood wastes, which are difficult to digest, and manmade substances, many or most of which are indigestible.

Living with Scum

If scum is not causing problems, it should be left alone. This is a good philosophy, but in practice we may not know whether or not scum will be a cause of problems until the generator is built and functioning. Sometimes then, we are left to wade through the trade-offs between the disagreeable situations which arise when we have made an investment (of time and money) and no problem appears, or we have a problem and no method of control was incorporated into our freshly built generator.

Test Generator

It is, therefore, practically indispensible to build a test generator. A test generator is usually made from a 55 gallon drum. (See Chapter 47.) Fill the test generator with the intended substrate and run it with the same parameters (insofar as possible) as you intend to use on the big generator. Find out what the response is. Be suspicious, as some possible problems, such as the gradual accumulation of grit or sand, will not show up rapidly. Before investing heavily in a large generator, experience with a smaller generator should be gained anyway.

On the assumption that at least *something* will float out of nearly any substrate, all generators will experience some form of scum accumulation, whether light and negligible, or heavy and undeniable. So, when is scum a problem, and when is it merely peacefully coexisting?

Evaluating Scum Problems

Scum causes problems in proportion either to its quantity or quality. When scum is abundant, it can: (1) decrease the effective capacity of the generator by filling up space with nondigesting materials; (2) interfere with heating systems by accumulating around heating pipes; (3) prevent the rise of gas; and (4) cause repeated periods of generator shutdown as the generator is taken out of service for cleaning (if this is a method chosen for control). When the quality of the scum causes it to be very closeknit and dense, then even if it is not abundant, it can cause problems.

The main problem caused by scum in small scale biogas production is the prevention of full gas production. Even without the seat-of-the-pants feeling (which can be developed with experience in biogas production) that some amount of biogas production is "enough," we can tell something about the health of the process by charting it. Simply put, we measure the amount of gas produced each day, and make a chart like the one in Chapter 8. If a heavy seed is used in the start-up of the test generator, the production of gas should climb rapidly to a peak in a week to ten days (if the generator is continuously heated). Gas output should remain at around one volume of generator per day for a time (at 7% – 9% solids, 35°C) and then gradually taper off.

When gas production slows down to about 10% of the peak rate, open the drum and examine the contents. A correlation of the gas production curve, which will have a sharp dip if scum problems have developed, and a visual inspection of the contents of the drum, will generally give you enough evidence of the probability of a scum problem to allow you to choose a method of handling scum in your generator. Poke something into the slurry to try to get an idea of the amount of scum and the degree of layer separation.

Remember that this test is run with a small generator, and the results gained in a larger continuous-fed generator may be different. Indeed, just as it may take time for a problem with grit or sand to appear, a scum problem may take time to show itself. Due to this fact, be extra suspicious.

Scum Control

There are several methods of scum control. The method chosen depends on the character of the scum—its composition and amount. High-rate mixing is supposed to prevent the formation of a scum blanket.

Drop Back and Punt

Some people seem happy taking care of scum problems only when they can no longer be ignored. In essence, no provisions are made in the design of the generator to discourage the slurry from forming a scum blanket. When the scum finally reduces the generator's volume or its gas production to a serious enough degree, the generator must be shut down, manually cleaned, and then restarted.

For those with a sufficient access to laborers who do not mind the thankless task of removing scum from a generator, this may be fine. However, it is not only disagreeable to remove scum manually, not only inconvenient to have a generator shutdown and substrate materials backlogged, but opening a large generator to clean it is very dangerous. Most of the fatal accidents which have occured in the history of anaerobic digestion have occurred because of explosions; and most of these explosions have occurred because a generator or digester was opened for some purpose and carelessness or chance caused a spark. (Sound effects: a huge roaring boom, followed by a dying rumble, like thunder.)

This method of dealing with scum is not recommended here. No generator should be designed so that it must be opened as a matter of course. This is simply poor engineering. It may be that in spite of careful planning and good preparation a generator must be opened due to some unavoidable circumstance, but this circumstance should not be planned into the generator.

Large Surface

Some generators are designed with a large surface to volume ratio, the thinking being that it will require more time for scum produced by a volume of slurry to cover the greater surface area. Generators of this type are often horizontal in their general orientation.

Small Surface

Studies on the ratio of volume to surface area, as reported by Mau (1956) showed that the opposite relationship of volume to surface area is, in fact, best—when using sewage.

If gas production is 50 cubic meters per square meter of surface area, (or 50 cubic feet per square foot of surface area, etc.), then the agitation produced by the gas as it rises through the slurry is, according to Mau, sufficient to prevent the formation of a scum blanket—

once again, for sewage digestors. No such research has been done on animal manure digestors, but municipal sewage has a long history of scum problems. The idea may well be applicable to manure-fed generators.

The gas rate Mau found to prevent scum problems was attained when the volume to surface ratio was 7.5 to 1 and when gas production was 0.67 volumes per volume of generator per day.

An approximation of the surface to volume ratio required in a generator which uses a different substrate can be calculated by dividing the expected rate of daily gas production by 50, and expressing the result as area.

For example, if a generator is expected to produce 50 cubic meters of gas per day, it should be designed, whatever its expected volume, to have a surface area of one square meter, if this method of scum control is desired.

Tiny Bubbles

The volume of gas which passes through the surface of the slurry can be artificially increased by gas recirculation, using the same 50:1 ratio to determine the volume of recirculated gas required. Biogas storage in large installations should be sufficient to provide for 5 to 10 minutes of recirculation, by which time the drawdown caused by the gas pump will be replenished through recollection.

No research, of which we are aware, has been done on the gas production surface to volume area ratio required for plant waste substrates—nor has research been done that suggests this method either would or would not work for plant-fed generators. Once again, you will need to find out yourself if you wish to use this method of scum control with a plant matter substrate. It seems clear that shredded, or small particle size substrates will respond better to this method of scum control.

It should also be remembered that the slurries on which such research was done were in the range of 6% to 9% solids.

Tiny Droplets

A further method of dealing with scum is liquid recirculation. Scum builds up gradually in most cases, and at first, the top layer is fairly low in its percentage of solids—around 15%. As scum builds up, new particles float up from the slurry and gas becomes trapped, which bouys the whole mass up. The top layer gradually compresses and drains, increasing to as much as 30% to 35% soilds.

The drier and more compact the scum is, the more trouble it will cause. Also, scum has the same peculiar property shared by manures, of being difficult to rewet, once it has had a chance to dry out. Giving this tendency a two dollar name, we call it *hydrophobic,* meaning literally "afraid of water," but in this case what we mean is that we can't get the darn stuff wet again (and therefore we can't mix it into the slurry to digest it). This isn't a sudden sort of change. In other words, there's not some point at which scum becomes suddenly and stubbornly hydrophobic, rather the tendency is there all along, becoming more pronounced as the scum becomes drier.

Liquid recirculation for scum control aims at keeping the scum moist and thus encouraging active digestion of anything digestible in the scum. The gentle flow of liquid washes through the scum, preventing a localized volitile acid buildup and continuously reseeding the floating mass. Sometimes a *surface wetting agent* is added to the recirculated liquid to overcome the scum's hydrophobic tendency. Any detergent will act as a surface wetting agent, but some of them are toxic. Basic H, or a chemical called sodium hexametaphosphate—$(Na\ PO_3)_6$ (Graham's salt)—or any organic detergent will serve. Sodium hexametaphosphate is poisonous to humans.

Liquid recirculation does not depend on force: it's not done with jets of liquid. Liquid recirculation involves a moderate flow of supernatent washing over the scum, either from enough outlets to cover the whole surface, or as few as one, near the center of the generator.

An alternative is to spray supernatent over the surface of the scum. Wisely (1941), reported that ordinary garden hose nozzles, set for wide spray, were used successfully in this tank and they did not clog. However, a filtration system of some sort might have to be installed if the supernatent had too high a solids content.

Heat Makes It Neat

Scum tends to dissolve into a generator with higher temperatures, and (possibly) with increased retention times.

Screen

One unusual approach to scum is found in Chapter 51. In essence, the slurry is contained inside a tube made of screen that can rotate within the generator, thus turning the whole contents of the generator over and breaking up the scum.

Chemicals

Chemically, scum in sewage digestors has been controlled by ammonium sulfate—$(NH_4)_2\ SO_4$, or alum (or sodium aluminum sulfate)—Na $Al(SO_4)_2$.

Just why the addition of ammonium sulfate should help the scum be digested is a bit mysterious, since all slurries have abundant ammonium ions in them. Some of the sulfate ion will show up as H_2S, depending largely on the pH.

The use of alum will tend to coagulate or clot the solids in the slurry so that they settle. Some of the sulfate will appear as H_2S, and the aluminum ion will make the effluent unsuitable for agricultural purposes if it is present in too great a concentration. The other methods of scum control are preferable, but if push comes to shove, you may want to try chemicals.

Terms

Colloidal gel: Like gelatin. A suspension of very fine particles which may behave either more like a liquid or a solid, depending mainly on the amount of moisture present.

Questions

None

Problems

None

33: Foaming

Foam is sometimes produced over slurries, and it seems to be related to two major factors: (1) the rate of gas production; and (2) the viscosity of the surface film of the slurry. Viscosity is a quality of fluids and a measure of how thick they are. Cold honey is very viscous, water much less so.

It appears that under different conditions and with different slurries, the surface film of the slurry begins to collect many large and complex molecules, in a concentration exceeding anything found under the surface, in the slurry itself. These molecules make this micrometer-thin surface-film quite viscous, and gas bubbles rising through the surface won't burst.

The particular conditions which cause this state of affairs are poorly understood, but different causes have been suggested. Cold, excessively high or low pH, hard water conditions, excessive loading rates, startup conditions, or unidentified bacteria have all been suspected, and certainly some of these, such as cold, are instrumental in setting up conditions for foaming.

The addition of lime to a generator with a fallen pH may have the same effect, because when the pH is more nearly neutral the methane-forming bacteria—if they were well established before the pH dropped—will start shufflin' and jivin', and a gas surge will be experienced. The higher the viscosity, the lower the rate of gas production that will produce foaming.

Because the causes of the viscous surface film are not well understood, the efforts directed to solving the problem have been of the trial and error variety.

The method which receives the most acclaim is that of Buswell (1929)—liquid recirculation, of the kind effective for scum. Spraying may also work.

Establishing proper digestion with moderate or no changes in substrate or parameters will probably insure that no foaming problems are experienced. Under such conditions foaming will only be seen—if at all—during startup.

If the problem is present and persistant, from 3 to 10 parts per million of chlorine (from pool supply stores) has also been shown to have an effect—but no one knows why. Apparently, with sewage-fed digesters, this dosage range of chlorine does not seriously disrupt gas production. For continuing problems, it has also been suggested that a complete clean out of the generator may be needed, based on the notion that mysterious bacteria are the cause. Experimenting with changes in parameters may be wise before such a drastic remedy is applied.

Terms

Viscosity: Thickness of a fluid.

Problems

None

Questions

None

34: Sediment

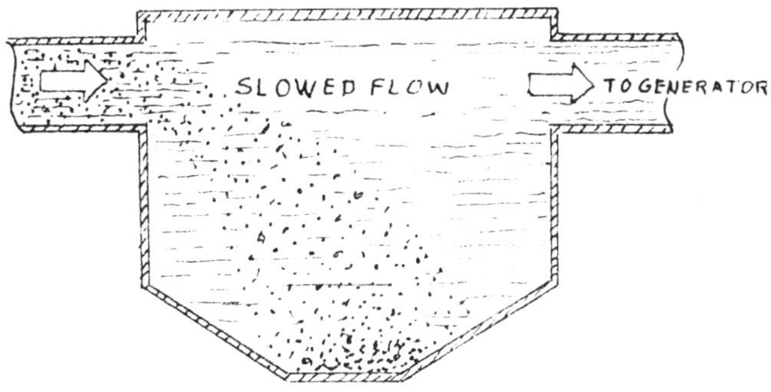

Fig. 34.1 Removal of Sediment (preloading)

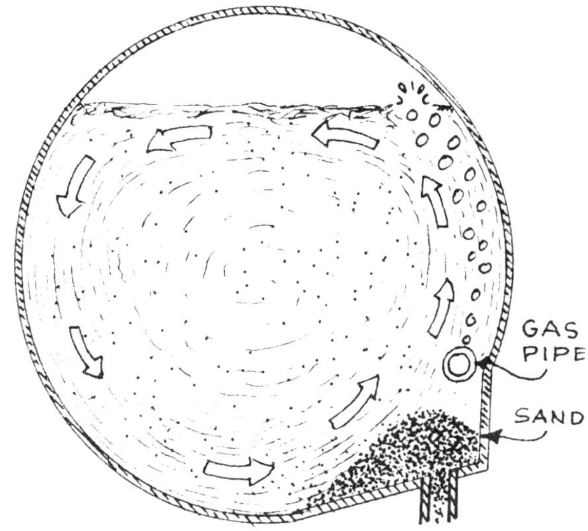

Fig. 34.2 Removal of Sediment (in the generator)

Sediment or grit accumulates in a generator at a rate which depends on the substrate and loading rate. Some substrates produce slurries which are grit-free; others include a certain portion of grit with each loading.

A little thought about your intended substrate, the method of its collection, and preparation, will show you whether you need be concerned about sediment.

The digestion of sewage-grown algae (for example) will not generally introduce sediment into a generator, but chicken manure will often introduce grit because chickens are fed grit for their gizzards. Horses eat hay and oats, but if their manure is collected off the ground dirt and sediment will show up in the generator fed such horse manure (and so on).

There are four methods of dealing with sediment: (1) separation; (2) high-rate mixing; (3) pumping; and (4) clean out.

Separation is recommended for any generator where a significant volume of sediment is expected. Some provision should be made for allowing the slurry to stand in, or move slowly through, a space where suspended grit can settle. The sediment which settles can be removed by gravity or by a pump. (If the slurry is heated before input to the generator, this should be done after sediment separation.) Separation can also occur in a pre-loading holding tank. (See Chapter 39.)

High-rate mixing will suspend smaller particles of dirt and grit, and this may prove all that is needed where very fine silt or clay sediments are involved.

Where the particles are larger, the velocity of the slurry as it is rolled by the gas recirculation will sweep the particles along the floor of the generator, and deposit them under the gas pipe outlets.

Pumping involves collection of sediment in the generator, generally on a sloping bottom. Effluent is often collected from the bottom of the generator, and when this is done, the sediment can be removed as well. If effluent is not collected from the bottom of the generator, periodic bottom pumping can help remove sediment buildup. Since pipe can be reasonably cheaply gotten, it's a good idea to install inlets and outlets at various places in a generator so that if a situation arises which requires access to a particular fluid level in the generator (e.g., supernatent needs to be recirculated, sediment needs removal, etc.), then this can be fairly easily done. Without having placed such pipes, a simple problem can create a lot of work.

As is the case with scum, it is here recommended that you not plan to deal with sediment by periodic shutdown and cleaning, but that such an approach be your last recourse.

Terms

None

Questions

None

Problems

None

SECTION VI

Design

In this section we will discuss questions related to the design of generators. The purpose of this section is to describe some of the options available for generator design, to give information relating to the economics of biogas generation, and to mention some points relating to the operation of generators.

35: Basic Generator Types

Probably the single most basic aspect of generator design is flow. How does the slurry move through the generator? There are several major types of generators, based on their different flow characteristics:

1. Continuous-feed, high-rate mixed (high-rate).
2. Continuous-feed, intermittently mixed (moderate rate).
3. Continuous-feed, unmixed (plug flow).
4. Batch-feed, mixed or unmixed.
5. Hybrid.

Continuous-Feed, High-Rate Mixed

Almost all large-scale manure substrate biogas generators are continuous- feed, high-rate mixed. As you no doubt remember, with a continuous-feed generator, its size depends on the HRT, and the volume of the daily load. If we feed a generator a liter each day and each liter stays in it about 10 days (a 10-day HRT), the generator will have to be 10 liters in volume. Since the rate of biogas evolution depends mainly on the temperature, and since the biogas is evolved faster at first and then ever more slowly, the idea here is to get as much biogas as possible out of each cubic meter of generator, which means optimum temperature (35°C, 95°F) and short (10-20 day) retention times.

In practical fact though, while a smaller, continuous-feed, high-rate generator will cost less than a larger generator of another type, it may require more supplementary equipment (for heating, agitation, etc.) and careful watching; this kind of generation can be more unstable than other kinds. This may cause the cost of the two types, geared to the same job, to be roughly equal.

High-rate generators— another name for continuous-feed, high-rate mixed generators—must be fed substrates which mix easily with water, and agitation will be difficult at or above 10% solids, depending to some degree on the percentage of VS in the substrate. High-rate generators generally avoid the problem of scum accumulation. These generators require an agitation system, a heating system, a pumping system for the slurry, and any of several kinds of monitoring systems, to keep track of the health of the biogas process.

A high-rate generator operates on a tightly knit balance of factors, including temperature, loading rate, percentage of solids, retention time, and sometimes the concentration of ammonia. Changes in these factors can upset the high-rate biogas process, and so care must be taken with high-rate generators to maintain a constant temperature, to not allow the load rate, percentage of solids, or concentration of ammonia to increase, nor to allow the retention time to decrease.

High-rate generators are not necessarily better or worse than other designs, but the short retention time for the slurry means that anywhere from 40% to 60% of the biogas which it is possible to generate from that slurry remains in it, and is not evolved. Of course, this last 40% to 60% is also hard to get, because it takes much longer to be evolved than the initial gas spurt.

Waste is a relative concept. The whole economic scheme of high-rate generators is based on whether we waste space, or waste potential biogas. Certainly the high-rate generator is among the best available for biogas production from intensively produced animal wastes, but it is less suitable for the small scale situation. Appendix 3 can help you make a sound decision, if you want a high-rate generator, about what trade-offs you may wish to make between retention time and volume on the one hand and rate and amount of biogas evolution on the other.

Among the alternative kinds of high-rate generators available, one which moves into the upper echelons of technology is the fixed-film generator. This is a generator which makes use of the fact that the biogas bacteria like to stick to solid surfaces. The generator is essentially a collection of tubes, through which the slurry is pumped. The inside surfaces of the tubes, once generation is well established, become coated with populations of bacteria, and these "react" with the passing slurry.

Continuous-Feed, Intermittently Mixed

These words describe a large number of different generators, as for example, many of the designs of Ram Bux Singh. These generators have less sophisticated agitation and pumping system designs, and use less energy. Many designs call for a gravity feed/displacement flow system, where slurry enters by gravity and displaces a portion of the digested effluent.

Because these generators usually have a longer HRT, they are larger than high-rate generators, and not as easily upset by changes in parameters. However, they are sometimes plagued by scum and sediment problems which can easily shut down a generator. (Any intermittently mixed design which fails to make a provision for scum and sediment accumulation will prove to be a costly experiment.)

Most municipal sewage digesters are of this type, with 30-day HRTs. Some municipal digesters have two separate tanks which can be used either one after the other, or as two separate, single-stage generators. For larger generators, this is an excellent means of insuring process stability, or a continuing supply of biogas if half of the generator should have to be shut down.

There is no overwhelming biological reason why a continuous-feed, intermittently mixed generator cannot operate with a greater-than-10%-solids-content slurry, although there may be mechanical reasons, depending mainly on the type of input (feeding) system—whether by gravity or pump, or some other method.

A further example of a partially mixed design is found further on in this book, where a continuous-fed design is shown that is reportedly capable of operating on plant waste substrates.

Continuous-Feed, Unmixed

These generators, usually long affairs which move slurry from one end to the other by gravity displacement, often have no auxiliary systems except for heating. The long ones are also known as plug-flow digesters, because they tend to operate like the digestive systems in our bodies, where small amounts of food move (without much mixing, or in other words, as "plugs" or discrete lumps) through a digestive process.

L. John Fry seems fond of horizontal, plug-flow, unmixed generators. His large South African "displacement digester" was of this kind, and the small generators described in *Methane Digesters for Fuel Gas and Fertilizer,* are of this sort.

It is a still simpler design than the two previously described, and suitable for farm-sized generators, as amply demonstrated by Mr. Fry. Some provision must be made for scum accumulation besides occasional cleanout.

Another noteworthy plug-flow design is the trench generator. This generator is constructed the way you might expect from the description. A long trench is dug in flat, gently sloping ground. Often, no provision is made for heating except what is inherent in the generator itself—that is, it may collect solar energy.

One way of turning such a trench into a generator is to lay a plastic tube (somewhat greater in cross-sectional area than of the trench) in it. The tube could be made out of 15-20 mil or heavier sheet plastic, heat sealed, or glued. All dimensions would depend on: (1) the volume of substrate (a mix-well); (2) the percentage of solids (below 10%); and (3) the retention time (30 to 60 days). More research should be done on this, as it may prove a very inexpensive means of construction for a large-scale generator. Climate (ambient temperature) and type of plastic (for UV degradation resistance) would much affect the economics of such a generator.

Another low-cost alternative which has stirred up much talk in biogas circles is the Chinese generator. This is an underground design, and the ultimate in simplicity. It is not, however, the ultimate in productivity. Whereas your average high-rate-generator will give one volume of biogas per volume of generator per day, the Chinese design will only give 0.2 volumes of biogas per volume of generator per day. Radical chic aside, these will not work well in cold climates. Construction starts with a circular wall, then the pit is filled with earth and a dome with a manhole in the middle is built on the mound in the pit. Finally the dirt inside is dug out and carried away through the manhole, and the floor is poured. The manhole is covered with a concrete plate, and the seam is sealed with moist clay. The clay is kept moist with a pool of water, and sometimes plants are grown in this small pool. It is claimed that there are more than seven million of these generators in existence, making this the world's most popular design. Further information can be gotten from the book *A Chinese Biogas Manual.* See the Bibliography for full information.

More suitable for the temperate north, and the U.S., where labor costs make the Chinese design impractical in most situations, are a number of generators being made out of what are, essentially, plastic bags. While the plastic provides leak-proofing, mechanical support for the generator is provided either by the earth (e.g., essentially the generator is underground) or by a reinforced plywood structure. Books on concrete form building will give you some idea of the stresses you are likely to encounter in a plywood support wall, and ways of counter-stressing the wall to resist the tremendous force of the liquid.

Among the kinds of plastic being used for the bags, the two most popular kinds are hypalon and butyl rubber. The various kinds of polyethylene, if they are kept away from sunlight (which will eat them up) can be used above the generator bottom as a gas collector. The use of this option creates a two-material bag, with the bottom of hypalon and the top of the more pedestrian polyethylene. William Jewell, who shows up in various places in this book and in the Bibliography, is one of the leading designer-testers of this kind of digester. A group active in designing, producing, testing, and selling similar generators is Perennial Energy Systems. They also give seminars in biogas production and generator design. Their address is Box 15, Dora, MO, 65637.

The two main movers and shakers behind Perennial are Ted Landers and David Jesse. They have come up with several interesting designs, not only for digesters, but also for a low-cost gas transport pump. We suggest you contact them.

Batch-Feed, Mixed or Unmixed

The simplest generators are batch-fed generators. They are filled up, they generate, they are emptied. The chief advantage of a batch-fed generator is its low cost, simplicity of design and construction, and its ability to operate well regardless of the substrate being fed. Batch-fed generators can even decompose twigs and unshredded leaves (given enough time), as you will see.

The chief disadvantages of a batch-fed generator? It generally requires manual labor to fill and empty, and it will only generate gas irregularly. The reason for this irregular gas production is that the whole batch of slurry is going through the biogas process at one time. After starting, biogas production will build to a peak, and at first sharply, then more gradually, diminish until it is barely perceptible. Sometimes gas production will continue for a very long time (years) at low levels.

To partially overcome these disadvantages, batch-fed generators can be grouped, so, for example, that while one is being filled, several are generating, and one or more are ready to be emptied. The modular batch-fed generator described in this book calls for small modules, made of 55-gallon drums, which are set up in groups of 1 to 20 or more. These modules can cost less than $4.00 each to build, which compares favorably (at 52 cents a cubic foot) with larger generators, many of which may cost 4 or 5 dollars a cubic foot.

Hybrid Generator

A hybrid is the offspring of two animals or plants of different varieties. Now obviously, the term is not used to mean exactly that here. But the generator described by this term is a mix of two basic generator types, batch and continuous, and of the two stages of biogas generation, acid digestion and gas digestion. Briefly, for the idea will receive more attention in the next Section, a hybrid generator has two stages, the stage where acid formation (or digestion) takes place, and the stage where methane formation (or digestion) takes place.

In the particular design we will discuss, substrate materials which must generally be used in a batch generator are loaded into the acid digestion stage. This stage of the generator is unheated. In it, the AF bacteria (acid formers, less affected by cold than the MF bacteria), break the complex organic molecules down into soluble molecules such as fatty acids—the main food source for the methane-forming bacteria.

These soluble molecules are pumped into the second stage and transformed by MF bacteria into high quality biogas. The second stage is, of course, heated.

So, a large cold batch-fed first stage produces fatty acids for a smaller, heated, continuously fed second stage—a hybrid generator. One principal advantage to this is the fact that less heat is required for the whole operation, resulting in a savings in energy.

A hybrid design need not be used only with a batch-loaded first stage. Mix-well substrates could be used, and the first and second stage continuously loaded. HRT would be the sum of the retention time for both stages, and loading rate would be a function of the total volume of both stages.

Research that has been done suggests that the continuous-fed hybrid design is very process-stable.

Terms

Hybrid Generator: Generator in which the acid-forming and methane-forming stages are separated.

Moderate rate: Laid back.

Plug-flow: A generator in which the flow occurs in such a way that each day's slurry addition does not mix too much with any other day's addition.

Trench generator: Simple plug-flow generator built in a trench.

Questions

None

Problems

None

36: Heating

This chapter is divided into two parts for better access. These are Heat Sources, where you will find short discussions of the sources of heat energy generally usable for heating generators and Methods, a series of similar short discussions on the various ways of getting the available heat energy into the generator. Even if you have already decided on your source and method, a reading of this chapter may introduce you to some general ideas that will prove helpful.

Heat Sources

It may be feasible to combine two or more sources, so that not as much of either is needed. We will discuss ten heating sources:

1. Ambient	6. Heat pump
2. Earth	7. Compost
3. Waste heat	8. Solar
4. Engine	9. Gas
5. Electric	10. Wood

This chapter does not cover the general mathematics of heating and insulation. For that information, see Chapter 28.

Ambient. The simplest heating source is none. This is called ambient, because the generator temperature adapts itself to surrounding environmental temperature. In the tropics, this is very feasible, even practical, but in many climates, biogas production will suffer too much from low temperatures. Indeed, winter is often the time of the greatest need for biogas—but with ambient "heating", it would be the time of lowest production. The trade-off, of course, is between the volume of the generator and the rate of biogas production. This peculiar process of anaerobic digestion, however, may not be easily pegged down by the rigid pluses and minuses of simple mathematics. Tests run by the author using one batch-fed 55-gallon module were stopped and the module was allowed to cool off to ambient after having been heated for some five weeks. Surprisingly, the module then continued over a period of time to produce good quality biogas, certainly not at the rate it had been while it was being heated, but above that rate at which the hard-edged graph would have led someone to predict.

The key may lie in three factors. (This is, of course, speculation.) (1) Generation had been well-established, producing needed extracellular enzymes for the breakdown of the substrate (fallen leaves) and a balanced population of bacteria. (2) The temperature of the module fell gradually, and this may have given the bacteria time to adapt. Upon reheating the module, gas production was easily and quickly reestablished at the higher rates. (3) The leaves were already fairly well-decomposed, and thus provided a fairly accessible food source even after the heat was turned off. In any case, it happens. The experts agree that a well-seeded material responds much better to the biogas process at any temperature than a poorly seeded material.

Earth Heat. Even in the coldest areas (excepting where, as at the poles, the ground may be solid ice many feet thick), the earth serves as a reservoir of heat. The temperature of the ground at a depth of a meter (a bit more than a yard), or at that depth under permafrost, will seldom be below 5°C or 45°F. Often, it will be as high as 12°C (54°F). Now, this doesn't fry eggs, to be sure, but on the other hand it's considerably warmer than the air temperature at International Falls, Minnesota in the dead of winter. The point is, that if you have to make the choice between heating slurry from 10°C to operating temperature, versus heating it up from 5 °C to operating temperature, you would obviously choose the former, and a knowledge of the temperature of the earth will help improve the picture in exactly that manner. Burying the generator or the slurry pipes, or piling earth up around the generator will help, if you have severe winters.

Of course, the opposite case is also valid. If you surround your generator with soil—particularly with wet soil—then it will be difficult to keep warm. Dry soil is a fair insulator.

Waste Heat. Heat is generally unrecognized as a commodity which has been paid for and can be recycled, so perfectly usable heat is often wasted, due to poor design practices. Of course, heat is abstract in the same sense that red is abstract. That is, we never have heat, we have something which is hot—possesses heat.

Most waste heat is wasted as hot air or water or—if you are very lucky—you may have waste heat as steam.

Engines. One special case of waste heat is the heat developed by an internal combustion engine. Heat can be extracted from the water which cools the cylinders, and additional heat is available off the manifold and from the exhaust. Because relatively cool water on one side and relatively hot exhaust gases on the opposite side can create stresses in metal parts (which may try to contract and expand on opposite surfaces), jury-rigging a water jacket cooling system on an ordinary manifold will probably lead to its cracking after short service. The use of glycerine instead of water will mean that higher temperatures can be tolerated in the manifold coolant (about 150°C, or 300 °F) but this may still not solve the problem, as the exhaust leaves at from 425° to 600°C (800° to 1,100°F). Another option is air cooling or steam cooling, but these are not generally low-tech approaches.

If anyone has information about manufactured or easily devisable liquid-cooled manifolds please write. The information will be included in subsequent editions. Fiat makes an expensive but apparently effective engine/generator combination which will run on biogas. It is called Totem. See Appendix 16.

Extracting heat from the exhaust after it leaves the manifold is easier since no special configurations are required. The exhaust manifold on an engine may have a very specialized shape, but any particular exhaust pipe heat exchanger can usually be adapted to different engines. The extremely important principle of countercurrent heat exchange (used by sea mammals to keep their internal organs warm in spite of a much cooler blood return from their flippers) can give us an idea for effective design.

In Fig. 36.1, the cooling fluid comes in from the right and exits to the left down. The hot fluid (to be cooled) comes in from the left and exits to the right. This way, the temperature differences between the two fluids are, throughout the exchange unit, as great as they can be. Naturally, this results in the greatest heat transfer possible. Countercurrent cooling should be used to extract other kinds of waste heat, as well as to put heat into fluids, such as in an instantaneous water heater. If both fluids were flowing in the same direction, the best that could happen is that they would exit at nearly the same temperature.

If waste heat from a stationary engine can be extracted with attainable efficiency—assuming, in other words, a well-designed, even if homemade, heat exchanger is used—one can expect that from 60% to 80% of the heat available in the biogas used to fuel the engine can be extracted. (Not this much is available as heat, of course. This is the overall efficiency of the engine plus the extracted heat.) This assumes scrubbed biogas, and/or an engine of good efficiency. Engines operating at poorer efficiencies (less than 20%) will have more of the fuel available as waste heat, but the engine will suffer for it.

A radiator is designed so that the engine can waste heat more efficiently. Ever think about that?

Electric. Electricity is expensive to use for heating, unless you have a wind/electric generator. And if you have a wind/electric machine, and you heat with wind-generated electricity, then you are making wind into biogas. There probably is not a job you want done which cannot be done better with electricity than with biogas if you have any such choice. Storage of energy is an exception, but don't try to store wind-generated electricity by making it into biogas. If you must, store it as heat (since this avoids the expense of batteries and inverters), but go no further. The low efficiencies of the transfers and transformations involved will tax your energy savings out of sight. (Heating and cooking are better done with biogas too, but the point is still valid.)

The basic value of electricity for heat is to start a

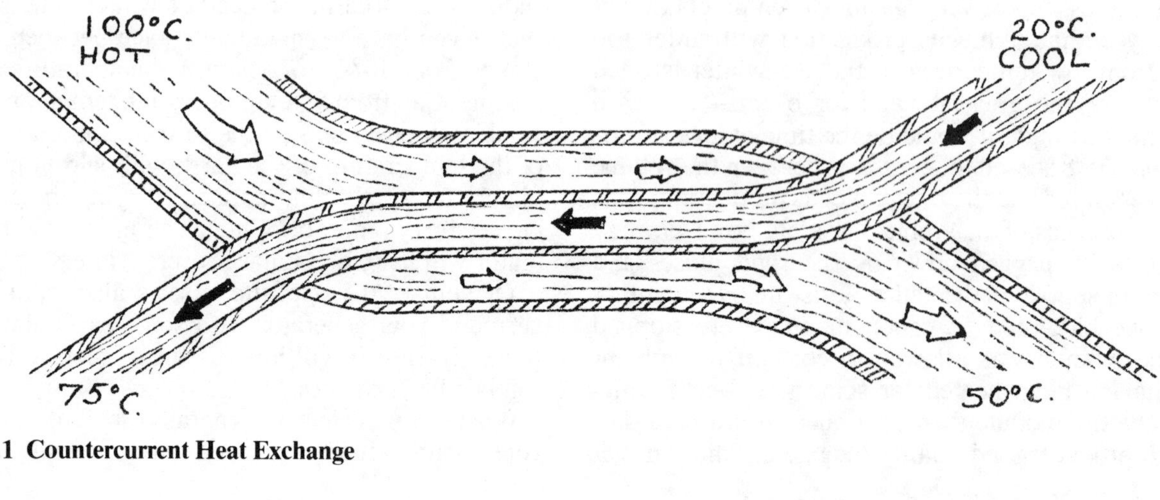

Fig. 36.1 Countercurrent Heat Exchange

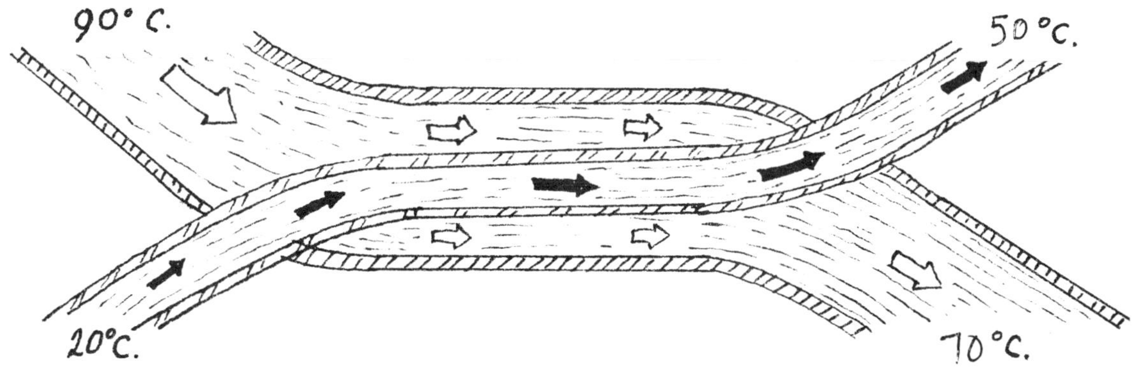

Fig. 36.2 Unidirectional Heat Exchange

generator which will thereafter be heated with some other source, or for purposes of experimentation.

Electricity is very high-grade energy, in the sense that it can do so many marvelous things. It's difficult to think of a general non-experimental, long-term situation where it would be necessary or cost-effective to use lovely juice to make lumpy heat. Nevertheless, if you do use it, on a large scale, the following is the best method.

Heat Pump. A heat pump is what the name implies —a pump which takes low-level and diffuse energy (from cool water or air) and transfers and concentrates that energy so it can be used for heating. A heat pump may also do the reverse, meaning that, instead of heating, it would cool.

A refrigerator, as one example, is a heat pump, since it pumps heat out of an (internal) air space and disperses it to the outside. If the heat pump has been designed to do so, the process can be easily reversed and both heating and cooling can be done with the same machine.

If you have heard about heat pumps, you may have heard that they are more than 100% efficient. This is nonsense. What is meant is that we will obtain from them more energy out as heat than we put in as electricity. To say, on that basis, that they are more than 100% efficient is also to say that the biogas process is more than 100% efficient, or the same of a solar greenhouse, since we get more energy out as biogas or heat than we put in as electricity in any of these examples. On this basis, a clothesline beats them all. A heat pump requires energy besides electricity in the form of available heat in air or water at ambient temperature, just as the biogas generator requires biologically collected solar energy (as plants or in the form of manure) in order to operate.

The efficiency of a heat pump depends on many factors, but chief among them is the temperature of that ambient air or water. A heat pump which operates on outside air must be designed to handle the largest temperature differential expected. So, for example, if, during a severe winter in your area, the lowest expected air temperature is to be 0°C (32°F), the heat pump which gathers its energy from ambient air will have to be large enough to find sufficient energy in that air, and hefty enough to concentrate that energy and pump it up to the required temperature—possibly 35°C (95°F)! During much of the year when such extreme air temperatures are not experienced, much of the pump's capacity is wasted. A better option, if available, is the use of a body of water as a heat source. This is because water stores more heat per unit volume than does air.

Another good option is well water, which has a very stable and generally higher (40° – 55°F, 5° – 10°C) temperature, year round, than either a body of water or the outside air.

Throughout this discussion, we've implied that we were referring only to electrically run heat pumps, but a heat pump could easily be run by an engine operating on biogas. In this case, the waste heat from the engine, as well as the heat collected by the heat pump, could be used to heat the generator. This combination is an interesting possibility for a larger generator, but since any engine sized to accept the output of a generator is usually more than adequate to heat the generator, a biogas-powered, engine run heat pump would only be useful in a situation where a lot of heat was required.

Compost. Aerobic compost typically can have an internal temperature of up to 70°C (160°F). This is a limiting temperature, in the sense that the process of composting might occur more quickly with greater heat but a higher temperature would damage the bacteria that decompose the compost, and generate the heat. If the excess heat were extracted, the temperature would not rise so much and the bacteria would be able to go full throttle. Naturally, the thought then arises: why not extract the heat from the compost pile and put it into a biogas generator? We would have happier aerobic, and happier anaerobic bacteria.

Well... controllable compost heat-extraction schemes (as for example, the thermostatically regulated circulation of water in flexible pipes buried in the compost) may conflict with the great need of compost to be turned in order to keep it aerated. And it is difficult to move large masses of compost without using machines powerful enough to damage the cooling pipe. Another interesting problem. Piling compost around a generator is one way to heat it, but it is quite labor-intensive, and there's no guarantee that sufficient heat would get into the generator past its walls.

Another scheme for compost heating is the aerobic/anaerobic generator. Materials are piled inside a large batch generator and an ordinary (aerobic) composting process is allowed to start. When sufficient temperatures are reached, water is flooded into the chamber to render the process anaerobic. The only source of heat is the biological heat generated by the aerobic half of the cycle. This has two drawbacks: (1) A poor quality and reduced quantity of biogas is generated; and (2) the heat generated is often insufficient to maintain good temperatures except in the tropics. If you try this scheme, do not use water for flooding, but the supernatent effluent left over from the last cycle. This will mean the faster establishment of an anaerobic population, if the effluent has been protected from exposure to the air. Further, this process would be better suited to high solids digestion of plant wastes than to dilute digestion of manure substrates.

Another possibility is to reverse this procedure and —in a modular generator—establish an anaerobic/aerobic cycle.

After each module has been generating gas for a given period, it is disconnected from the gas collection system and air is forced through it. Theoretically, this should cause it to go aerobic and begin producing heat. Most likely, this expectation is another fantasy. The probability is that too much water in the module would steal heat from the organic matter; it would be difficult or impossible to distribute the air to all parts of the soggy mass; and there would be a loss of nitrogen in the effluent.

Admittedly, the possibilities for using compost to heat your generator are not good. However, see Wiley (1957). Isman, the French pioneer in early biogas digestion studies, also is a source of ideas on this possibility.

Solar. Solar heating a generator, on the other hand, is a lovely idea. After all, the biogas we expect to generate is solar energy, biologically collected, and a little more solar energy—for heat—can't hurt.

Solar heat can be gathered directly by the generator. Of course, the simplest way of accomplishing this is just to put the generator out in the sun. A more efficient way of using direct solar heat is to put glass, fiberglass, or Filon around the generator. This stops the immediate reflection of a good portion of the energy collected, and cuts down on the heat loss from the wind.

What's involved here is the process which occurs when you park a car in the sun. High-frequency radiation enters through the windows, is absorbed by the interior (seats, carpets, etc.) and then it is reradiated at a lower frequency. Reradiation is different than simple reflection, of which it could be said, "It's all done with mirrors." When energy is reradiated, it is first absorbed, then the molecules of the substance that absorbed the incoming radiant energy are excited, and they in turn begin to radiate energy, albeit at a lower frequency. This lower frequency energy does not easily pass through the window glass, and so its exit from the car is blocked. It shows up as heat.

The same principle is used in greenhouses, and one elegant way to heat a biogas generator is to put it in a greenhouse. The greenhouse acts as a solar collector, the mass of the generator can store heat for the greenhouse, and the burning of biogas can provide CO_2 for the plants as well as supplemental heat for the generator and greenhouse. However, David Beiter (1979) informs us that CH_4 is toxic to plants. The only missing ingredient is insulation.

Steve Baer, solar whiz kid and main force behind Zomeworks, is promoting an idea one of his students had to insulate large areas of glass. It's called the *bead wall.* Whether a bead wall is small or large, it's made the same way. Two sheets of glass or fiberglass, separated by 1 or 2 inches, form a window or a wall. Styrofoam beads are blown down from the top into the space between them whenever insulation is desired. The beads are vacuumed out from the bottom (and stored for the next insulation cycle), whenever it is desirable to collect or release heat.

Plans for the bead wall are available from Zomeworks, P.O. Box 712, Albuquerque, New Mexico 87103.

An earlier Baer idea is called *drum wall*—indeed the two ideas could be combined. In the drum wall, 55 gallon drums filled with water are stacked directly behind glass or fiberglass. The sun heats them and they store heat for the structure in which they are located. (The drum wall can be used for cooling as well. In this application, the drums are exposed at night and they lose heat. During the day, heat from the structure soaks into them, providing cooling.) The drums are insulated, whether by a bead wall or by a movable insulating wall, when necessary.

A modification of the basic idea is to have the drums in the back of a greenhouse (rather than up against the glass). With the addition of some means of moving the barrels (pully or winch) a modular batch-fed digester could be used instead of water-filled drums. This brings us back to the idea of a few paragraphs ago—a generator in a greenhouse. Drain the liquid from the drums before moving them, and put in the water or supernatent effluent, only after they are up in place.

Indirect solar heating is accomplished by using flat plate collectors or concentrating collectors. The former simply accept the incoming energy and heat water or air flowing through them, and the latter focus the sun's radiant energy on a smaller area before heating the heat transfer medium. (Concentrating collectors must be used if steam is to be generated.)

Once the air or water is hot, it can be used to heat the generator. Quality manufactured flat plate collectors run 10 to 20 dollars a square foot, although some of the newer ones are running as low as 5 dollars a square foot. The size of your collector (and therefore its cost) will depend on your climate, latitude, and generator—both its size and its insulation.

Since this is not a book about solar energy—another fascinating field fully as complex as the study of biogas—we will not go into further detail on this subject. For more information, see Appendix 16.

Gas. Because the rate (but not the final amount) of a biogas evolved from any quantity of slurry depends on the temperature, it is often a good option to heat the generator with its own gas. Depending on the heating method and its efficiency, the climate and the quality and amount of insulation on the generator, as well as the design temperature (mesophilic or thermophilic), anywhere from 15% to 100% of the gas evolved will be used for heating.

Wood. Wood heat is also a source to consider. The amount of energy potentially available from a cord of wood is remarkable, averaging about 5 million Calories. However, as most books on wood heat will tell you, the final figure will depend on the kind of wood and its dryness.

Wood stoves differ remarkably in their ability to deliver this potential energy. In almost all wood stoves, a great deal of energy is wasted up the flue. While some of this energy is useful in the sense that it is required to heat the air so that it will rise out of the flue and therefore pull fresh air into the stove to continue the burning process, still this function might be better done by a small electric fan, if the stove and chimney were so exceptionally efficient that the exhaust gases were not hot enough to pull air into the stove.

One's first thought upon realizing the considerable heat loss up the stack is to consider ways to extract that heat, possibly via a simple heat exchanger—and several are on the market. What the manufacturers will often fail to mention is that if too much heat is extracted from the flue gases, problems will soon appear. These hot gases are not only composed of CO_2 and water (the results of complete combustion) but as well, unburned but vaporized tars and resins from the wood, and small particles of ash (better known as soot). When the heat is stolen from this mixture, the tars and resins condense on the stove pipe or the exchanger surface, and soot immediately sticks to it. As this process continues (layer by layer), the flue becomes blocked by deposits and of course the surface of the heat exchanger becomes effectively insulated. Fire danger increases.

If you want to recover substantial heat from the stove flue, then you must have a downdraft stove. In essence, the gases which leave the stove are first forced to go through the fire. Even if you don't recover wasted flue heat, efficiency is immediately increased by the fact that the residual tars and resins have for the most part been burned. The fire, of course, must be burning at the point of stove exhaust, with no unburned fuel between it and that area of exhaust. This will not eliminate the problem of tars and resins, but it will lessen the importance or the speed with which residue accumulates. Different types of wood, seasoned differently, will also give off widely varying amounts of tars. Until you have experience with your stove and wood type, check the flue every month or sooner.

If you go to wood heat, then several things are recommended:

1. An airtight stove, preferably with a thermostatically controlled damper;
2. Downdraft burning;
3. A heat exchange process or device for hot exhaust gases;
4. Or, if the above are too expensive and wood is plentiful, get the cheapest stove you can. Convert a 55-gallon drum, for example.

If you have or use a wood stove, and if you don't go whole hog on a wood-heated generator, you might consider supplementing the heat available to the generator by circulating water around your stove and using that hot water for generator heat, or making steam on the stove and using that.

Heating Means

We will discuss eight means of transferring heat into the generator. They are:

1. Heating incoming slurry
2. Slurry circulation (heat exchanger)
3. Adding hot water
4. Steam heat
5. Heating the generator skin
6. Hot air circulation
7. Hot water circulation
8. Submerged combustion

Heating Incoming Slurry. In continuous-fed generators with good insulation, it may prove adequate to provide heat to the generator by heating the incoming slurry. The advantages to this method are that it may simplify the design of the generator (thereby lowering construction costs) and it may prove an easier system to maintain and repair (thus decreasing downtime when there is a heating system failure, and decreasing operating costs).

Some disadvantages are that it may necessitate calculation each time slurry is added to the generator, to insure that the temperature in the generator remains constant, and it may make it more difficult to regulate temperatures effectively if he can be added only when slurry is added. This method is not suitable for batch digesters, except under unusual circumstances where, for some reason, pumpable substrates are batch-digested and temperature regulation is not considered a high priority.

For well-mixed slurries and ordinary purposes, it may be assumed that the heat in the incoming slurry and the heat required by the generator are directly relative to the temperatures involved. Thus:

$$\Delta ts\ Vs = \Delta tg\ Vg$$

Where:

Δts = temperature difference between desired generator temperature and the incoming slurry temperature.
Vs = volume of incoming slurry
Δtg = temperature difference between desired generator temperature and actual generator temperature.
Vg = volume of slurry in the generator, or the volume of old (cool) slurry in the generator which will still be there after the incoming slurry displaces its own volume.

As a practical matter, unless the slurry is pumped through a pipe at a fair rate of speed (something over 0.6 meters—2 feet—per second), the slurry cannot be in contact with a surface of over 60°C (140°F) or it will begin to cake onto the pipe.

At the velocity mentioned above, it may be possible to achieve 70°C (158°F) without much caking, but any deposits (even a very thin crust) will slow down the transfer of heat tremendously, and thus require more pipe surface for a similar amount of heat transfer. High temperatures work against us as much or more than they work for us.

Another option would be steam injection, discussed below. With injected steam, temperatures in the incoming slurry could be made to exceed the 60° to 70°C limit placed on any surface-to-surface heat transfer method.

In principle, the idea of heating the incoming slurry is not a bad idea, since the major heat requirement of any well-insulated, continuous-fed generator is the heat required to bring incoming slurry up to the design temperature.

Slurry Circulation. With the addition of a pipe or a pump, the slurry could be recirculated, and the above methods could be used, albeit at a somewhat higher equipment cost. This method is often used in municipal sewage digesters. In these large installations, the slurry is pumped through a heat exchanger, similar in principle to the heat exchanger discussed in Chapter 28. The principles are the same as those involved in a radiator.

Adding Hot Water. In the early history of municipal sewage disposal in Germany (ca. 1915), digesters were often heated by the addition of hot water. Percentage of solids is one among several factors which affect the efficiency with which the space in a generator is used. Obviously, adding hot water will reduce the percentage of solids, thus reducing the amount of biogas available from each unit of generator volume. However, the system of adding hot water for generator heat has its advantages. It is simple and inexpensive, and, unlike slurry, water will not cake on a surface heated above 60°C. Since we are interested in adding heat to the generator, and not merely water, it is to our advantage to add the water as hot as possible. For example, a liter of water at 100°C can heat 10 liters of slurry from 23°C to 30°C (a gain of 12°F). This assumes that the specific heat of slurry = 1.00. In other words, it takes exactly the same amount of heat to change the temperature of slurry as it does to change the temperature of water.

If the original percentage of solids in this 10 liters of slurry was 12%, the one liter of added water would drop that to 9.2%. Since 9.2 is 23% less than 12, this drop is really a 23% per unit volume drop in solids. Referring to Chapter 8 on Temperature, we find that this drop in percentage solids would be compensated by an approximate 44% increase in the rate of gas production. The overall effect (assuming that no slurry had to be dumped to make room for the hot water), is a 44% increase in gas production rate, because even though the generator volume is not as efficiently used, all of the VS which were in the generator before are still there.

If an equal volume of slurry had to be displaced to make room for the water, there would still be a 40% increase over the gas production rate of the original volume of slurry, but there would be a decline in total gas production, because of the change in the HRT or, more accurately, a change in the VS retention time.

Unless a generator using such a heating system is well insulated, and unless entering slurry is preheated, the expense in generator volume and slurry dilution would soon render this method impractical.

The changes in percentage solids which occurs as a result of adding hot water is shown in Table 36.1. The amount of water added is given as a percentage of the original slurry volume (%H_2O added). The resulting change in percentage of solids, and the percent volume loss are both given. Percent volume loss simply means that if the hot water displaced an equal volume of slurry, it is as if the volume of slurry were less, by the percentage shown. Of course, in reality, the slurry has the same volume, it is just more dilute. The real loss is in volatile solids, some portion of which will get washed out with the exiting slurry. Loss in volumetric efficiency is given only once for each column of "% H_2O added."

% Added H_2O	% Volume Loss	35	30	25	20	15	Original % Solids 12	10	8	7	6	5	
2	− 2	34.3	29.4	24.5	19.6	14.7	11.8	9.8	7.8	6.9	5.9	4.9	final % solids
4	− 3.8	33.7	28.8	24.0	19.2	14.4	11.5	9.6	7.7	6.7	5.8	4.8	
6	− 5.6	33.0	28.3	23.6	18.9	14.2	11.3	9.4	7.5	6.6	5.7	4.7	
8	− 7.4	32.4	27.8	23.1	18.5	13.9	11.1	9.3	7.4	6.5	5.6	4.6	
10	− 9	31.8	27.3	22.7	18.2	13.6	10.9	9.1	7.3	6.4	5.5	4.5	
15	−13	30.4	26.1	21.7	17.4	13.0	10.4	8.7	7.0	6.1	5.2	4.3	
20	−16.7	29.2	25.0	20.8	16.7	12.5	10.0	8.3	6.7	5.8	5.0	4.2	
25	−20	28.0	24.0	20.0	16.0	12.0	9.6	8.0	6.4	5.6	4.8	4.0	
30	−23	26.9	23.1	19.2	15.4	11.5	9.2	7.7	6.2	5.4	4.6	3.8	
40	−28.6	25.0	21.4	17.9	14.3	10.7	8.6	7.7	5.7	5.0	4.3	3.6	

Table 36.1 Changes in Percentage Solids

A 15% addition of water would result in a 13% loss in volumetric efficiency, regardless of the original percentage of solids.

Steam Heat. Heating with steam is similar to heating with added hot water, except that the amount of heat which steam gives is very large compared to the volume of (condensed) water added, thus resulting in less dilution for the same amount of heat. No pump needs to be provided to put the steam into the generator since it will inject itself into the generator. (Of course, hot water can be pumped into the generator merely by raising it above the top of the liquid in the generator.)

The heat provided by the steam comes not only from the fact that its temperature is greater than that of hot water, but also from the heat of vaporization. If you need to know more about heat of vaporization please read Chapter 25.

If we know the temperature of the water, and the quantity consumed as steam (and injected into the generator), then we can calculate the approximate amount of heat which we have put into the generator.

$$Ht = \Delta t\, W + L\, W$$

where:

Ht = total heat available
Δt = the difference between the original temperature of the generator and the boiling point of water
W = the weight of water converted to steam
L = latent heat of vaporization of water (539 Cal per kilogram)

For example, suppose we evaporate 3.5 kilograms of water and inject it into the generator as steam. If the generator was at 20°C when we started, then the heat available is:

$$Ht = (3.5)(100 - 20) + (539)(3.5)$$

$$Ht = 280 \qquad + 1{,}887$$

$$Ht = 2{,}167 \text{ Calories}$$

(If you want to figure out the units in terms of kg^{-1}, °C and that sort of thing as per Chapter 3, don't forget to put in the specific heat of water 1 Cal° C^{-1} kg^{-1}, or 1 Btu° F^{-1} lb^{-1}). This formula gives conservative approximations.

No heat is lost once the steam enters the generator, so direct steam injection can be a very efficient form of generator heating. Further, not as much elaborate equipment is needed to set up a direct steam heater as, for example, the equipment needed for circulation of hot air or water. Not as much water is needed either. For example, where 3.5 liters (1 liter = 1 kilogram) of water injected as steam can give about 2,170 Calories to a generator at 20°C, it would take about 30 liters of hot water to impart the same heat energy to the slurry if hot water (at 100°C) were added, and still more if the hot water were circulated through the slurry in pipes.

Steam heat is an excellent option. The steam generator should be small, as more heat will be lost in trying to heat a large amount of water up to steam temperatures.

Heating The Skin. For very small generators, it is sometimes easiest to directly heat the generator skin with a biogas flame. If this is done, for example on a demonstration 55-gallon drum generator, a layer of sand, about 5 cm (2 inches) thick should be put on the bottom of the drum (before the slurry is put in) to spread the heat from the flame and prevent caking of the slurry.

When larger generators are heated directly by the sun, as in a greenhouse situation, this also is skin heating. Because slurry conducts heat so slowly, if the generator has much volume, it will need agitation to distribute the heat.

The heat transfer capacity of the generator skin can vary widely, depending on such small things as what paint is used, whether or not the inner surface is rusted and so on.

Reference to Chapter 28 and Appendix 9 which

lists the K values of various materials will allow you to determine approximately the amount of heat imparted through your generator's skin by warm air.

Heating the skin via hot water, the same formula will work tolerably well, but it will not serve so well for calculating the energy available to the generator skin from the radiant energy of the sun. For this information, please consult a book on solar energy.

Hot Air Circulation. A heating option not often discussed is the circulation of hot air. Such a means of heating, whether the air is circulated through the generator in pipes, or blown around the generator skin in some manner, might serve in situations where the heat source was either solar energy (flat plate hot air collector) or the energy found in the gaseous exhaust products of biogas combustion.

The chief disadvantage to hot air circulation is the volume of air required to transfer a given amount of heat, as compared with the circulation of water. In general, at the temperatures we are concerned with, multiplying the volume of water required to transfer the desired amount of heat by a factor of 300 will give the approximate volume of air (at the same temperature) required to accomplish that same task.

Further, since the transfer of heat by air must necessarily be done through a surface, air temperature cannot exceed 60°C.

For further information, see Appendix 11.

Hot Water Circulation. The subject of hot water circulation through pipes can be either simple or complex depending on whether we are trying to measure the heat which has been put into an existing generator (simple) or design an adequate water circulation system for a proposed generator (more complex). As we have already found out, the amount of heat in water is related to its temperature, its weight (or volume), and its specific heat.

If we want to find out how much heat has been put into a generator via water circulating through pipes in that generator, we need to know the amount of water we have put through the pipes, the beginning (inlet) temperature, and the final (outlet) temperature. In another form:

$$Ha = Vw\ \Delta t$$

where:

Ha = heat available
Vw = volume of water
Δt = temperature difference between inlet and outlet temperatures

Since volume can be expressed in terms of 3 dimensions, or an area (2 dimensions) and a length, and since length can be related to velocity, then:

$$Vw = Ap\ F\ T$$

where:

Ap = cross sectional area of the pipe, or $\frac{\pi D^2}{4}$
D = the diameter of the pipe
F = flow rate of water, per unit of time
T = time elapsed

(For example, if the cross-sectional area of the pipe is 300 square centimeters, and the flow rate is 30 centimeters per second, and the time elapsed is one second, the volume of water is (300 x 30) 9,000 cubic centimeters.

So, when we know the pipe size, the flow rate, and the inlet and outlet temperatures, we can calculate the heat available:

$$Ha = \frac{\pi D^2\ Fw\ T\ \Delta t}{4}$$

where:

Ha = heat available
π = 3.1416
D = diameter of the pipe
F = flow rate of water
T = time elapsed
Δt = temperature difference between inlet and outlet

Both flow rate and diameter must be measured in the same units, be it inches, feet, centimeters, or meters. Flow rate and time elapsed should be measured in minutes or hours.

Just for you tired pencil pushers, $\frac{\pi}{4} = 0.785$.

Simple. Merely a matter of multiplication. The formula will give us answers for any situation where we can measure input and output temperatures and so on.

However, how do we estimate the needed pipe size, length, etc., when we want to design a generator? This problem is more complex. For rough estimates, Fry and Merrill (1973) report that a square foot of pipe surface per 100 cubic feet of generator volume (one-third square meter per 100 cubic meters) is adequate. (Formulas for figuring the surface area of a pipe are found in Appendix 13.

The heat output of a heating pipe depends not only on its surface area and temperature, but as well, its flow rate. This is, in essence, a problem in heat transfer, and it requires formulas beyond the scope of this chapter to explore. However, a quick and dirty method is found in Appendix 11.

Submerged Combustion. This is mentioned as an option to consider in large-scale operations because it is a fairly efficient way of making use of biogas energy. However, it is impossible to completely burn biogas without providing a surplus of air, and therefore of oxygen, which will contaminate the generator contents. To make use of submerged combustion, both biogas and air will need to be compressed. This rapidly becomes a high-tech operation, but it is possible. For those interested in pursuing the subject, information found elsewhere in this book will provide a basis for experiments, or consult references found in the Bibliography.

Terms

None

Questions

Unless, for some reason, we have an abundant supply of hot water (such as that from an engine cooling system) and barring solar energy, and where it is feasible to produce higher temperatures, what method of heating would generally be cheapest?

Problems

None

37: Agitation

Agitation is a fairly general term. It can mean anything from what happens to a batch-fed unstirred generator during an occasional mild earthquake, to the rapid and continuous pumping of a slurry through a high-rate generator.

Each kind of agitation (or non-agitation) has its place depending mainly on the substrate used and the design of the generator. As before, these factors differ in each situation, and there are advantages and disadvantages which become emphasized differently in each situation.

There are several—indeed many—different kinds of agitation. First, we have what might be called the natural agitation methods: loading, gas evolution, and heating.

Loading

In a continuous-fed generator, the act of loading can stir the actively digesting slurry. Fresh slurry is pumped into the generator, and imparts its motion to the rest of the slurry. Generators designed to make use of this simple principle usually have the end of the inlet pipe near the bottom of the generator, positioned so that a vortex or swirl is created.

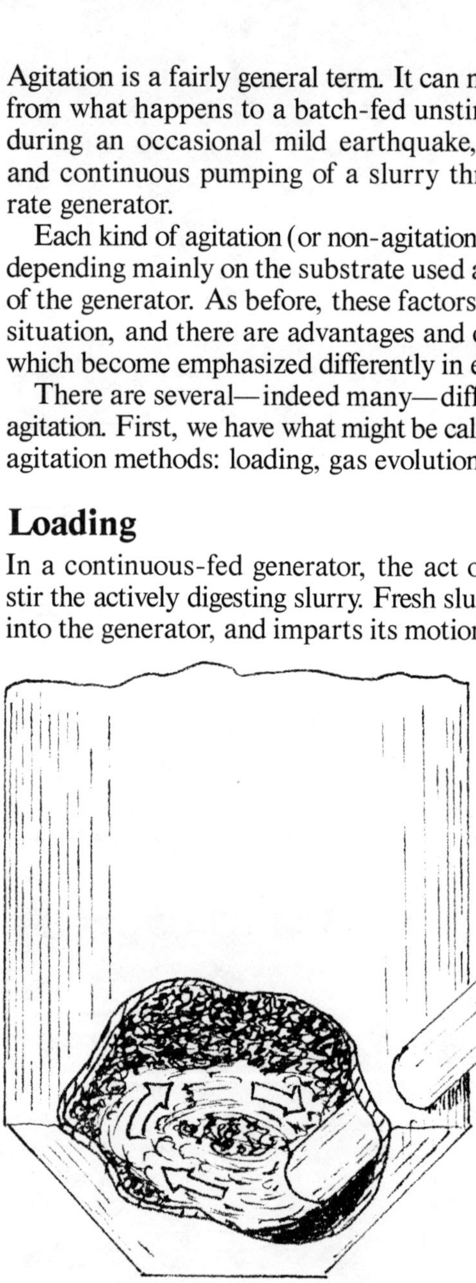

Fig. 37.1 Loading a Continuous-Fed Generator

The disadvantages of this kind of agitation are its slow and intermittent quality, and the fact that if there is a buildup of a sediment layer in the bottom of the generator, its effectiveness is even further decreased. The advantage is that it's cheap, and may be all that's required.

More Tiny Bubbles

Another kind of natural agitation occurs as biogas is formed in the sludge layer, at the bottom of the digestor. The gas forces sludge particles to rise to the surface, where they are released, and, theoretically, the particles should then fall back to the sludge layer. In practice, this kind of mixing does not occur except at the higher mesophilic temperatures, and possibly not until a thermophilic digestion is taking place. Generated gas agitation will be more effective in deeper generators, or narrow generators.

Convection

The third and last "natural" agitation occurs in heating the sludge. Heat rises, and hotter slurry will tend to rise within a body of cooler slurry. We spoke of this before—it's called convection. So, to some degree, heating the bottom of a digester will cause some of the warmed sludge to rise. Again, in practice, this will not work well enough to justify its use for many situations. Slurry is very sluggish, and it does not convect very well. Mix-well slurries of low-percentage solids will have the best convective behavior.

The main advantages then, of all three of these natural agitation methods are their low cost and simplicity. For small, low-cost, simple generators, they may be perfectly adequate. The methods of "artificial" agitation are numerous.

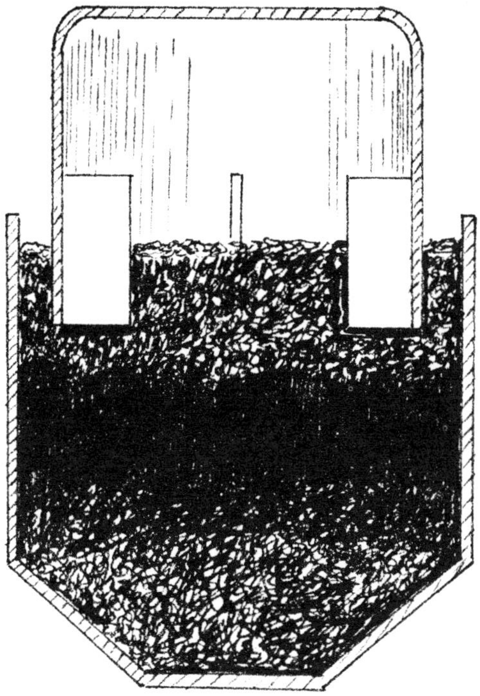

Fig. 37.2 **Agitation by Vanes**

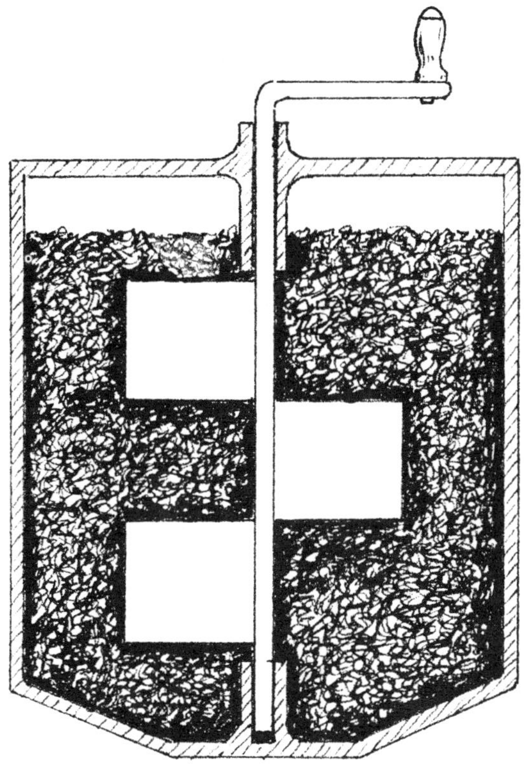

Fig. 37.3 Hand Agitation

Vane Ideas

The simplest way to agitate the slurry is to stir it. Sometimes vanes are attached to a bottomless tank which floats above the generator itself, and when this gasholder is rotated, it stirs the slurry.

However, this method will not stir the slurry vertically (from top to bottom), because it will not take sludge from the bottom and directly mix it with higher layers. As well, the depth of the mixing vanes will change depending on how high the gasholder is riding on the slurry. This method is ineffective for large or wide generators.

Further Vaneity. Another method is to attach vanes to a rod which can be moved from the outside of the generator either by hand, or possibly by a wind machine. One prime candidate for such a wind machine is a Savonius rotor.

Plate

Another simple stirring device is a flat plate, moved up and down by a rope or rod. Notice in Fig. 37.4 that the rope is surrounded by a pipe. This is done so that gas cannot escape from the generator through the top hole. Clearly, unless the holding pipe is below the surface of the supernatent, gas will be able to escape. By this means of agitation, then, the scum layer, which floats above the liquid, cannot easily be broken up to allow the escape of gas.

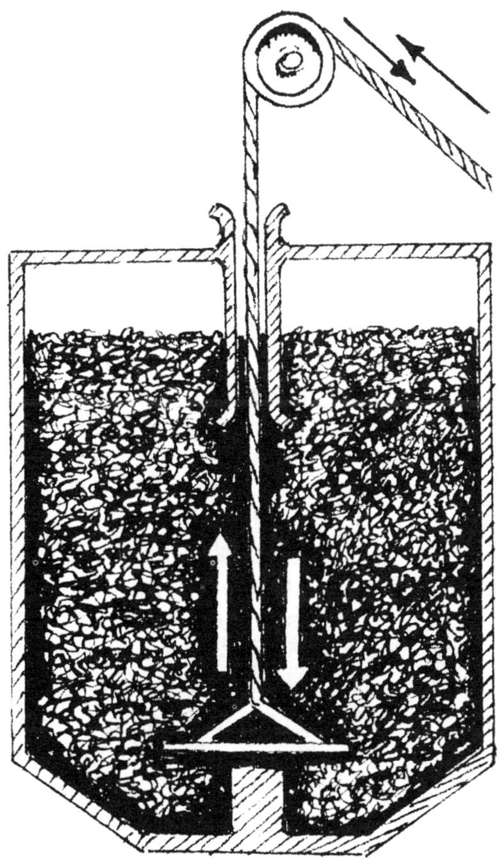

Fig. 37.4 Rope Agitation

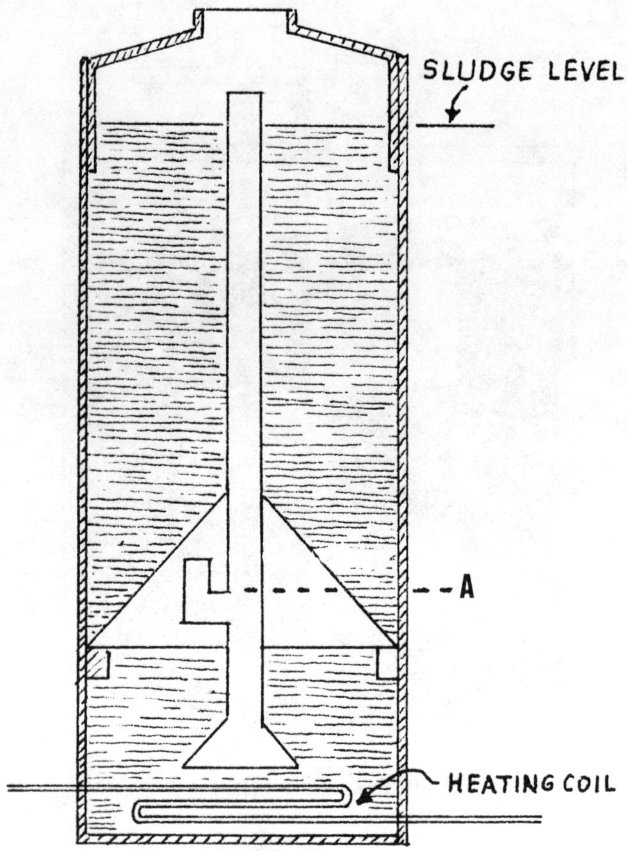

Fig. 37.5 Biogas-powered Agitation

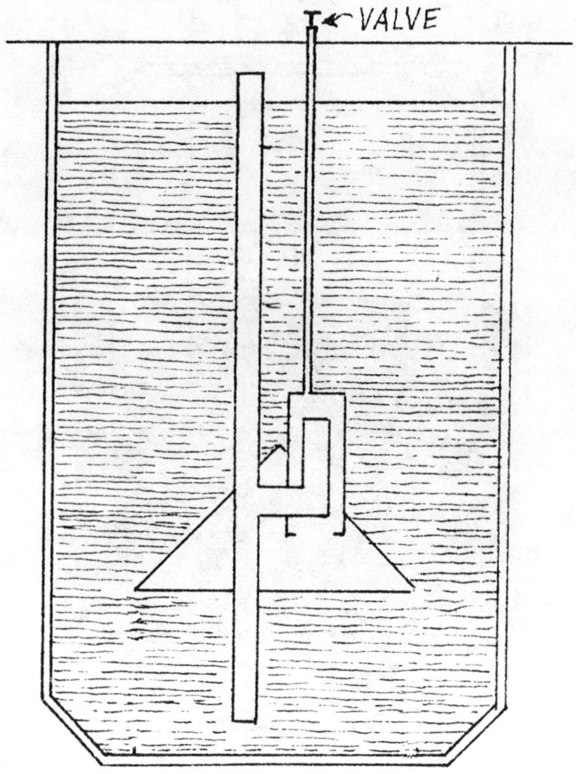

Fig. 37.6 Modified Gas-powered Agitation

Biogas Power

Some interesting gas-powered agitators were briefly tested at the Joint Disposal Plant of the Los Angeles County Sanitation Districts. As reported by Rawn (1931) these experiments were done using small generators.

The principle used in the agitators was percolation, as in coffee pot percolation. Gas built up under a hood near the bottom of the generator, and was suddenly released. Flowing up through a pipe, it pushed slurry out the top of the pipe. Two designs were found successful.

The principle of both is similar. In Fig. 37.5, the gas collected under the hood gradually drives the liquid down to level A. When this happens, a surge of gas is released up the main (vertical) pipe, and the slurry above the rising column of gas is forced up and out. Rawn reports that when the top of the generator was removed, the device "threw sludge from ten to fifteen feet into the air." Since the temperature in all parts of these generators was the same, and since there was no formation of layers (supernatent, sludge, sand) Rawn concluded that they did the job satisfactorily. As is usual in such experiments, Rawn was working with municipal sewage slurry.

The second automatic percolation agitator (Fig. 37.6) resulted from tests which were undertaken to make the device operate with a greater hood area and thus a higher rate of gas collection. Enlarging the previous design was unsatisfactory, because the hood failed to fill rapidly enough and would not create the necessary surge—rather, it dribbled small bubbles up the pipe continuously.

The modification shown eliminated that difficulty by providing a smaller chamber on the side into which large bubbles collected by the full hood could surge. The first device (Fig. 37.5) operated about once every two hours, the second (modified) device (Fig. 37.6) operated about once every fifteen minutes. Heat was applied immediately below the hoods. The second hood was the same length as the generator, but not as wide.

Gas Recirculation

Either mixing or high-rate mixing can also occur via pumping the slurry, or recirculating the gas. These two are very viable options, since gas recirculation is reported to increase biogas production and recirculation is not too difficult to achieve. It would be a stirring experience.

Gas recirculation can be done directly, by pumping the gas into a pipe which has had holes drilled into it. Rates adequate to agitate sewage sludge hover around 35 to 40 liters of biogas per minute for every 1,000 liters of generator volume (35 to 40 cubic feet per minute per 1,000 cubic feet).

Experience will show that if the recirculation pipe

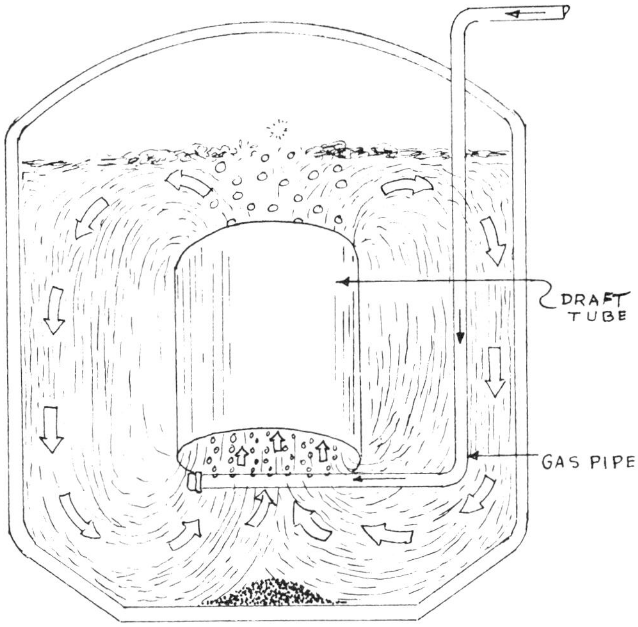

Fig. 37.7 Agitation by Draft Tube

is long, and if all the holes in it are of uniform size, more gas will come out of the first few holes than from any others. If a long run is needed, a better option is to split the pipe into two or more lengths and feed the gas from the middle with a T-connection. The holes should start smaller and become larger the further they are from the gas pressure source. Gas can be recirculated intermittently as, for example, for five to ten minutes an hour, or 10 to 20 minutes every two hours. An arrangement similar to that in Fig. 37.7 or Fig. 37.8 can be used. The draft tube improves circulation of the slurry.

Liquid Recirculation

In generators which use fibrous plant substrates (leaves, for example), the whole slurry cannot be recirculated, but a liquid will drain off the soggy mass and that liquid can be pumped hither and yon. For the mix-wells, supernatent or slurry can be recirculated.

Recap

The methods mentioned are agitation via:

1. Loading slurry
2. Natural gas production

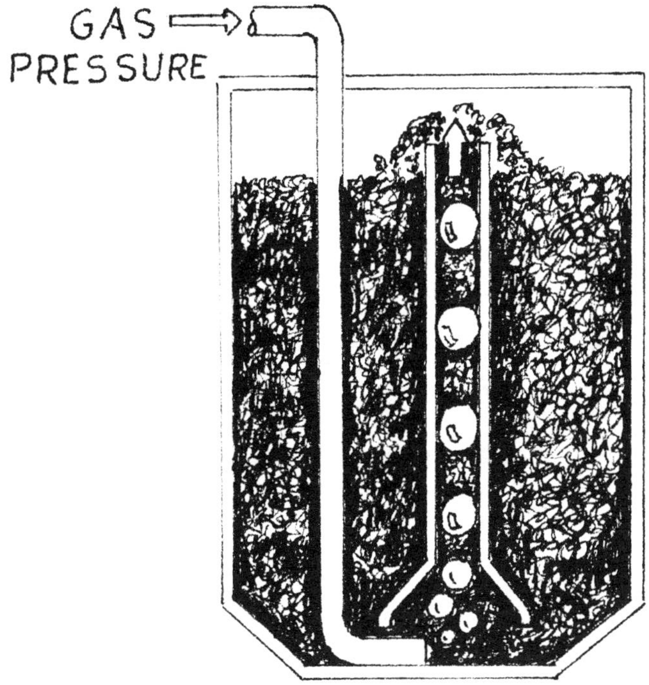

Fig. 37.8 Bubble Pump

3. Convection
4. Stirring with gasholder vanes
5. Stirring via wind- or hand-powered crank
6. Movable plate
7. Automatic percolation
8. Bubble pump
9. Gas recirculation
10. Liquid recirculation

Terms

Bubble pump: A pump which pushes liquid up a tube or pipe by bubble power.

Draft tube: A large diameter cylinder, such as a 55-gallon drum, open at both ends, up through which gas is bubbled to provide agitation.

Questions

None

Problems

None

38: Gas Handling

This chapter will discuss gas storage, sizing gas storage, and gas transport.

Storage

Storing the biogas produced by a generator can be done in a number of ways.

Gasholder. In low-tech situations, one of the most widely used methods of gas storage is displacement storage, under a gasholder. A gasholder is simply a large bottomless container floating in water or oil, under which gas collects. These gasholders can be configured in the shape of a can with holes in the bottom.

As gas comes into the gasholder, the liquid is displaced (hence, "displacement storage"), and the gasholder rises. Some gasholders are positioned over the generator itself ("floating cover"), and some are set up as a separate unit. A hole should be placed very low on one side of the gasholder so that gas can escape if it accumulates in too great a quantity.

The gasholder, placed on rollers or guides of some kind, will move up or down, depending on the amount of gas available to it, and the pressure of the gas. The pressure of the biogas under a gasholder is a function of the weight of the holder and the area of its horizontal cross-section.

For example, if our gasholder is a cylinder (the best shape) with a radius of 1.8 meters and a weight of 100 kilograms, then a horizontal cross-section would be a circle with a radius of 1.8 meters, and an area of about 10 square meters. Gas pressure will be (100 ÷ 10) 10 kilograms per square meter. Reference to Appendix 14 will show that this is the equivalent of only four-tenths of a centimeter of water on an open-tube manometer. Barely enough for a hiccup.

This gasholder would need to be weighted to bring the pressure up, since such low pressures are often not enough to force the biogas quickly out of the gasholder, through pipes, to the point of use. Further, low pressures make it harder for jets on stoves or similar devices to mix sufficient air into the biogas so that it will burn well. A general guide is that the ratio of weight to horizontal cross-sectional area should not exceed 200 in the SI system (kilograms per square meter) or 40 in the American system (pounds per square foot). For heavier gasholders, counterbalance weights are used. While there is no hard and fast minimum pressure, 10 centimeters (4 inches) of water, as measured with an open-tube manometer, is about as low as gas feeding and safety would prefer.

Gasholders can be constructed of fiberglass, ferrocement, steel, or any other material which is at hand, strong enough, and which can be protected against the corrosive power of biogas. (See Chapter 4.) It is not necessary to feed the gasholder with a pipe coming through the bottom of the water holding tank (assuming that the gasholder is not a "floating cover"). Connecting the generator directly to the distribution pipes will work just as well, if there is, somewhere in the system, a T-connector which goes to the gasholder. Care should be taken, however, to ensure that flashback will not happen. See Chapter 43.

Plastic Bags and Inner Tubes. Biogas can be stored in plastic. Ideally, if you can find seconds (cosmetically damaged merchandise) at a water bed factory, you will have the basis for an excellent modular storage facility. Waterbed-type bags can also be constructed out of vinyl and glue, or any strong plastic which can be glued. Such plastic should be 10 to 12 mils thick, at a minimum. Some plastics can only be heat-sealed, which requires special tools and expertise to achieve the proper weld on a seam. Unless you have access to these tools and that expertise, stick with glue (no pun intended) or high quality tape. Tape can also come unstuck easily, particularly when exposed to sunlight. Indeed, most plastic is exceedingly sensitive to sunlight, and should be kept away from it where possible.

But there's no point in making something if you can find something suitable. Any elastic and airtight container with sufficient wall strength can be used—inner tubes, for example. Such containers can be grouped together into a single system for greater capacity. Weights should be placed on top of these containers to achieve a counter pressure, so the gas will flow out of them when it is needed. One of the drawbacks of such a system is that the pressure varies so much from what it is when the bags or tubes are full to what it is when they are nearly empty. Thus, this is not only a low-cost, but also a low-performance system, unless it is interfaced with some means of keeping the pressure constant.

Pressurized Gas Storage. For larger generators, pressurizing the biogas may prove economical as compared with the construction of a larger low-pressure storage facility.

If the gas is to be used in tractors or for other mobile uses, pressure storage will be a requirement. The volume of gas which can be stored under pressure can be roughly approximated by multiplying the volume of the pressure container(s) by the input pressure, divided by standard atmospheric pressure. In other words:

$$Vs = \frac{Ps\ Vc}{Pa}$$

where:

Vs = volume of storage
Ps = pressure under which the gas is stored
Vc = volume of the pressure canister
Pa = atmospheric pressure

If the pressure of storage is 10 times normal atmospheric pressure (10 atmospheres, 147 psi, 4,072 inches of water, 10,333 centimeters of water, 7,600 millimeters of mercury, etc.), and if we have one volume of storage capacity, it will hold 10 volumes of biogas. Ordinary paint spray compressors usually operate in this range.

Brown and Smith (1975) made an interesting and valuable report on the feasibility of constructing a large continuous-fed generator on a dairy farm. The study is a good example of the design process in action, as applied to a specific situation.

For gas storage, they concluded that for the expected daily output (about 1.2×10^3 cubic feet per day—34 cubic meters), sufficient storage (50% of daily output) could be gotten using three 1,000 gallon propane tanks. (This is 134 cubic feet, or 3.8 cubic meters, assuming that propane tanks are measured in liquid and not dry gallons—there is a difference.) The pump they were using is capable of 240 psi (16.3 atmospheres).

This provides us with an example to use in the storage equation:

$$Vs = \frac{(240\ \text{psi})(3)(134\ \text{ft}^3)}{14.7\ \text{psi}}$$

$$Vs = 6{,}563\ \text{ft}^3$$

Brown and Smith estimated the propane tanks and fittings at $4,500 and the compressor at $1,000. This means that the estimated storage capacity of 6560 cubic feet will cost $5,500, or about 84 cents per evolved cubic foot of biogas stored. Naturally, once it is compressed, each evolved cubic foot will have much less volume—about one sixteenth the original volume. As well, if the biogas is scrubbed before storage, then the volume of storage is used with greater efficiency.

It is not required that heavy cylinders be used for low-pressure storage. We could use inner tubes, for example. Using pressure storage in inner tubes, the volume of biogas stored will depend on the final (expanded) volume of the inner tube and pressure should probably not exceed 50-60 psi (3 to 4 atmospheres). (See Appendix 13 for information on calculating the volume of an inner tube. This doughnut shape is called a torus).

CO_2, H_2O and (possibly) H_2S scrubbing should generally be practiced before pressurizing biogas.

Liquefaction. Methane is notoriously hard to compress to the point where it becomes a liquid. This means to do so requires complex and expensive equipment, and requires a lot of energy. The heart of a system capable of liquefying biogas is a 3-to5-stage pump. Each stage (after the first) accepts pressurized gas from the one before and compresses it still further. Because of the high pressures involved, the process can also be dangerous, and various safety devices are attached to such compressors—or they certainly should be. One further drawback of liquefaction is the weight of each of the storage cylinders, about 45 kilograms or 100 pounds. This is for cylinders with an internal volume of 0.05 cubic meters, or 1.8 cubic feet. Such a cylinder will store biogas at 160 plus atmospheres (2,400 psi).

Because liquefaction of biogas or methane is so expensive and because very few biogas installations will produce a sufficient volume of gas to make this a feasible option, we will say no more. If you wish to know more about this subject, consult a compressor manufacturer, or the library.

Other Storage Options. In storing biogas, we are essentially trying to store energy. While it will not be true in all cases, we should remember that the energy in biogas may be stored economically in some other form—perhaps as heat. Such means of storage must be relative to a particular situation, but look sharp! Maybe you'll find a better way for yourself.

Another option for storage is the storage of volatile acids—although this would be more likely to find use for longer term storage. See Chapter 49.

Storage Capacity Requirements

The volume of gas storage required will depend on the balance between production and use. Storage is required chiefly to provide a buffer between the steady production

Time	Hourly Use	Total Use	Total Production	Difference
9 PM–10 PM	17	17	29	12
10 PM–8 AM	17	187	319	132
8 AM–6 PM	17	357	609	252
6 PM–7 PM	109	466	638	172
7 PM–8 PM	109	575	667	92
8 PM–9 PM	109	684	696	12

Table 38.1 Production and Storage Example

of biogas, and the intermittent use of biogas.

For example, suppose a generator is built to handle the manure of chickens and the resulting biogas is used only to run a fairly large engine for two hours each day. As it happens (indeed, as it should be planned)—the engine requires a full 24 hours of production to run for 2 hours.

In such a situation, the storage capacity should clearly be at least 22 hours of production. (Question: If the chicken farmer takes weekends off, and doesn't have automatic equipment, what will happen to the biogas?)

In another example, a couple plans to use biogas for lighting in the evening and space heating in their three room house. Suppose that the lights require 540 Calories per hour and are generally used between 6 and 9 PM (*e.g.*, for 3 hours) each night. The bedroom is heated between 9 PM and 8 AM at a cost of 100 Calories per hour and the other two rooms are heated between 8 AM and 9 PM for 100 Calories per hour. Biogas production is 700 liters per day, at 65% CH_4, or about 170 Calories of available biogas energy per hour. Translating everything into volume of biogas, we have:

average Calories per liter = 5.85 calories
hourly rate of production = (170 ÷ 5.85) = 29 liters
lamp, use per hour = (540 ÷ 5.85) = 92 liters
heat, use per hour = (100 ÷ 5.85) = 17 liters

The average daily biogas use and production pattern would look like Table 38.1. Notice the time.

Production outstrips use all during the night and until the next evening, when the lights go on. With both lights and heat on, use is occurring at (109 − 29) 80 liters per hour faster than production.

Storage capacity in this situation should be about 260 liters, which is a bit above the greatest per day gas surplus (at 6PM). If these people do not have regular habits, storage capacity will need to be greater, or gas will be wasted.

Storage capacity requirements can be estimated according to the above method if you have an idea of the heat requirements of the different uses to which you will put the biogas, the average production of the generator, and the heat value of the biogas.

(If that all seems like too much bother, then estimate that somewhere between 20% and 40% of the per-day gas production will serve for most situations.)

Intermittent uses—once a week, or once in a while—can be treated similarly, except that the gas surplus needs to be averaged over a longer period. In the above example, there is a theoretical surplus of 12 liters per day. If we want to carry this surplus over a week, then we will need (12 × 7) an extra 85 liters capacity. The total storage capacity will then become (260 + 85) 345 liters.

We get a lot more use out of our storage, however, if we realize that it is more like a shock absorber than actual storage, in the same way that a refrigerator should be like a shock absorber, rather than storage.

If we truly store the food in our refrigerator (and don't use it), then soon we will need another refrigerator for use, and that's energy and dollar expensive. Similarly, if we expect to generate and store biogas during the summer and use it in the winter, then we have stepped over the line.

Therefore, if our storage facility is designed for average per-day use and per-day production, we will be ahead in most all cases. For intermittent uses, cutting back on other uses to build up the surplus needed is generally better than constructing a huge storage facility.

Gas Transfer

Remember that the gas from a warm generator will inevitably be saturated with water vapor, and that water vapor will condense to a liquid in a cooler pipeline and seek the lowest level available. Water is one thing that will always level with you.

Also remember that biogas can be fiendishly corrosive, especially in the presence of water. Armed with these two memories, you are well equipped to design a biogas pipeline. It should slope to a low point where the water can be drained, and if possible, it should be made of plastic pipe—which is cheaper than galvanized or copper pipe in any case.

In the absence of a transit—those useful telescope-like gadgets surveyors use—there are inexpensive levels which can be put on a long string to obtain the proper drop in the pipeline. Ask at your hardware store. (Transits, however, can often be rented at rent-all places.)

At the lowest point of the pipeline, put a moisture

Cubic Feet/Hour of Biogas Flow	3/8	1/2	3/4	1	1-1/4	1-1/2
5	0.28	0.07	0.01	—	—	—
6	0.40	0.10	0.02	0.01	—	—
8	0.71	0.18	0.04	0.01	—	—
10	1.11	0.28	0.05	0.01	—	—
15	2.5	0.63	0.12	0.03	0.01	—
20	4.5	1.11	0.20	0.05	0.01	0.01
30	10.0	2.50	0.46	0.12	0.03	0.01
40	—	4.5	0.82	0.20	0.05	0.02
50	—	7.0	1.28	0.32	0.07	0.03

pressure drop

Table 38.2 Pipe Sizes and Pressure Drop

collector—which might be a large pipe—and a valve to drain it, or, if there's enough room, put in a manometer out of which water can drain naturally. The end of the manometer open to air pressure should be below the level of the transfer pipe. This device cannot be used for higher pressure pipes, and check it often so that it doesn't dry out and leak gas.

A pump can provide the pressure required to transfer the gas, or the ordinary storage pressure can be used for that purpose.

Table 38.2 shows the pressure drop in inches of water, for the flow of a given quantity of gas through a pipe of the diameter shown, per 1,000 feet of pipe length. (Specific gravity of gas = 0.60)

Terms

Gasholder: Container inverted over water, or over the slurry of the generator to form a floating cover, which collects gas and stores it under low pressure.

Questions

None

Problems

1. If we have a continuous need for 300 liters of biogas per hour, and a need for 3 hours in the evening of 600 liters of biogas per hour over and above our continuous need, what minimum volume of storage will be required, assuming production averages 9,000 liters per day?

39: *Slurry and Effluent Handling*

Preloading

Prior to loading a substrate into a generator, it may have to be gathered, prepared (for example, by shredding), diluted with water, mixed, and so on. The methods you may want to use to handle the slurry and substrate will be largely situational. They will depend on your resources, substrate, parameters of digestion, and generator type. For small scale generation, the answers will be fairly obvious once you're out there working with the situation.

For larger scale generators, however, more preplanning is necessary. Because of the scale a greater quantity of substrate will be loaded than can be manually handled. If a tractor is used to carry the substrate, it will be more elegant if the tractor is biogas-powered.

A loading area should be established where the substrate is brought, with a large sink-shaped mixing and loading tub. The tractor can skip-load the tub, or if it has a scraper blade, push the substrate up a ramp into the tub. If necessary, the tub can be constructed so the tractor can drive right over it, dumping substrate through a grid or between tire-wide beams, spanning the tub.

If any substrate is nitrogen-poor, supernatent liquid can be recycled into it to provide water, nitrogen, and innoculation, but take care to keep as much air as possible away from the effluent supernatent.

Loading

Loading can be done by gravity feed, by pump, by hand, or by an auger, otherwise known as an Archimedes screw. The type of loading selected will depend first on the substrate and parameters chosen, and second on the scale involved. (The modular batch generators mentioned throughout this book can be loaded either by hand, or by hand and by pump.)

Unless you are dealing with a mix-well substrate (and even then) it is wise not to use anything but straight pipe. Pipe blockages, when they occur, generally occur at bends or constrictions in the pipe, and they are most difficult to clear out of these places.

If it's not feasible to avoid bends in the pipe used to transfer slurry inside the generator or used to carry effluent away from it, the size of the substrate chunks can be regulated by screens, filters, or easily accessible pipes, smaller than the internal transfer or effluent pipes, used for loading. These screens, filters, or smaller pipes will serve to insure that whatever can get past them will very probably not get caught in any other part of the system's piping. Wherever possible, if elbows are necessary, use less than 90 degree (45 degree or 22.5 degree) elbows to provide more gradual changes of direction, or bend the pipe around in a gradual curve.

For continuous-fed generators, higher loading rates can be tolerated if the loading of slurry is more or less continuous. Since cleaning the barn or moving the substrate from the place where it's produced to the generator may be more conveniently done once a day or once a week, it is good practice to put a holding tank between the mixing/loading tub and the generator. Depending on the schedule visualized, the holding tank should have a capacity slightly exceeding the daily or weekly substrate production plus the volume of make-up water required to dilute the substrate to the desired percentage solids.

The generator should be loaded from the holding tank by a constant feed, low-volume pump at a rate which will, over a period of 24 hours, equal the design loading rate for the generator.

This setup has the added advantage of allowing the slurry time to become anaerobic so that no oxygen is introduced into the generator, thus further increasing process stability, and increasing the relative percentage of methane in the resulting biogas.

Effluent

Effluent handling methods will be determined by factors similar to those which determine loading systems. High-rate digestion of animal manure leaves a fairly stable (no pun intended), but quite liquid effluent. Wheelbarrows will seem a bit out of place for handling such an effluent. Dry digestion of leaves will generally result in two effluent phases, one a moist residue, the other a dark liquid. The liquid can be pumped or drained, the soggy mass will need to be shoveled or dumped.

Gravity handling of effluent requires either a generator aboveground, or a slope falling away in front of the generator, so that, at some point downslope, a pipe coming from the bottom of the generator will be aboveground.

If our substrate is low in nitrogen, or if water is scarce, or if for some reason, we want to increase the percentage of solids in the effluent, we may wish to consider an effluent holding tank, where the effluent will be allowed to settle into a supernatent and sludge phase. The effluent supernatent can then be recycled into the generator to provide nitrogen, water and bacteria, or it can be used for irrigation, and the effluent sludge can be used as fertilizer.

Terms

None

Questions

None

Problems

None

40: Pumping

In many places we have mentioned pumping. Information included in this chapter is designed to enable you to better understand some of the basics of pumping. Covered briefly are both liquid and gaseous pumping.

Liquid Pumping

Because basic information on pumping is easily available from various pump manufacturers, the focus here will be on the special and peculiar aspects of pumping anaerobic slurries. However, some general information also follows so that you can find your way around a pump manufacturer's literature, and answer the questions salesmen may have.

Because many operations around farms and feedlots require manure pumps—it is easier to pump manure than shovel it—and manufacturers, recognizing a market exists, have designed and now sell many different kinds of pumps for this purpose.

Pumps designed for large particle vegetable slurries are few and far between, because such fluids are often handled by different means and they are difficult to pump. They either clog most ordinary pumps, or they simply will not move into the suction, feed, or inlet side of the pump.

The kinds of pumps used for manure slurries or thick fluids are many; inventors continue to invent, as they should. Among these kinds, we will discuss a few, since some of these will come to our attention shortly.

One pump we have already discussed is the Archimedes screw, or auger. At least one company manufactures a version of this pump for sewage sludge transfer. Another kind of pump used for sewage and manure slurries is the diaphragm pump. It operates in much the same way most aquarium air pumps operate (See Fig. 40.1). Sewage and sludge are also transferred using plunger or piston-type pumps. (See Fig. 40.2.)

Slurries can often be successfully handled by a positive displacement pump. One variety used in pumping slurries is the helical rotor, which operates similarly to the auger pump. (See Fig. 40.3.)

Last there is a class of pumps known as centrifugal. (See Fig. 40.4.) These pumps sling or throw slurry out through an impeller. If you take a piece of plastic tube and whirl it around your head, essentially the same thing will happen. Air will enter the bottom of the tube and be thrown out the whirling end. If you place the bottom in water, water will come out. Some centrifugal

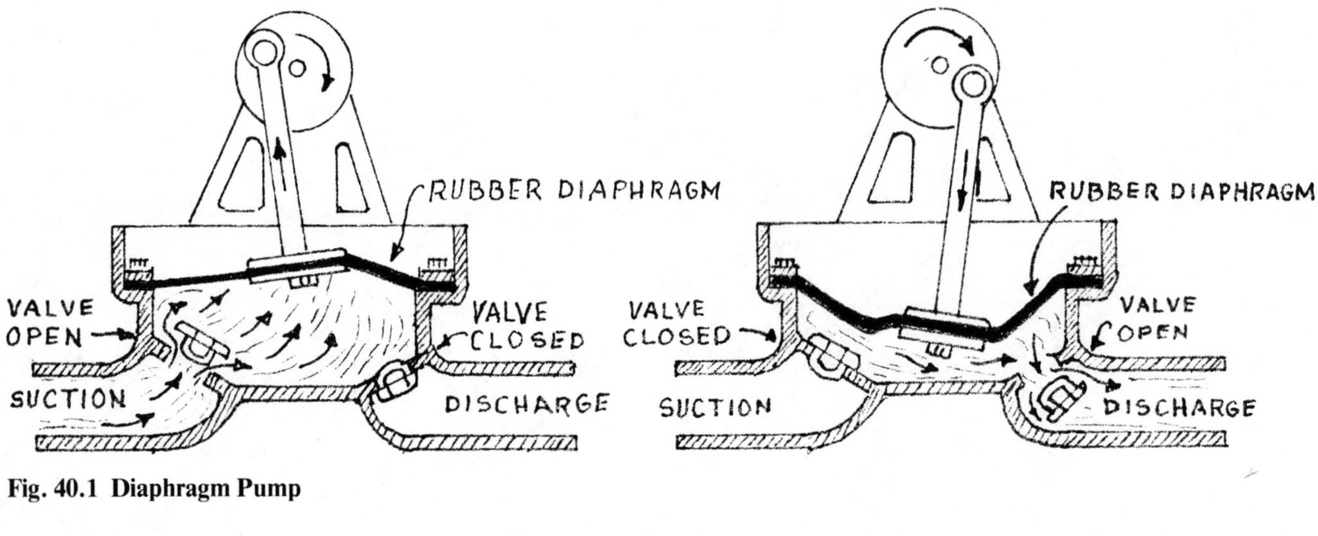

Fig. 40.1 Diaphragm Pump

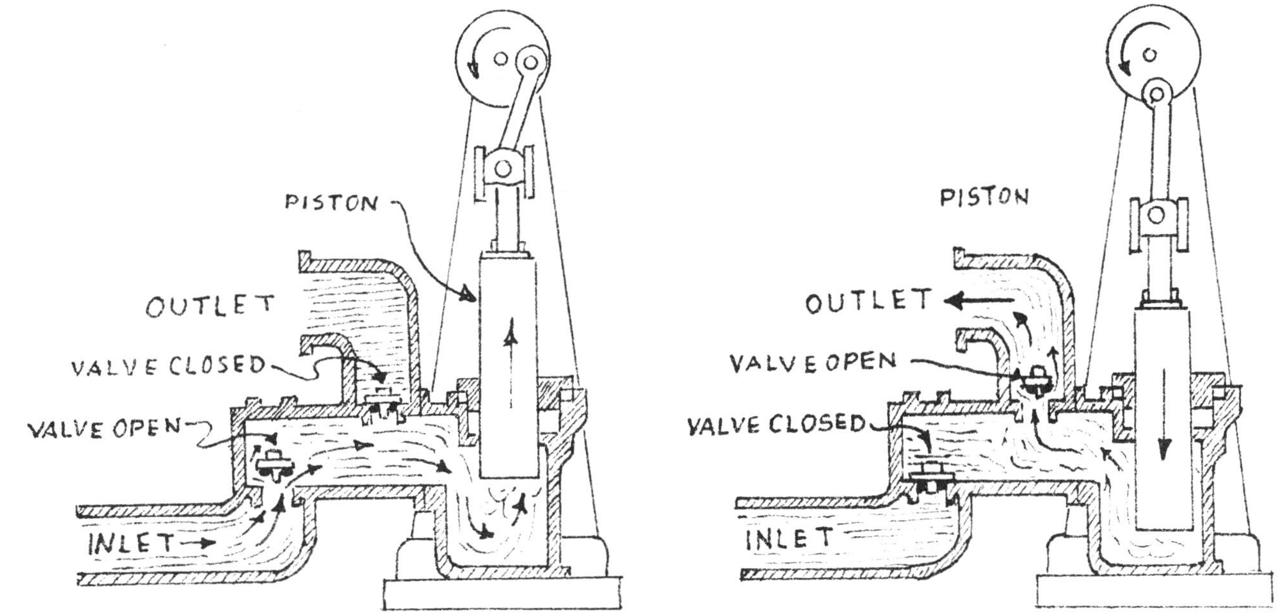

Fig. 40.2 Piston Pump

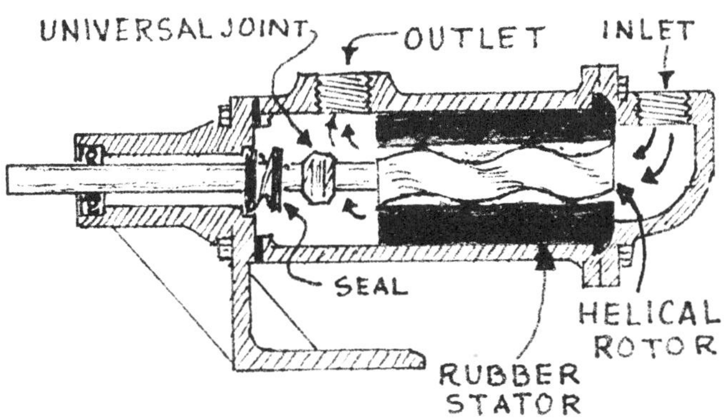

Fig. 40.3 Positive Displacement Pump

pumps are modified to chop or cut any materials in the slurry which might otherwise clog the pump. This feature is useful when pumping manure slurries containing straw or feathers, for example.

A pump should be efficient (in terms of work delivered for power expended), durable, and of a simple design for easy maintenance. Parts should be easily available —and this aspect needs very careful investigation before purchase.

Each pump operates best under the conditions for which it was designed. You may lose efficiency, damage your pump, and void your warranty by using a pump in a situation for which it was never designed.

Liquid pumping requires an energy input to overcome what is called the "dynamic head." Head is a word which refers to the vertical distance a liquid must be lifted above its original surface, and dynamic head refers to that same height, plus losses in pipe friction and the like. Since the loss of power can be expressed in terms of additional vertical distance the pump must overcome, head plus other losses experienced in pumping are expressed as a total vertical distance—dynamic head.

Because friction increases dynamic head, the number of bends, valves, and restrictions in the pipes through which the slurry moves, all of which increase the friction, also increase the dynamic head. For this reason, it can help reduce both initial costs (since a smaller pump may serve) and operating costs (since there is less dynamic head to overcome) if the piping is made with as few of these bends and restrictions as possible. Full flow gate valves are practically mandatory. Sharper bends and more acute angles increase friction. For bends of 90° or less, the friction encountered will decrease as the sharpness of the bend decreases, until the radius of the bend is made as large as 5 to 10 times the radius of the pipe itself; beyond this, there is no practical advantage in friction reduction to be gained by increasing the radius of the bend.

Friction is also proportional to the length of the pipe; it increases as the square of the velocity; it is inversely proportional to the diameter of the pipe; and abrupt changes in size or direction of the pipe will increase the friction. Mathematically:

$$fh = \frac{f\,L\,V^2}{2\,D\,g}$$

where:

Fh = loss of friction (head)
f = roughness of coefficient
L = length of pipe
V = velocity of flow of the liquid being pumped
D = diameter of pipe
g = value of gravity

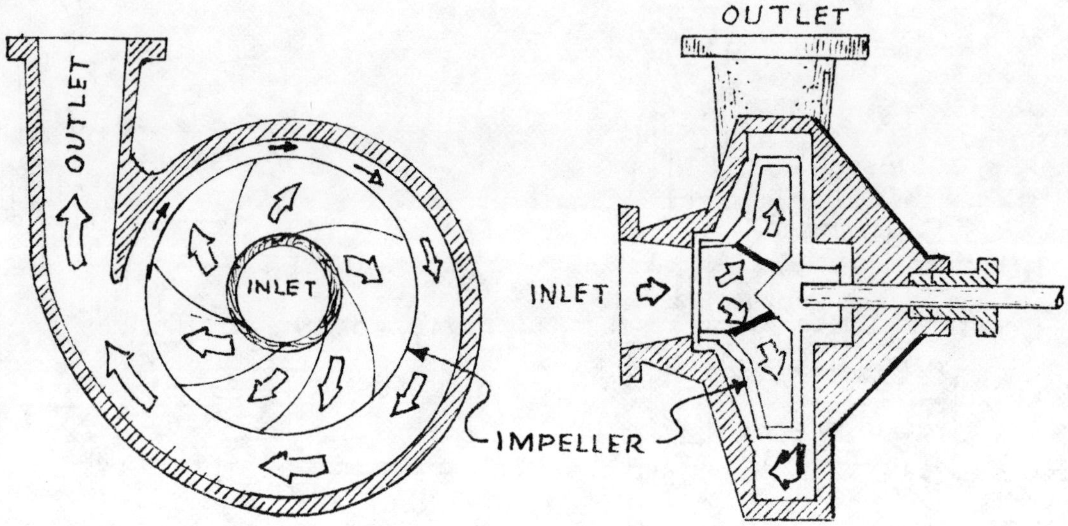

Fig. 40.4 Centrifugal Pump

Charts listing f can be found in numerous places, and most of them disagree with each other. The roughness coefficient varies according to the velocity of the flow, the size and condition of the pipe, and the viscosity of the fluid being pumped. You may be disappointed that it is not here listed—or you may be relieved, since now we can't throw a problem at you—but there would be no real purpose served in such a list, since further formulas and lists would be required (for figuring friction losses in bends, valves, etc.) to actually determine pipe friction. For you mathematical diehards, f varies between about 0.04 and 0.02 for smaller diameter pipes at ordinary velocities—that is, for water. For slurries, f is largely unknown.

The purpose of the formula is conceptual, to show you the relationships involved. Notice that because the velocity is squared, the effect of flow rate or velocity on friction is greater than any other component. Faster is harder, squared. (This tends to legislate against recirculation of slurry for purposes of agitation.)

Another consideration which affects the size and power requirements of pumps is the nature of the fluid being pumped. Is it thin or thick? Is it uniform or chunky? Is it full of grit, straw, feathers, twigs, comic books?

For each particular set of circumstances—total dynamic head, characteristics of the slurry, and flow rate desired–a different pump will prove most suitable. The difficulty comes in choosing one, since most manufacturers test their pumps—even those specifically designed for manure handling—with water. Faith, hope, and charity play a large part in this choice. It's like choosing an off-road vehicle based on how well it drives on a level freeway.

Fortunately, tests with cow, pig, and poultry manure slurries of various solids contents, using several different pumps, have been made, so we don't have to depend entirely on faith, hope, and charity—though it would certainly improve the world if all of us did anyway.

Hart, Moore, and Hale (1966) for example, tested five pumps: the Parma number 2 (centrifugal manure service), the Vaughan chopper model (centrifugal manure service), the Gorman-Rupp 13 A2B (centrifugal sewage service), the Moyno 1L10 (positive displacement helical screw), and the Gorman-Rupp 3 D (diaphragm pump).

These five pumps represent three general kinds often used in pumping slurries, as mentioned above; the centrifugal, the positive displacement, and the diaphragm types.

They found that the three centrifugal types were sensitive to pressure, and therefore, were not the type which should be chosen where a large dynamic head was experienced. However, these centrifugal models were less expensive, and had greater capacity (ability to pump large volumes) than either the diaphragm or the helical screw positive displacement type. Efficiency in terms of work delivered for a certain horsepower input was best in the Monyo, but the diaphragm pump was able to pump the thickest slurries of any of the five pumps tested. Because their test results are rather complete, it is recommended that anyone interested in more detail read the article. For a summation of their results on the viscosity of various manures at various percentages of solids (that is, with different amounts of water added), see Chapter 10. These viscosity differences will affect the ease with which different manure slurries of different percentage solids can be pumped.

Pumping Biogas

Pumping biogas presents us with many fewer problems, since the fluid we are considering is of a much simpler composition. The biggest problem encountered will be the condensation of water and possible corrosion due to carbonic and other mild acids dissolved in this water.

Corrosion problems can be overcome by the use of suitable materials for construction of the pump.

Beyond this, the size of the pump will depend upon the capacity (volume per minute) desired, and the pressure required.

For gas recirculation, the volume of gas required will depend on the volume of the generator and the pressure required will depend on the liquid depth to which the gas is to be pumped (see Appendix 10). For purposes of storage or transfer, the volume requirements will depend upon the rate of gas evolution. Storage pressure is chiefly a matter of the cost of tanks and a pump, and transfer pressure depends largely on friction losses and the desired pressure at point of use, which generally need not exceed 5 to 10 centimeters of water for biogas appliances.

Terms

Centrifugal pump: see Fig. 40.4.

Diaphragm pump: see Fig. 40.1.

Dynamic head: Sum of all factors (such as friction and pressure) which the pump must overcome to push fluids through a pipe.

Head: The vertical distance which a pump must push the liquid being pumped.

Piston pump: see Fig. 40.2.

Positive displacement pump: see Fig. 40.3.

Questions

1. You have a 7% solids, small particle size, manure slurry. Which of the pumps discussed could you use?
2. Which one should you use?

Problems

None

41: Construction Materials

Due to the peculiarities of the anaerobic environment, special construction materials are sometimes required. The anaerobic environment of a generator consists of two distinct phases, the liquid and the gaseous. It is the gaseous phase, (and the interface between this phase and the liquid phase), which is most corrosive. Almost any material which can resist being corroded by water can resist being corroded by slurry, but many materials do not do well when exposed to biogas.

Gas Transfers

Galvanized pipe, for example, which is often used for gas transfer systems, can be plagued by troubles, particularly where water collects in it. Hazeltine (1933), reported that a pipe, buried without concern for leveling, became completely blocked in one of the low spots over a period of two years, due to corrosion and the growth of sulfur-loving bacteria.

Plastic pipe is cheaper, easier to work with, and (when buried or otherwise protected) will last (in ordinary service) as long as galvanized pipe, but any pipe used to transfer gas should be laid on a slight slope, to allow condensed water to drain away. PVC plastic pipe will give off toxic chemicals in very low concentrations, but unless you are planninng to drink the slurry, it shouldn't matter much. Clay pipe is also a good possibility for slurry or effluent transfer, as it will not react to the mild acids in slurry the way galvanized pipe sometimes does.

Some other metals also react badly to exposure to biogas. In tests by Young (1948), it was shown that Monel, Everdur, and Phosphor bronze screens were severely corroded over a period of two years exposure. Stainless steel showed no sign of corrosion.

Gasholders

Corrosion of metal is particularly a problem with gas holders, as these float in water, and the high CO_2 content of the biogas plus the pressure of storage combine to make the water rather strongly acidic. Some protection can be gotten by the use of bituminous or acid-resisting and waterproofing paints, such as are used to seal concrete foundations, but these break down sooner or later. Another solution is to adjust the pH of the water with alkaline chemicals, which may temporarily have the advantage of making the gas holder do double duty as a gas scrubber.

Fiberglass, however, does not react with biogas, and this material is suitable for a gasholder. Various kinds of plastic are also used, quite successfully, for gasholders.

Main Tanks

Many generators are constructed partly or wholly out of cement or concrete. Portland cement—a popular and particular kind of cement—is degraded by mild acids, such as water containing dissolved CO_2. Increasing the amount of the carbonate ion (CO_3, from, for example, calcium carbonate, $CaCO_3$) in the water to 10 or 20 parts per million by weight will reduce this kind of cement degradation.

As well, water or liquids which contain sulfates seriously harm the cement by forming chemical compounds which have a much greater volume than the original molecule, thus breaking the cement apart by internal pressure so that it actually becomes soft and mushy. This kind of reaction occurs wherever water vapor and H_2S are present.

Portland cement can be protected from sulfate attack, according to Hammerton (1944) by adding the sulfate ion (SO_3^-), to the cement prior to mixing it. This will need to be added at the rate of 7.4 percent by weight. This addition causes the above described sulfate reactions to take place prior to the setting of the cement so that the cured cement is immune to attack. This is similar to type V Portland cement.

Another approach is to use a high aluminus cement with a water content of not more than 60 percent by weight of the cement used, in a mixture of one part cement, 2 parts sand, and 4 parts aggregate (1:2:4). This should not be overmixed, and the curing temperature must be below 25° C (77° F). Coatings on concrete which is otherwise liable to corrosion will only generally delay, but not stop that corrosion.

Other materials suitable for the main tank of the generator are plastic or butyl rubber storage tanks, surplus steel tanks, or glass-fused-to-steel tanks, such as those used to store silage. Each of these has been reported as having been used, and each will have suitability in different situations, depending mainly on the cost per unit volume.

In the U.S., more and more attention is being paid to using plastic and/or rubber bags to contain the slurry and biogas, with support provided by the earth (e.g., the generator is put into a hole), or by plywood forms of the sort used for concrete forms. Perennial Energy Systems , P.O. Box 15, Dora, MO 65637, has a great deal of information available on this low-cost alternative. Please contact them for further information and prices. Everyone has to make a living, so please include two or three dollars and a self-addressed, stamped envelope if you want them to spend some time answering your question(s).

Terms

None

Questions

None

Problems

None

42: Sizing Generators

The size of a generator will depend on several factors. The first is generator type. If the generator is a batch-fed type, the minimum size required will be estimated by simply multiplying the expected volume of slurry (organic materials + water), by the desired number of days those materials are to be kept in the generator for biogas production. Add 10% to 20% to give the biogas room to accumulate.

The volume of a batch-fed generator can be calculated using this equation.

$$\text{Vgen} = 1.1\,(\text{Vs})(\text{T})$$

where:

Vgen = volume of the generator
1.1 = gives a 10% increase in volume for gas accumulation space
Vs = volume of slurry produced or available daily
T = number of days (time) per batch cycle

Of course, it may be difficult to measure or estimate the volume of slurry. Much easier, generally, is the estimation of the weight of the substrate (manure or plant matter) available daily. What is needed is some means of relating the weight of the substrate to the volume of slurry. Unfortunately, this information is not easy to find, and must be developed for each situation. A liter (volume) of water will weigh 1,000 grams. A liter of manure will weigh, on the average, within 10% of that figure.

The ratio between the weight of a substrate and its volume is called "density." Lead is more dense than water; thus, given the same volume of each, the lead will weigh more. Air is less dense; it weighs less for a given volume.

You may wish to determine the density of the substate you are using yourself. This can be done simply by weighing a standard volume of your substrate, or, if you wish to find out the density of your slurry (substrate + water) then you can weigh a standard volume of slurry.

$$\text{Den} = \frac{\text{W}}{\text{V}}$$

where:

Den = density
W = weight
V = volume

If the generator is to be of the continuous-fed type, the volume will depend on several factors. We can approach the problem from several angles. Most simply, if we know the weight of VS produced per day per animal load, and the number of animal units, and if a loading rate is chosen from Table 16.6, then we can calculate the minimum generator volume using this equation.

$$\text{Vgen} = \frac{(\text{Wvs})(\text{Nau})}{\text{Lr}}$$

where:

Vgen = volume of generator
Wvs = weight of volatile solids produced per day per animal unit
Nau = number of animal units
Lr = loading rate chosen

For instance, according to Table 16.5, 450 kilograms of pigs produce 2.7 kilograms (2,700 grams) of VS per day.

Let's assume we have a 450 kilogram (1,000 pound) pig in the back yard, and that our chosen loading rate is 3.5 grams of VS per liter of generator per day. We need to convert everything to the same units, so that kilograms become grams below.

Then:

$$\text{Vgen} = \frac{(2{,}700\ \text{gm. a.u.}^{-1}\ \text{day}^{-1})(1\ \text{a.u.})}{3.5\ \text{gm. liter}^{-1}\ \text{day}^{-1}}$$

$$\text{Vgen} = 770\ \text{liters}$$

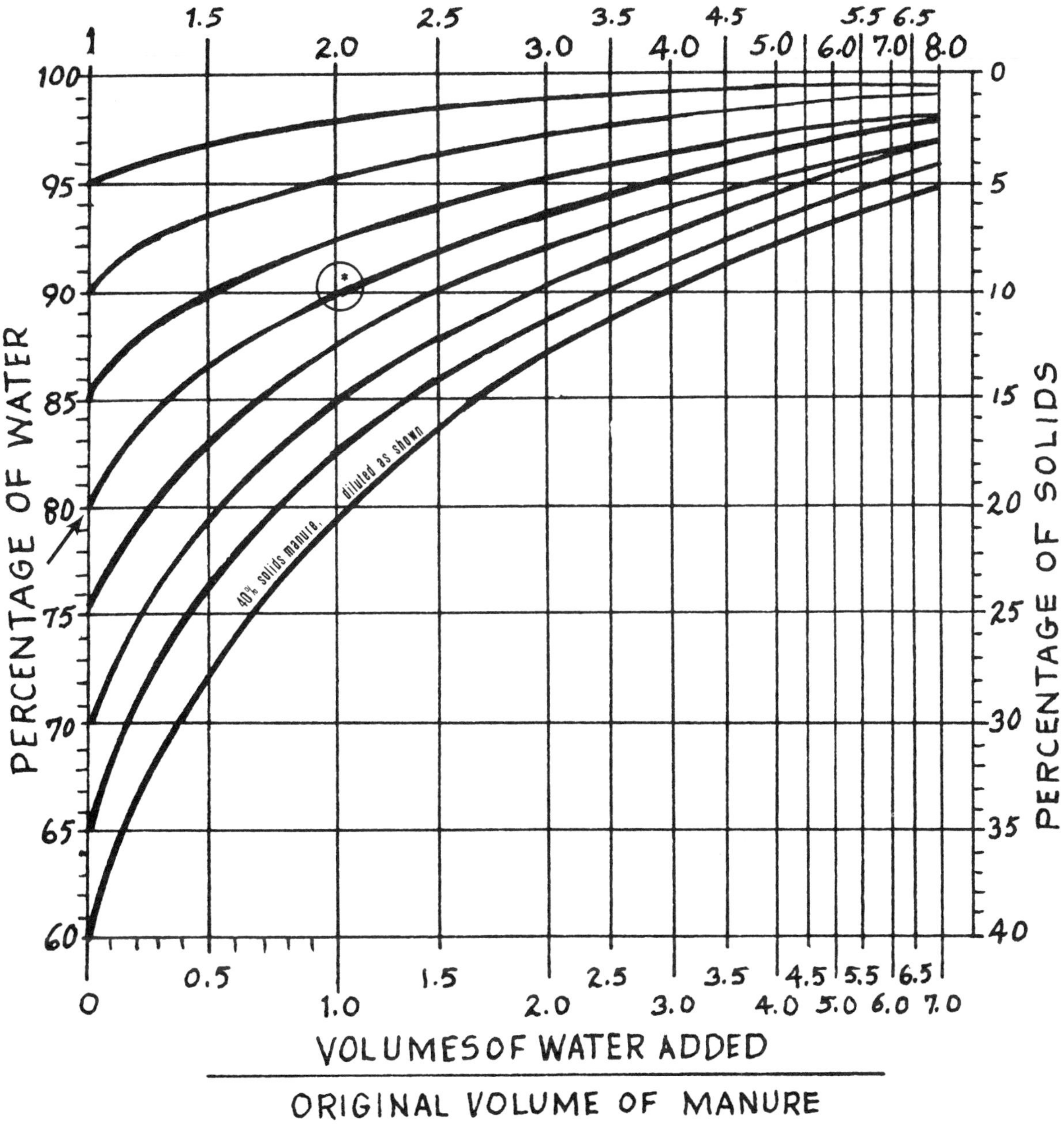

Fig. 42.1 Dilution Factor

(The a.u. refers to "animal unit"—450 kilograms of pig.) Notice that this says nothing about the dilution of the manure slurry—that is, we're merely loading a specified amount of VS per liter per day, regardless of how much water comes along for the ride. To figure out the HRT in the absence of dilution, we need only divide the volume of the generator by the volume of slurry we produce each day if we know that volume. On the other hand, if we wish to take dilution into account, the equation below will tell us the hydraulic retention time:

$$HRT = \frac{Vgen}{(Vm)(Df)}$$

where:

HRT = hydraulic retention time
Vgen = volume of the generator
Vm = total volume of manure produced per day
Df = dilution factor

This equation assumes that adding a liter of water to a liter of manure (as one example) will give us equal liters of slurry. (In other words, Vm times Df equals the volume of slurry produced each day.) In fact, it might not, since the manure might have air in it which would be forced out by the water, resulting in less than two liters of slurry. Nevertheless, the volume of slurry will be quite close to two liters if we start with fresh manure. If, on the other hand, we are starting with old manure, dried manure, or plant materials, it will be best to make tests to determine the final volume of the slurry.

Alternately, we may wish to find out the dilution factor we need to use to get a certain HRT (given the same assumption about how the volume changes); the equation below will help.

$$Df = \frac{Vgen}{(Vm)(HRT)}$$

Where those symbols are as described just above. Fig. 42.1 gives the dilution factor (Df).

As an example of using this figure, we can see that when one volume of 80% H_2O manure is diluted with one volume of water (Df = 2), the resulting percentage of H_2O is 90% (See asterisk in Fig. 42.1.)

So as you can see, the dilution factor is merely:

$$Df = 1 + \frac{Vs}{(Vm)}$$

where:

Df = dilution factor
Vs = volume of slurry
Vm = volume of manure before adding water

Again, the assumption here is that 1 + 1 = 2, volume-wise.

With the formulas above, we can choose either the Df or HRT we wish to use, and then obtain the other by calculation. Suppose, for example, that we want less than 10% solids in our generator slurry, and we don't care too much about the HRT, as long as it exceeds 10 days. If our pig manure (from the example above) starts at 80% H_2O, then the minimum acceptable Df (from the charts) is 2.0. If our pig produces about 28 liters of manure a day (from Table 16.5), this means:

$$HRT = \frac{770 \text{ liters}}{(28 \text{ liters day}^{-1})\ 2.0}$$

$$HRT = 13.6 \text{ days}$$

Or, working from the other direction:

$$Df = \frac{770 \text{ liters}}{(28 \text{ liters})\ 10}$$

$$Df = 2.7 \text{ liters}$$

This means that HRT can vary between 10 and 13.6 days, while Df varies between 2.7 and 2.0. (When Df is 2.7, the percentage of solids—from the chart—is about 6%. Remember that as dilution decreases and HRT falls, the cost of the generator generally falls too.)

There are several other ways to size a continuous-fed generator, and other formulas for finding various parameters. Some examples follow:

$$HRT = \frac{(\%VS)(\%TS)(Den)(Df)}{(10^4)(Lr)}$$

$$Vgen = \frac{(Vau)(Nau)(\%VS)(\%TS)(Den)(Df)}{(10^4)(Lr)}$$

$$Vgen = (HRT)(Vau)(Nau)(Df)$$

where:

HRT = hydraulic retention time
%TS = percentage of total solids in the wet manure
%VS = percentage of volatile solids in the wet manure
Den = weight of manure per unit volume of manure density
Df = dilution factor
Lr = loading rate
Vgen = volume of generator
Vau = volume of manure produced per day per animal unit
Nau = number of animal units

Of course, all units should be in the same system, and must be of the same kind. In other words, all the units must be either American or S.I., and all references to weight, if we are dealing in S.I., should be grams or kilograms, and not a mix of the two.

The per-day volume of biogas produced by a manure substrate generator can be roughly estimated by using the formula below, and Table 16.5.

$$Vb = (Nau)(Gau)$$

where:

Vb = estimated volume of biogas
Nau = number of animal units
Gau = estimated gas production per animal unit per day

In the case of our enormous pig, Vb = 1.6 cubic meters per day.

Terms

None

Questions

None

Problems

1. You have 3 cows, weighing 550 kilograms each. Take the needed information from Table 16.5 assuming a loading rate of 4 grams VS per liter of generator per day, using Vgen = (Vau)(Nau)(Df), what is Vgen?
2. In the above situation, what is the volume of biogas produced?

43: Safety

The chief danger in biogas production is explosion, and therefore safety in biogas production has a great deal to do with avoiding explosion. Such a purpose has two related efforts: avoiding explosive mixtures and eliminating sparks, and we'll discuss both of these below. However, regardless of what design changes may be made in the name of safety, regardless of how consumer-proof your biogas system is, the primary way you're going to avoid accidents of any description is by simple awareness. If you've damaged yourself very often in minor accidents, then you know how easily they can happen, and how often they are due to mere carelessness, of the kind that most often shows up when you are tired or preoccupied. So, when dealing with biogas production, the point is to do so safely: always remember that biogas is very dangerous, given the right circumstances. Scattered throughout the periodicals dealing with sewage digestion, you can find reports of many accidents and explosions that have occurred. In many cases, the people who were close to the explosion when it occurred were never found. Buildings, strong and well built, virtually disappeared.

Avoiding Explosive Mixtures

Air leaking into a small space filled with biogas, is much more dangerous than biogas leaking into a large space filled with air. This is the case because a small amount of biogas is only explosive when mixed with the proper small amount of air, but not when it is diluted with a large amount of air. It is generally more dangerous to have air leak into the small volume of the generator than it is to have biogas leak out to the surrounding air. This is why the first biogas safety rule is: *maintain a positive pressure in the whole biogas system at all times.* If this is done, any leaks which occur will not allow air into the system. If there is air in the generator, in the proper amount, you may never find out until you light your stove (or whatever), and then stand there wondering whence the flame disappeared, as the burning mixture races back down the pipe to the doomed gas-holder. (BOOM). So, maintain that positive pressure.

Frankly, any generator will leak to some degree—all we can really try to do is avoid serious leaks. This is mostly a matter of exercising care when constructing a generator. It should be obvious that it's much more likely that a leak will develop at the juncture of two pipes than it is that a leak will develop in the middle of a good piece of pipe. Saving money is fine, but using worn or damaged parts can be foolish. And so on. Secondly, that positive pressure mentioned should not be too extreme. Five to ten centimeters (2 to 4 inches) of water is plenty, and any greater pressure will magnify leaks and reduce gas output.

The best way of detecting leaks, when there is some pressure in the system, is by the use of a soap and water mixture, brushed or sprayed on the likely spots. If a leak is present, bubbles will continue to appear at that spot. Whenever smells indicate a leak, and at regular intervals as part of a safety program, use this means of assuring that there are no leaks in the system.

That Stinking Sensation

The smell of biogas depends on the parameters of digestion and the substrate used. Most biogas has a quite tolerable, musty odor. Some biogas smells of rotten eggs, for the reasons outlined in the chapter on scrubbing, and rarely, it smells worse. In any case, it always has a detectable odor. The only possible exception to this rule is well-scrubbed biogas, since methane itself is odorless. While it is unlikely that low-tech scrubbing processes will completely remove all odors, they might make odors more difficult to detect.

The human nose is a marvelous instrument, however, and if yours works, you should use it and rely on it to help you determine when biogas is leaking. This is the first line of defense against explosive mixtures outside the generator. If you have a large scale operation and you wish to put your trust in transistors, you might consider purchasing an electronic gas leak detector. The name and address of one manufacturer is listed in Appendix 16. Natural gas detectors will generally work in detecting biogas leaks, but just to be sure, ask the manufacturer.

Ventilation

If your generator is enclosed, as for example in a greenhouse, you must take care in arranging for ventilation. Even if biogas is leaking, as long as the methane component does not climb up to a critical percentage, no explosion can occur. Thus, if we can dilute the methane by replacing the air around the generator at regular intervals, we increase our margin of safety. Of course, one of the primary reasons for enclosing a generator is to keep it warm, and ventilation will work against that by introducing cool air into our enclosed space. However, safety is definitely first here. (Don't put your gasholder in any enclosed space.)

For each unit volume of air surrounding the generator, 0.07 volumes of biogas will need to be present, per unit volume of air, to achieve an explosive mixture. This assumes that your biogas is 75% CH_4. If we allow ourselves a safety factor of 3.5, then we will allow no more than 0.02 volumes of biogas to co-exist with each unit volume of air.

Armed with this information, we can then make some reasonable decisions about ventilation:

$$0.02 \text{ Vsp Cf} => \text{Vbp Lf}$$

where:

Vsp = volume of the space around the generator

Cf = change factor. This equals the number of air changes in the space per day plus one. No change of air, Cf = 1. If 50% of the air is replaced with fresh air every day, Cf = 1.5.

$\geqq$ = this symbol means "greater than or equal to"

Vbp = volume of biogas per day

Lf = leak factor as a decimal percentage. If 10% of the gas leaks out of the generator, Lf = 0.10.

Suppose, for example, that we have a space of 10 cubic meters, biogas production is 1 cubic meter per day, there is very little ventilation (*e.g.*, Cf = 1) and no more than 10% of the daily biogas production leaks out into our space.

$$(0.02)(10)(1) \geqq (1)(0.10)$$
$$0.20 \geqq 0.10$$

This reads "0.20 is greater than or equal to 0.10." A true statement. This means that, given the conditions assumed, we will not exceed our safety factor in one day. However, if the air in that space is truly stagnant and the biogas does leak at 10% of daily production, we will exceed our safety factor in two days! For this reason, replacing the air fully, at least once a day, in any enclosed space which contains a generator, is excellent insurance. In the example given, this would mean a mere 0.417 cubic meters of fresh air an hour—about 7 liters a minute—should enter the room. Generally, this will not radically increase the heat load of the room. Notice that in our formula, a lot depends on what we assume the amount of leakage is. It is well not to underestimate this factor.

If, in the above example, biogas production is 3 cubic meters a day, then we have a different story:

$$(0.20)(10)(1) \geqq (3)(0.10)$$
$$0.20 \geqq 0.30$$

This reads "0.20 is greater than or equal to 0.30." Not true. Cf needs to be beefed up to at least 1.5 to avoid danger:

$$(0.20)(10)(1.5) \geqq (3)(0.10)$$
$$0.30 \geqq 0.30$$

Okay?

Baby, Don't You Light My Pyre

Avoiding explosion also means avoiding sparks and open flames. Much common electrical equipment, particularly light switches and motors, routinely produces sparks. (Light switches which use sealed mercury contacts are all right.) Heating directly or indirectly with biogas means open flames, and so on. If you're going to rub sticks together, make sure they're not dynamite sticks, and if you're going to have a generator enclosed, show a spark of intelligence. Smokers beware, and beware of smokers.

Last Chance

The very last line of defense is a flame arrestor. This is a device put into the biogas supply line which, at least theoretically, stops any flame front burning its way gleefully down the supply pipe to the gasholder. There are two basic designs. One is a cooling design, where, as mentioned in Chapter 22, a fine mesh of wire blocking the pipe, cools the burning gases down below the flash point. The other kind of design relies on some kind of lock in the supply pipe. A simple version of this would be a device which bubbles the biogas up through water. The idea is that any flame which flashes back to that point, will be stopped, since it cannot travel through the water.

Terms

None

Problems

None

Questions

Name 3 gadgets around your home that could set off an explosion (given the proper carelessness on your part) around a biogas system.

44: Economics

For a generator to make sense economically, its annual dollar benefit must exceed its annual dollar cost:

Anb = annual net benefit
Agb = annual gross benefit in fuel and labor
Ac = annual costs of maintenance, repair, and overhead
Ec = initial equipment and construction costs
If = interest factor (see Table 44.10)

The formula assumes a 10 year life on the equipment, and 10 years to pay off the loan taken out to build the generator at different annual interest rates as reflected in Table 44.1.

In an environment of inflation, even 23% (or more?) may be the prevailing rate of interest. The conclusion may be to build with non-borrowed money, or to do it while interest is merely painful, rather than insupportable.

% Interest	*If*	% Interest	*If*
6.5	.139	18.5	.227
7.0	.142	19.0	.231
7.5	.146	19.5	.235
8.0	.149	20.0	.239
8.5	.152	20.5	.243
9.0	.156	21.0	.247
9.5	.159	21.5	.251
10.0	.163	22.0	.255
10.5	.166	22.5	.259
11.0	.170	23.0	.263
11.5	.173	23.5	.267
12.0	.177	24.0	.272
12.5	.181	24.5	.276
13.0	.184	25.0	.280
13.5	.188	25.5	.284
14.0	.192	26.0	.289
14.5	.195	26.5	.293
15.0	.199	27.0	.297
15.5	.203	27.5	.302
16.0	.207	28.0	.306
16.5	.211	28.5	.310
17.0	.215	29.0	.315
17.5	.219	29.5	.319
18.0	.223	30.0	.324

Table 44.1 Interest Factor

Generator Costs

Information on the costs of generators as reported in the literature, is often vague and conflicting. For example, Morris, Jewell, and Casler (1976) estimated that a high-rate mesophilic generator would cost about $270.00 per animal unit. This is about 170% greater than the reported cost of the high-rate generator designed and built by Smith and Brown (1975) at $160.00 per animal unit. It seems as though, in abstract studies, the motivation is to be rather conservative, since nobody minds building a generator which then turns out to cost a great deal less than expected, but cost overruns are anathema.

Capacity and Scale

Scale is also a factor. Costs drop, per animal unit or per unit volume, as size increases; unless that is, the generator is very small. Cost per unit volume, however, is not a reliable measure of generator costs. For example, Smith and Brown's generator was estimated to cost $4.24 per cubic foot, about $150.00 per cubic meter. Auerbach (1973) built a small generator for $4.94 per cubic foot, $174.00 per cubic meter, but his generator, due to a overly long HRT (90 days) will only handle 19% of the amount of substrate materials that the Smith and Brown generator will handle.

Thus, while Auerbach's generator is but 17% more

expensive per unit volume, it is more than six times as expensive (616%) in terms of amount of substrate materials handled. Dilution of substrate and loading rate will also affect the efficiency with which a generator functions. For these reasons, generator costs per unit amount of substrate handled can vary dramatically. The pivotal values of HRT, loading rate, and percentage of solids must therefore be carefully chosen, as these will determine the size of the generator, given a fixed supply of substrate materials available per day.

Smith and Brown who outline the design process for a diary operation, were working under the design assumptions of a 17-day HRT, 3.5 grams VS per liter of generator loading rate, and a 10% solids slurry.

Due to the difficulty of agitating a slurry of over 10% solids, high-rate generators are almost never designed with any greater percentage solids than this. Thus we cannot, given the state of the art, expect to make the generator smaller by increasing the percentage of solids. However, several studies have shown that a loading rate of 3.8 grams VS per liter and an HRT of 10 days can produce a stable digestion process when using dairy cow manure. Under these conditions, the generator of Smith and Brown could have been nearly 50% smaller, at an estimated savings of $28,000 (assuming percent reduction in volume equals percent reduction in cost).

This is not intended as a criticism of Smith and Brown's work—indeed, their report, "Process Feasibility Study," is a recommended reading for anyone wanting to build a large-scale, high-rate digester. They are among the very few whose work is clear and careful enough that such examples can be extrapolated. The design parameters chosen insured stability of the process and even a savings such as that outlined would not justify a generator which experienced repeated process failures. However, the illustration serves to indicate the dramatic cost difference which can be made when seemingly small adjustments in parameters are made. Also, the longer HRT means increased gas production per unit TS or VS.

Since any generator designer faces two conflicting goals—the need for cost reductions and the need for process stability—it is obviously worthwhile to run tests on the intended substrate. Here is where the tremendous variation on values reported for various aspects of the biogas process reaches a point of frustration. A 5% difference could be accounted for, accepted, and planned around—but a 50% or 500% difference is more difficult to justify. The best way to resolve these difficulties is with your own tests. (For simplicity, calculate your loading rate as a function of TS rather than VS—see Appendix 1).

Let's take a couple of examples to show the process in action. L. John Fry (1974) reports that he built a plug-flow generator to take care of his pig manure. The cost in 1958 is reported as $10,000, including auxiliary equipment. He estimates that the plant could have been duplicated in 1974 for three times that cost. This may be conservative, as his original labor was very inexpensive. (His farm was in South Africa.) Running the generator, he estimates, saves him 56 farm/hours per week in labor, and he reports producing about 230 cubic meters per day in methane. (Biogas production was, of course, greater.) He estimates 4.5 metric tonnes (5 tons) of nitrogen, 4.2 tonnes (4.5 tons) of phosphates, and 0.9 tonnes (1 ton) of potash in fertilizer value per year.

In many studies of the economics of the biogas process, the value of these nutrients is thrown into the pot and added to the benefit of the biogas process. However, in regard to calculating the value of these nutrients, it sould be remembered that they do not simply appear. They were in the manure to begin with and then were put through the generator. The anaerobic process provides an economic incentive in regard to its fertilizing value, only as compared to (a) composting, and (b) putting it in the fields without composting it. This in turn introduces several other factors. More nitrogen is lost in aerobic composting than is lost in anaerobic (biogas process) composting, but it seems that plants respond better to the nitrogen in compost than they do to the nitrogen in effluent. At least most of them do. Soil composition will also have an effect on crop response to effluent versus compost versus raw manure. And there are other factors as well. (See Appendix 3.)

Because the economic benefit of the nutrients cannot be manhandled into a simple formula, we will simply view them as being essentially unchanged by the biogas process. In other words, unless the farmer is presently discarding his manure, and there is something about putting it through the biogas process that suddenly causes him to regard it as so valuable that he then begins to use the effluent on his crops, then he is probably getting the benefit of its nutrients anyway, and whether or not he makes the manure into biogas will not change this. Of course, since the effluent is liquid rather than semi-solid, the farmer may find it easier to pump it out to the field, and may on that account begin to use it. This is, however, a special case. In most cases the manure will be used regardless of whether it passes through a generator or not. Therefore, we will not here consider the dollar amounts of nutrients as being a part of the economic benefit of the biogas process.

Dollar Benefit of Biogas

As regards the value of the biogas, this is relative to the net amount available (after heating needs are accounted for), the value of the fuel it replaces, and the conversion efficency of the replaced fuel as compared with the conversion efficiency of biogas in the job or jobs being considered.

For example, in Mr. Fry's case, the biogas was mainly used to fuel a modified diesel engine to produce electricity. Assuming the engine had a conversion efficiency of biogas into mechanical energy of 22%, and the generator it ran had a conversion efficiency of 60%, the total conversion efficiency of biogas into electricity would have been (0.60 x 0.22) 13.2%—which, by the way, is an excellent efficiency for such a set-up.

This means that Mr. Fry's 230 cubic meters of methane becomes 30 cubic meters of methane energy available as electricity, or about 278 kwh of electricity available each day. Since Mr. Fry heated his biogas generator with engine waste heat, this energy need not be subtracted. It is accounted for when we assume a 22% engine conversion efficiency since some of the 78% of the wasted heat will show up in the water from the cooling system, which can be used to heat the generator. However, in some other situation, this may need to be taken into account.

If electricity is available, and assuming its price to be 2.5 cents per kwh, this is \$6.95 net energy per day. If electricity is not available, then the energy value of the methane must be compared with whatever is available, and the conversion efficiency of that fuel in the job required.

Let's run the economic equation once with the net value of the biogas as electricity, and once with the net value as natural gas. If we assume that this too would need to be converted to electricity for use on the farm, then we can run the second comparison without regard to conversion efficiency, since natural gas will behave very similarly to well-scrubbed biogas in any application. The labor savings, as reported by Mr. Fry, will in either case be equal, 2,900 man hours per year. At \$2.75 per hour, that's \$8,000.

For electricty:

Fuel benefit per year = (365)(\$6.95) = \$2,540
Labor benefit per year = 8,000
Total dollar benefit
per year = \$10,540

Note that such a figure does not include savings in equipment costs, if any, for equipment which would otherwise be required to handle the manure.

Assuming, then, a loan for the \$30,000 (new, estimated) construction cost, an interest rate of 9%, and a maintenance, repair and overhead (insurance, tractor repair and fuel, etc.) cost of \$2,500 per year, we have:

Anb = 10,540 − 2,500 − (30,000)(0.156)
Anb = \$3,360 per year

Remember that this is the net annual benefit, and the equation assumes an amortization period (pay back on the loan plus interest) of 10 years. In other words, after making an average year's payments on interest and principle and subtracting the yearly overhead, more than \$3,000 in benefit are still available.

For natural gas, the dollar benefit of energy gained would be \$10.91 a day for the 230 cubic meters of methane, as energy, compared to that amount of energy as natural gas, if we assume that natural gas costs \$9.92 per million Calories. This is \$6,646 per year.

Therefore:

Anb = 14,640 −2,500 − (30,000)(0.156)
Anb = \$7,460 per year

The pivotal factors, as you can see, are:

1. the cost of the loan;
2. the net amount of gas produced;
3. the cost, availability and suitability (conversion efficiency) of other energy sources;
4. the savings in labor, handling, and equipment costs.

Remember that even if we can only show a net annual benefit of one dollar, the generator will pay for itself in 10 years. (Thereafter we will gain, per year, the amount we formerly paid out to the loan.) However, from the point of view of classic economics, the net annual benefit should be at least 10% of the construction cost, since many other investments will yield at least this much return.

With inflation running 12%, an unchanging annual net benefit of 10% will actually lose 2% of real value per year. But the generator is not generating dollars, it is generating biogas, saving labor, replacing fuel. If we assume these things to be worth a certain dollar amount, and then run our calculations on this dollar amount, we miss an important point. That is, these real things will increase in their dollar worth as inflation continues to drive up prices. Trends indicate that the inflation in energy prices will be more rapid than inflation in general. Thus, dollar wise, we have a

Fuel	Dollars per million Cal	Dollars per million Btu
Electricity	29.07	7.33
Fuel oil	8.84	2.20
Gasoline	17.15	4.31
Natural gas	9.92	2.50
Propane	21.64	5.45

Table 44.2 Energy Costs

new ball game each year. To say, on the basis of the present dollar amount of the annual net benefit, that payback will be in eleven or fifteen or six years is misleading, precisely for this reason.

The formulas herein, however, do not take the inflationary rise in energy prices into account primarily because it is difficult to be certain that the assumptions we may make about gas production, heat requirements, and construction costs are accurate. It is wise to go slowly and carefully and try to approach the problem in several different ways.

The value of the energy from the biogas can be roughly calculated from Table 44.2 of the costs of other kinds of energy. It should be remembered that if the generator provides all its own energy requirements (such as electricity for pumps), then the surplus energy is not subject to Arab oil embargo, coal strikes, power line failure . . . or rate hikes. This, as well, is difficult to measure in terms of dollars.

The expected amount of biogas can be roughly calculated from the information and tables in Chapter 15 and the heat value, or energy produced (assume 55% CH_2) calculated from the formulas in Chapter 29. Table 44.2 assumes:

1. Electricity at 5 cents per kwh.
2. Fuel oil at 10 Calories per liter (150,000 Btu per gallon), 26.4 cents per liter ($1.00 per gallon).
3. Gasoline at 8,600 Calories per liter (130,000 Btu per gallon), 31.7 cents per liter ($1.20 per gallon).
4. Natural gas at 8,900 Calories per cubic meter (1,000 Btu per cubic foot), $1.24 per 100 cubic meters (35 cents per 100 cubic feet, or 35 cents per U.S. therm of 100,000 Btu).
5. Propane at 6,100 Calories per liter (91,800 Btu per gallon), 19.5 cents per liter (75 cents per gallon).

Based on the above values, the amount of energy in units of either a million Calories or a million Btus, are given for a number of energy sources in Table 44.3.

Using local prices, the value of the energy in biogas can be approximated. For example, if propane is 38 cents per gallon, a million Btu would cost:

$$(10.9)(0.38) = \$4.14$$

There is further information in Appendix 16 where the answers to questions and problems are found. If economics is of further interest to you, it is recommended that you look at those answers.

Terms

None

Questions

None

	10^6 Calories	10^6 Btu	Units
Biogas (at 55% CH_4)	228,000	57,460	liters
	228	57.5	cubic meters
	8,050	2,030	cubic feet
Electricity	1,164	293	KWH
Fuel oil	100	25.2	liters
	26.5	6.7	gallons
Gasoline	116	29.2	liters
	30.5	7.7	gallons
Natural gas	39.7	10	U.S. therms
	1,000	252	metric thermes
	112	28.2	cubic meters
	3,970	1,000	cubic feet
Propane	164	41.3	liters
	43.31	10.9	gallons

Table 44.3 Heat Value of Fuels

Problems

1. Construction costs on a generator run $35,000, the loan can be obtained at 9% interest, annual costs are estimated at $4,800, annual dollar savings in labor and handling are estimated to be $4,200, and 300 cubic meters of biogas is the expected daily production. A farmer wants to build a biogas plant to use the energy to replace propane in a grain drying facility. Should he go ahead with the plan?
2. What about if he is going to use the energy to replace electricity for motors and lights?

45: Design Process

The whole purpose of design is to come up with rational, concrete, economically feasible answers to questions. Unfortunately, this purpose is often subverted and "design" becomes a word which describes the process of merely modifying some valueless but traditional archetype. Good design depends on being able to see a situation clearly and that means "as if for the first time," as well as with the benefit of whatever you have learned from experience.

Therefore, be creative. Much of the information previously presented, for example in the Parameters Section, is there so that you can decide whether or not a particular design will work.

Following the general chart is a discussion of its application in one particular situation; that of a manure substrate, large-scale operation. The focus of this process is economics, and reliability of function, which are the primary factors in large-scale and therefore expensive generators. Small scale generation is more casual and is considered in the latter part of the chapter.

Large Scale Generators—Design Flow Chart

Below, a step by step process for designing a larger scale generator is outlined.

1. **Substrate:** Assess the quantitiy and quality of the available substrates.
2. **Basic type:** Choose among the types listed below in two columns for the two general kinds of substrates.

Float-much or combination	Mix-well or combination
batch	high rate
continuous fed fibrous	moderate rate
hybrid	plug flow
dry	dry

3. **Sizing:** Situation-specific, run the formulas.
4. **Auxiliary systems:** Choose methods of dealing with these aspects of biogas production.
   Heating (source and means)
   Agitation
   Biogas storage
   Primary biogas use
   Preloading (shredding, mixing, chemical addition, holding tanks for continuous feed?)
   Configuration (dual tanks?)
   Scum and sediment systems
   Parameter measurement and control
   Effluent handling systems
   Effluent use
5. **Time, materials, cost:** Now get more specific.
   Construction materials (of main tank(s), etc.)
   Construction logistics (Will concrete truck get stuck in the mud? Is there room on the north side of the barn?)
   Estimate costs or get bid.
6. **Economics:** Run the formulas from Chapter 44. This may lead you back to the second or third step.

Sample Flow Chart for Large Scale Situation

1. **Substrate:** Manure, as mentioned.
2. **Basic Type:** The safest and best researched generator type for manure substrates is the high-rate, continuous-fed generator.
3. **Sizing:** Run the formulas in Chapter 42.
4. **Auxiliary systems:**
   Heating source: In the choice of a heating source, biogas is the most obvious. If you are running an engine on the output of the generator, however, you will be better off using engine waste heat for a heat source. Solar energy can be used as an auxiliary source, if more heat is needed.

Heating means: Hot water circulation is recommended. Steam heat, where a higher technology approach is being used, will work.

Agitation: Gas recirculation is the best means of agiation for the manure substrate generator.

Biogas storage: Low pressure compressed storage, in the United States at least, where compressors are easy to come by, will probably be least expensive. A gasholder, is a second, if much less satisfactory choice.

Primary biogas use: Generally speaking, with the amount of biogas generated from a large scale generator, heating air or water is the most likely use. Conversion to electricity through a biogas-powered ICE/generator, is likely, but in fact, depending primarily on the cost of other sources of energy, you might use the biogas as a replacement fuel for any number of gases.

Preloading: Use if sediment removal is required.

Loading: A holding tank is recommended to enable continuous feeding.

Configuration: Dual tanks will give the best process stability, where they are designed to operate in tandem, alone, or sequentially.

Scum: As mentioned, scum should not be dealt with by periodic shutdown and removal. Rather, high-rate intermittant agitation and some provision for a gentle spray are better options.

Sediment: Bottom pumping is a necessity where any sediment may accumulate. While the high-rate agitation recommended above will help suspend sediment for removal, preloading settlement should be the first line of defense.

Parameter measurement and control: These auxiliaries are not required, although they may make life easier. Use if desired.

Effluent handling: This is situation-specific, but some provision for recycle is recommended.

Effluent use: As above.

5. **Time, materials and cost:** Situation-specific.
6. **Economics:** Run the formulas.

Small Scale Generators—Design Flow Chart

Small-scale design is not principally economic, but rather situational. Money is part of that situation, but, with inexpensive design and low-cost materials, it can remain only a small part.

Among the factors which may affect your decisions about design, and whether or not to build at all, are these:

1. **Knowledge**
2. **General assessment**
3. **Skill/experience**
4. **Money, time**
5. **Space**
6. **Tools and equipment**
7. **Construction materials**
8. **Available substrates**
9. **Other resources**
10. **Other impediments**
11. **Needs**
12. **Economic analysis**

We will discuss these factors below, relating them to a chain of decisions about the design of small scale generators.

1. **Knowledge:** Obviously, before we can build a biogas generator, we must know that the possibility exists. Then we must further understand the theory and practice of, and the options available in, generating biogas. This is what the book up to this point has been about. Finally, we must be to some degree familiar with a different body of knowledge—knowledge of construction, building or fabrication—if we are going to be able to design and build a low-cost, low-maintenance, and long-lasting generator. The knowledge we possess will, to a large degree, determine the design and success of our generator.

2. **General Assessment:** Next, ask "Why do I want/need to build a generator? What do I expect to gain from it?" These are very important questions, since we have to have accurate answers to these questions before we can really understand what this is worth. The cost of a generator in terms of money, sweat, blood, frustration, time, and work must be balanced or exceeded by its results in terms of dollar value, a better life, peace of mind, fertilizer, methane, increased time through fewer hours of work, knowledge, satisfaction, pride of ownership—whatever. Some of these can be assigned a dollar value, some cannot.

In small scale situations, the intangible factors can be allowed to become more important. It is in these situations that decisions are less clear cut, because so much depends on what the internal weather report says—sunny and mild, or overcast and stormy. If you know what your motives and expectations are, then you will be a giant step towards knowing if a small scale generator is for you.

3. **Skill/experience:** You have it, or you don't.

4. **Money, Time:** With enough money, we can remedy almost any lack except a moral one. Money can replace mechanical knowledge, skill, or experience,

but it cannot create or replace love, patience, understanding, or wisdom. With enough money, we can even throw our assessment out the window, but most of those who have any money, generally have acquired or retained it as a result of very careful and considered use of it as a tool. To some degree, in many small scale situations, time can replace money, as we can often find or produce ourselves what we need for a particular task, given enough time. In a nation like America, bent as it seems on the ultimate consumption of the known universe, durable, and useful tools and materials are every day discarded for something new and more expensively produced, and the careful and considerate scavenger can find and use many an unappreciated item. However, take care, for a certain mentality inevitably develops in the one who lives on someone else's garbage midden. Also free, and deeply to be prized are the sun, the wind, love, and God's grace. These, no one can throw away, but only fail to use.

5. Space: Have you enough room? For example, one of the easiest ways to heat a generator is to heat its outside skin and shield it from contact with the cold. Have you the space for this (e.g. your living room)? Can you squeeze your generator between the house and the vegetable garden? Do you have the space to dispose of or adequately use the effluent? Space can also be a limiting factor.

6. Tools and Equipment: Some designs for generators require welding. Either you must have a torch or you need someone else to do this. To clear the ground for a foundation in a poured concrete generator, you may need a tractor, etc. These tools can be rented, given that you have enough money. Otherwise, the design must be modified to fit your limitations—or borrow and barter.

7. Construction Materials: Happy is the person who has whatever he/she needs at his/her fingertips. Again, with limited funds we must make do with what we have. Ingenuity can make the difference between success and failure of a generator design based on the availability of construction materials. A study of books and articles on construction with indigenous materials (stones, rammed earth, etc.) may provide ideas. Further information on materials has already been information on materials has already been presented.

8. Available Substrates: If you haven't met up with a limit yet, an inventory of available substrates may provide one. With any kind of energy, we can do just about whatever we want, if we have enough energy. However, in biogas generation, we will only get out of it what we put into it. With a limited amount of available substrates, we will end up with a limited amount of biogas. The Substrates Section can help us make choices.

For the ordinary citizen—house, car, leaves in the yard—biogas generation is necessarily on a smaller scale, and may come under the heading of hobby, rather than under the heading of business. This is no reason not to generate biogas. Indeed, the same situation exists in the suburban garden, and gardening is a hobby because it is difficult to trace the economic benefits of healing work and healthy vegetables on an accountant's ledger sheet.

9. Other Resources: The position or design of a generator may depend on other resources, such as sunlight, waste heat in the form of hot water or air, the possibility and practicality of wind-powered agitation, the possible use of a gravity fed system (a generator downslope from a barn) or other unique and unobvious advantages and resources. Some of these, only careful, detailed thought will reveal. A slow walk through the future—mentally—will help in discovering these pluses.

10. Other Impediments: Building codes, zoning laws, neighbors better suited to the planet of the apes, tired blood, or other obstacles to success may arise or already be present. Take time to be pessimistic, then make suitable plans or calculate the odds.

11. Needs: An assessment of energy needs, based on what can be learned from the section on Uses, should be made. Winter needs often vary from summer needs, and priority for biogas use should be assigned to these needs on the basis of how essential to life and comfort they are (unless that need can be more easily met in another way), the quantity of biogas required to power a device which meets the need, and the efficiency of conversion which serves as a general guide to whether or not that need is being met effectively by the use of biogas.

For an example of this last criterion, we may look at someone who is comfortable in a cabin on a mountain, warm and well fed, but who feels a need for some music. Biogas might well be a poor answer as a source of energy (to run an alternator to provide electricity to operate a radio or record player) to fill this need. Unless other elements of this situation make biogas necessary and desirable, a wind-electric system would probably require less maintenance, or a small radio operated by a battery charged by an old pick-up truck (on a simple, auxiliary charging circuit) on trips into town, or learning how to play the guitar and sing—without electricity. All three of these options have great advantages over the fairly large biogas generator

NOTES ON MY GENERATOR

SUBSTRATE – COLLECT MANURE FROM BESSY.

GENERATOR TYPE – SIMPLE CONTINUOUS FEED.

SIZING – BESSY PRODUCES ABOUT 4 kg VS/da

CHARTS IN CHAPT. 16

VOLUME = ABOUT 38 l/da

LOADING RATE SHOULD BE ABOUT 3gm/l/da

$$V_{gen} = \frac{(W_{VS})(N_{au})}{L_r} = \frac{(4\ kg)(1\ au)}{3\ gm/l/da} = 1300\ \text{LITERS}$$

ADD ½ AGAIN AS MUCH WATER = SLURRY OF 7%–8% SOLIDS

$$HRT = \frac{V_{gen}}{V_m\ D_f} = \frac{1300\ l}{(38\ l/da)(1.5} = 22\ \text{DAYS}$$

AUXILIARY SYSTEMS – GENERATOR IN GREENHOUSE

MAIN HEAT FROM BIOGAS

ADD. HEAT FROM SOLAR + SIMPLE HOT H_2O COIL

AGITATION: NATURAL

STORAGE: INNER TUBES

TIME, MATERIALS + COST

NEED: 180 GAL. TANK ← ZOE'S JUNK YARD?
ABOUT $100

OTHER PARTS 80.

REFILL WELDING TANKS 100

$280 – TOTAL COST

PLUS LOTS OF MY TIME

required to give us even a modicum of electricity at the low efficiencies involved.

In other words, separate needs from desires, and be honest in you general assessment. if you simply want a biogas generator, build one, but don't imagine that unrealistic goals will be met simply because you didn't plan for realistic ones.

12. Economic analysis: In situations where small-scale generation is possible, economic analysis often consists of two questions, "Do I have the extra money needed?" and more importantly, "Do I have the time and interest needed?" Small scale generators can be inexpensive, easy to construct, require little attention, and accept the variety of substrates likely to be found in these situations.

Now, go back to the first design flow chart and make the decisions as listed for your situation and desires. Bon voyage.

Terms

None

Questions

None

Problems

None

SECTION VII

Plans

The generator plans presented in this book are for small scale generators, meant to be constructed and used in the U.S. One of the reasons this book was written was that there was a lack of such designs available.

The author's original intent, before this book was well started, was to establish a solid basis in information (e.g. write a small book), and then proceed to experiment with these designs. However, as the project developed, this book became the focus, and time was not available for further experiments; thus, only the "test generator" and the modular batch generator were built and tested.

All through the book, solid information has been the goal: we see no reason to stop at this point. So, the above is said not in terms of "confession," but rather simple fact.

But you may wonder why untested designs have been presented. Well, the purpose is the same one presented in the very first chapter—grass roots research. This is a handbook, precisely because it is intended to be used to carry you further on a journey; but you must decide the goal and take the steps.

Designs for larger generators can be extrapolated from this section, and from the preceding section.

46: Pipes and Drums

Drums

Several of the generator designs presented here call for the use of 55-gallon drums as the basic container. This is because such drums are everywhere. Because of the importance of 55-gallon drums or barrels in the design and construction of small scale generators, we will discuss these drums in some detail.

Any industry which is involved with paint, oil, or chemicals probably has its share of 55-gallon drums. The drums come in two basic styles, the tin-can types, which are constructed like a large tin-can with several holes or bungs on top, and the open-head type, which have a top or lid that is removable. The open-head lids generally have one or two bungs in them also.

Because the bungs on the tin-can type are no more than 2 inches (5 centimeters) in diameter and because these bungs are the only access to the barrels, these barrels are unsuitable for most small-scale generators; they won't accept leaves, straw, and other float-much substrates.

Some open head drums are made of cardboard-like materials (called fiber drums) and usually these will not work for us either, both because of their permability (they get soggy) and their low mechanical strength (we can't bang them around). If you have an abundance of fiber drums and want to try them, soak them in sealant material, such as is painted on concrete to make it waterproof.

For the most part, however, we will be using metal, open-head barrels. These barrels or drums come in several different configurations. For example, the thickness of metal of which they are made varies from 24 (U.S.) gauge (0.025 inches, 0.635 millimeters) thickness to 16 gauge (0.625 inches, 1.588 millimeters) thickness. (Common sizes are 24, 20, 18, and 16 gauge.)

The lids have two holes most commonly, with either two 2-inch holes, or one 2-inch hole, and one 1-inch or ¾-inch hole (2.5 or 1.9 centimeters). Less commonly, the lids will have one 2-inch hole, and rarely, three holes: two 2-inch holes and one ¾-inch hole. Some of these different kinds of lids look like Fig. 46.1. Notice the placement of the holes.

For the most part, open top barrels have no bungs in them. Occasionally, however, one will be found with a 12-inch bung near the bottom, as in Fig. 46.2.

The last important difference between various open

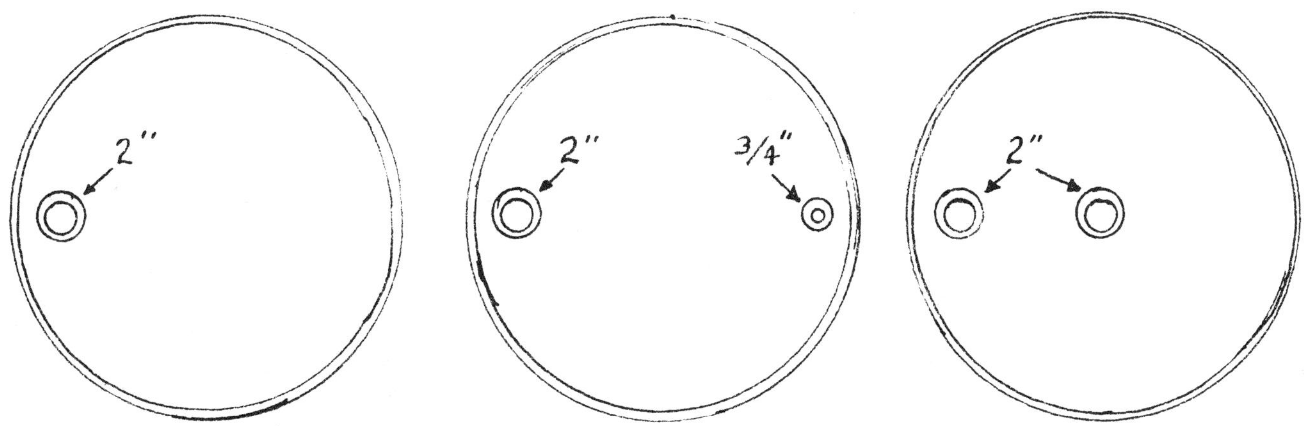

Fig. 46.1 Barrel Lids

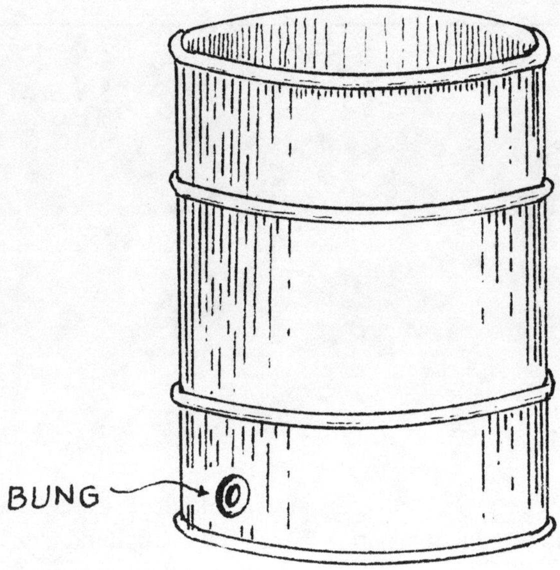

Fig. 46.2 Barrel

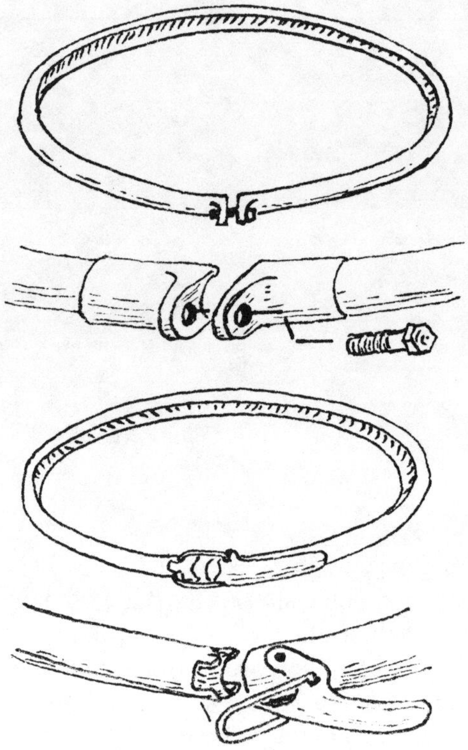

Fig. 46.3 Lid Sealing Methods

head drums is the means used to secure the lid onto the lip of the barrel. The only two means which merit our consideration are the lever lock ring, which operates the way a lot of suitcases do, and the bolt ring, this being a strong ring which is tightened by a large bolt around the lid. They are shown in Fig. 46.3.

The bolt ring takes longer and is more difficult to put on, but it makes a sturdier and possibly a more airtight seal than the lever lock ring. For a barrel that will be in use (without opening) for a long time, try to find bolt rings.

These differences in barrels and lids are important, particularly the differences in bung numbers and sizes, since if we want to use them as generators, we will always be trying to find a way to stick pipes into them, and we can do that most simply by using the holes that are available, rather than make new ones.

Sorry, You're Not My Pipe

Before we go further, let's pause and discuss some elements of plumbing. In what follows, we talk a lot about various plumbing fittings, and so we'd like to stop for a moment and show you some of them.

Notice that we're talking about both plastic and galvanized metal pipe fittings, and that two kinds of plastic pipe—PVC (polyvinyl chloride) and flexible black poly— are shown.

These fittings and these kinds of pipe are widely available in the U.S. This is one good reason for using them. They are also fairly cheap especially if gotten from a discount store, and they require only a few tools to put together and use. In what follows, we'll use the flexible poly fittings mostly to get us across certain gaps (you'll see what we mean) and the galvanized pipe and fittings mostly to carry steam, which is the main suggested means of heating these designs.

More Holes

Other than welding, the simplest way to put another hole in a barrel lid or bottom is by the use of a floor flange. If a rough hole can be punched in the lid or barrel bottom (it has to be a flat surface), then the floor flange can be put on with a bit of inner tube or some other gasket material under it. A single flange (outside only connection) and a double flange (outside and inside conections possible), are shown.

The use of flanges will probably be more expensive than welding, so if possible, find a commercial welder and compare the costs of welding a short piece of pipe (or a straight line connector) into the barrel with the cost of using flanges. Tell the welder specifically that the weld should be airtight, and the threads unharmed. Come prepared to check it out before you pay for it.

If a friend has a welding outfit, you might try welding, but take care that the weld is strong and reasonably airtight. Some asphalt sealer on the inside will help stop leaks and prevent rust. Make sure when welding that the drum has no explosive fumes or flammable residues in it, and weld only in a well-ventilated area, since poisonous gases can be produced from the hot

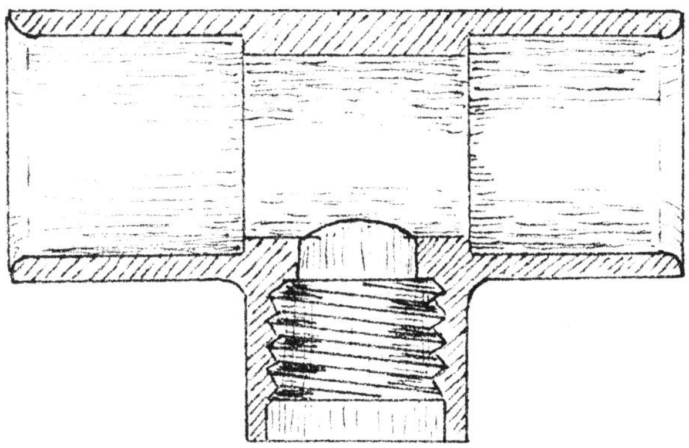

Fig. 46.4 SxSxT (slip by slip by thread)
PVC reducing T (cutaway view)

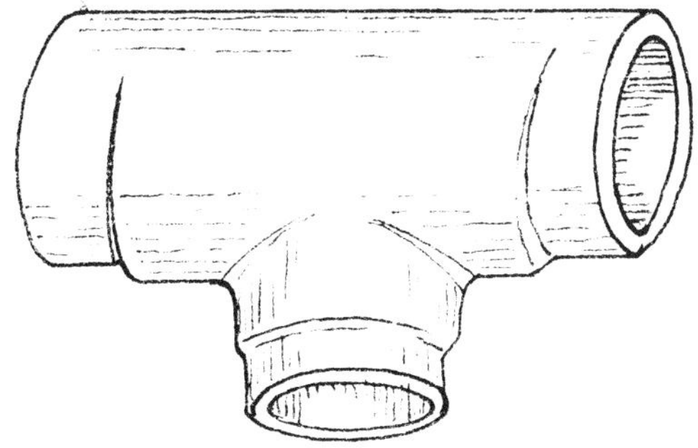

Fig. 46.5 PVC T

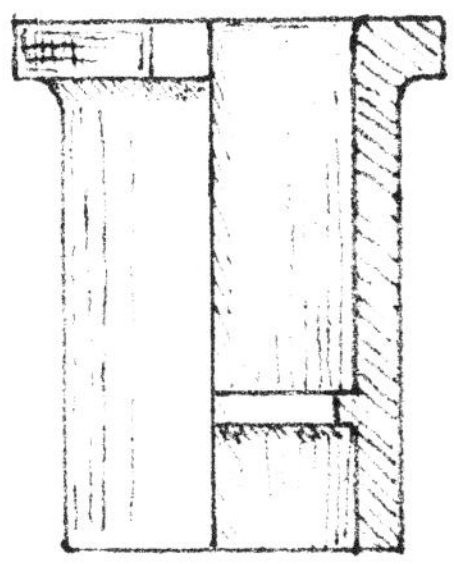

Fig. 46.6 SxS (slip by slip)
PVC reducing bushing (cutaway view)

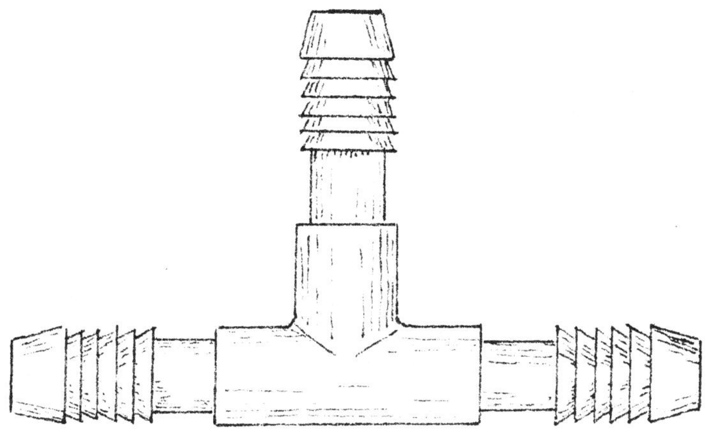

Fig. 46.7 SxSxS (slip by slip by slip)
Flexible Black Poly T-connector

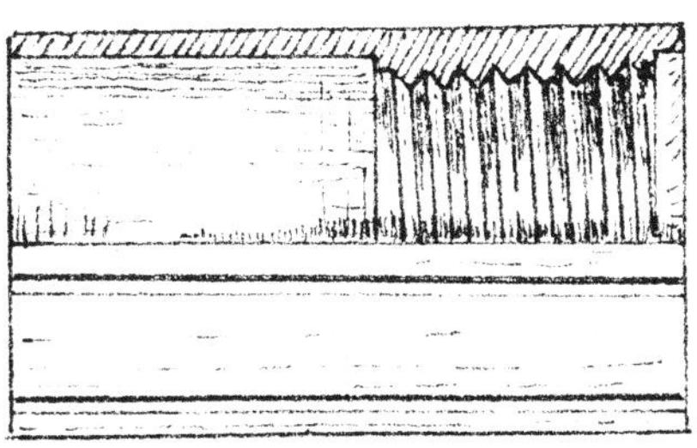

Fig. 46.8 SxT (slip by thread)
PVC Adaptor (cutaway view)

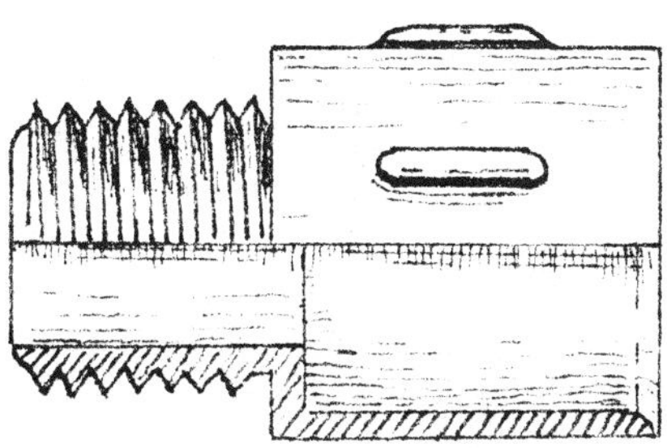

Fig. 46.9 SxT (slip by thread)
PVC Adaptor (cutaway view)

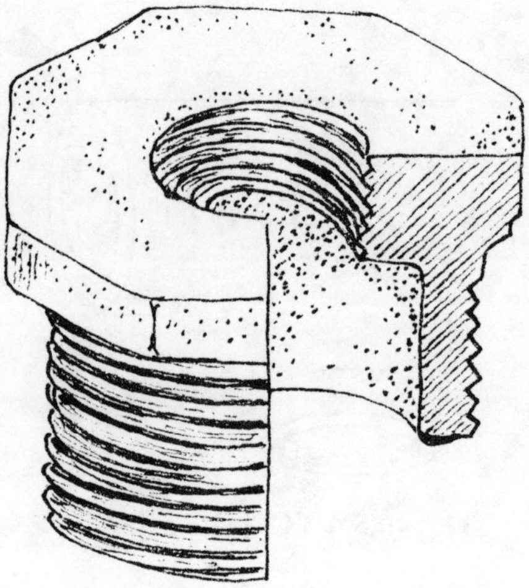

Fig. 46.10 Galvanized Pipe Reducing Bushing 2" to 3/4" (cutaway view)

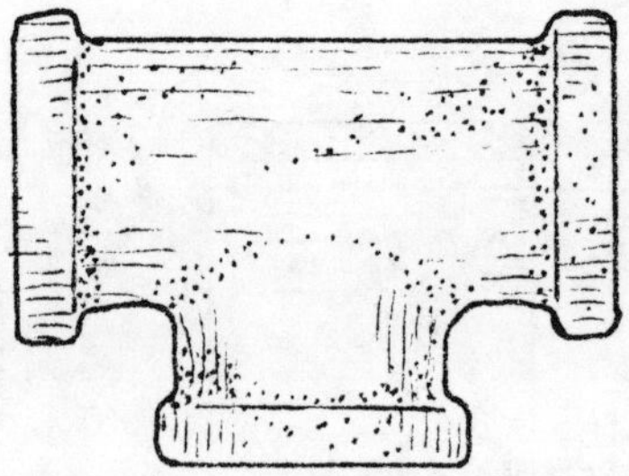

Fig. 46.11 Galvanized pipe T

Fig. 46.12 Galvanized Pipe Cap

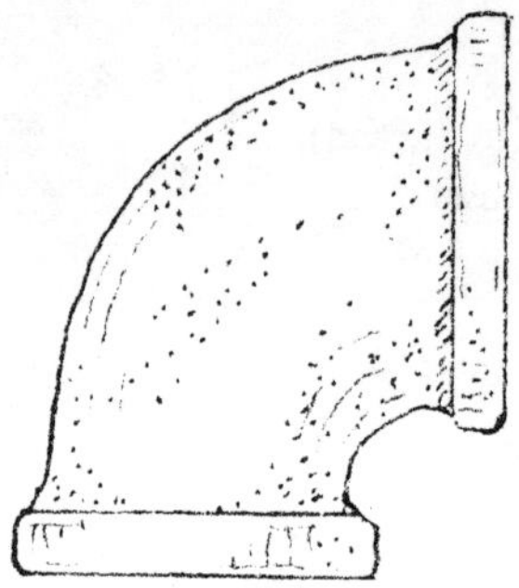

Fig. 46.14 Galvanized Pipe L (elbow)

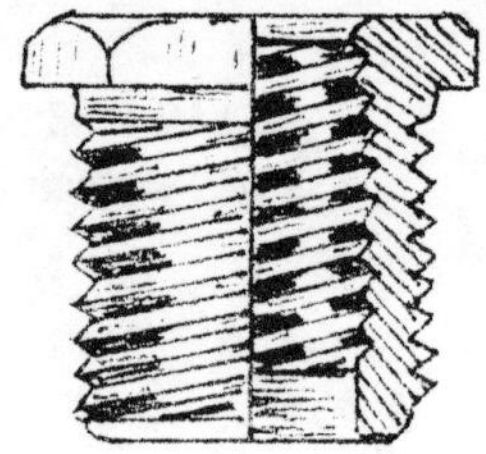

Fig. 46.15 Galvanized Pipe Reducing Bushing 3/4" to 1/2"

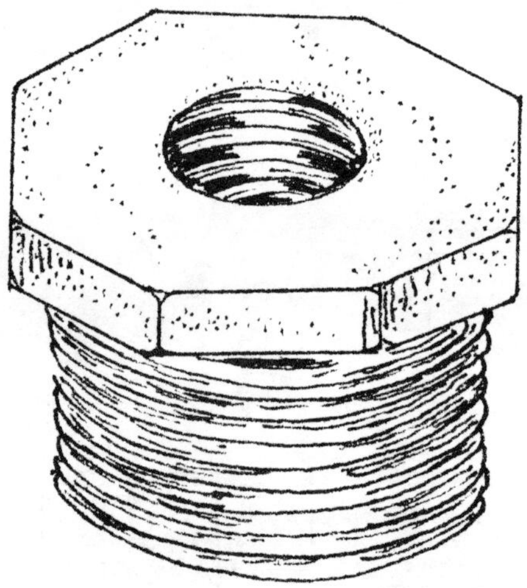

Fig. 46.13 Galvanized Reducing Bushing

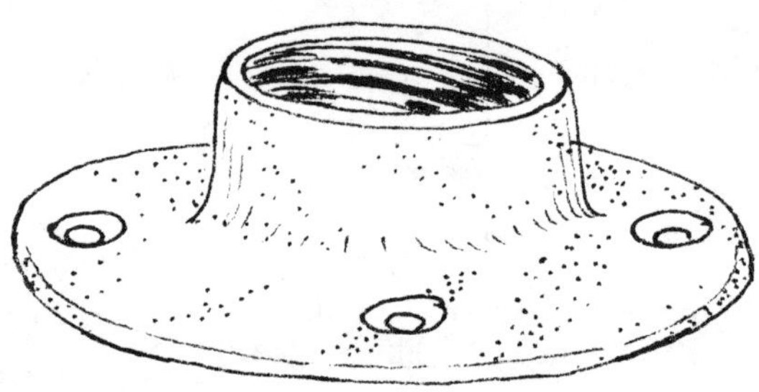

Fig. 46.18 Galvanized Pipe Floor Flange

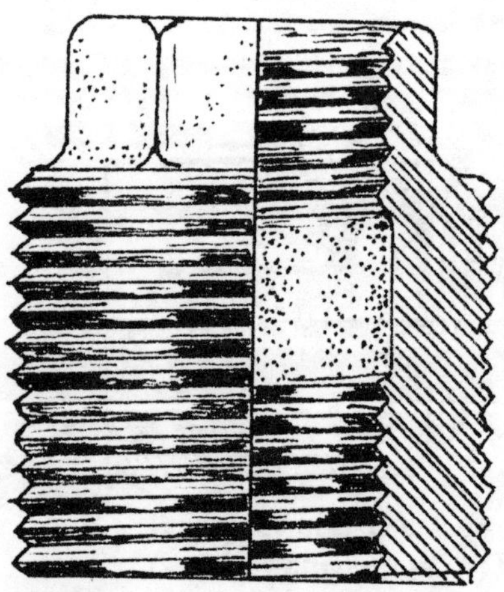

Fig. 46.16 Galvanized Pipe Double Tap Reducing Bushing 2" to 1"

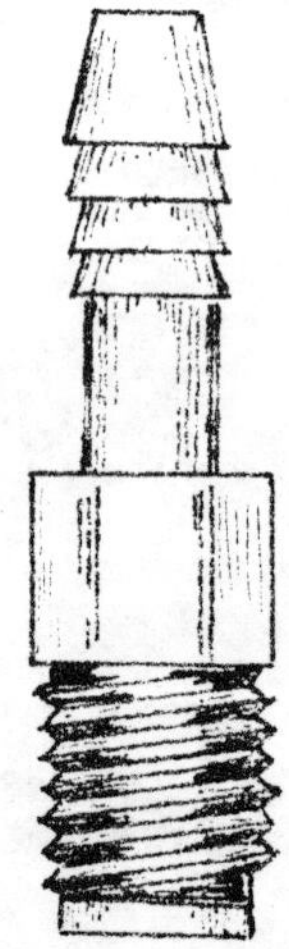

Fig. 46.17 SxT (slip by thread) Flexible Black Poly Connector

Fig. 46.19 Galvanized Pipe Coupler

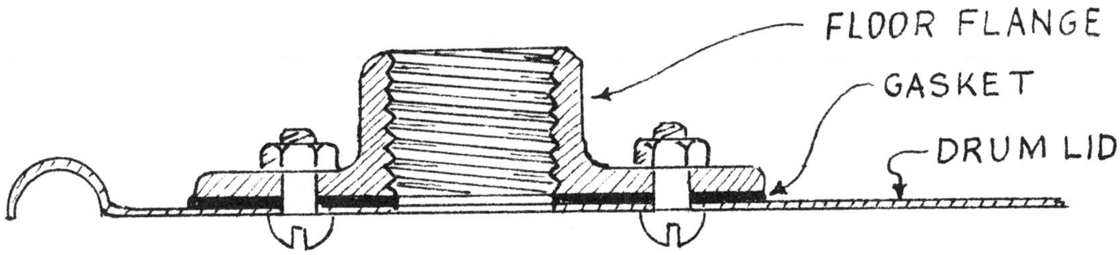

Fig. 46.20 Single Flange

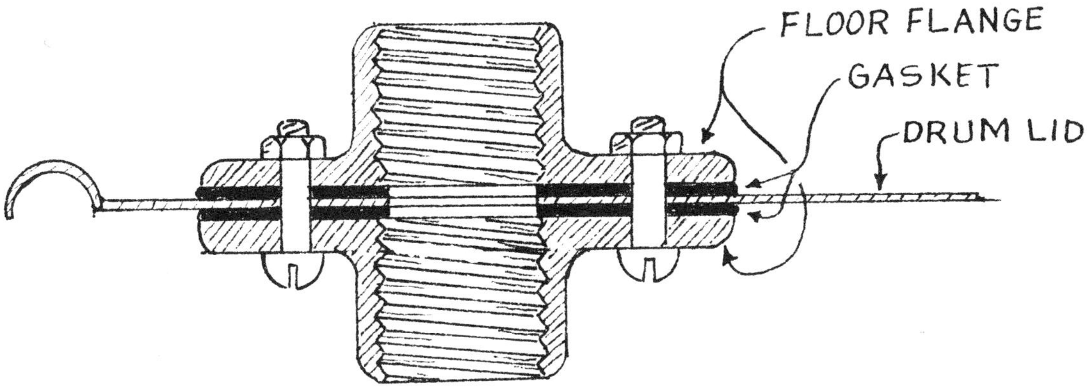

Fig. 46.21 Double Flange

galvanized pipe. The heat of welding can warp the barrel or lid, and you might find after much effort that you can't use what you've got, so consider ways of preventing this warping before starting.

Getting Barrels

Cheap barrels are used barrels, and the cheapest barrels, are uncleaned, rusty, or dented. There's no point paying a Rolls Royce price for a Volkswagen generator and the cost of the drum is the main controllable expense of the generators constructed out of these barrels. If you can afford one, and pay more than the going rate (assuming your barrels are not free), then you effectively jack up the price for someone else. Bad karma.

These barrels are used in many places and for many things, and you can find them by asking around (your friends may know where some are), by calling gas stations, painting contractors, asphalt driveway contractors, fiberglass factories, chemical supply houses, pool supply houses, ans so on. If all else fails, you can call the nearest manufacturer and find out who he's selling them to. Some of his customers may have used or leftover drums.

The going price ranges from free to $5.00. At around $5.00 and up, you should be able to get reconditioned barrels (used and cleaned) and at $10.00 and up, you should be able to get new barrels, or even barrels made to your specifications. As far as we're concerned, more than $5.00 is much too much.

Because the barrels are so widely available you can generally afford to be choosy. Barrels with badly dented lips or barrels that are badly rusted (to the point of leaking) canot be used. Dented lips will allow the escape of gas, and severely rusted barrels may break open in use.

Look for 2 or 3 hole lids, bolt rings, side bungs (on-the barrel), clean barrels (or easily cleaned), and complete drums (rings, lids, barrels). A little rust, a dent here or there—this we can live with.

Cleaning

Nearly anything which comes in 55-gallon drums can be either washed out, or burned out. Better, of course, is washing, since burning uses more energy and it can be very unecological. But necessity is the mother of excuses.

Those barrels which are hardest to clean are also often the cheapest, so burning is sometimes inevitable. Fill the barrel with twigs and small branches for burning (or paper—which is less suitable). If you must use a petroleum derivative to start the fire, don't use gasoline. It's explosive. Diesel is better but harder to start. Kerosene is also useful and not as expensive. Keep the whole circus away from anything flammable because a 55-gallon drum fire, going full bore, may prove difficult to control. Burning barrels can be very dangerous, so take the time to consider what might go wrong and what to do if it does.

If at all possible, simply wash the barrels. You may

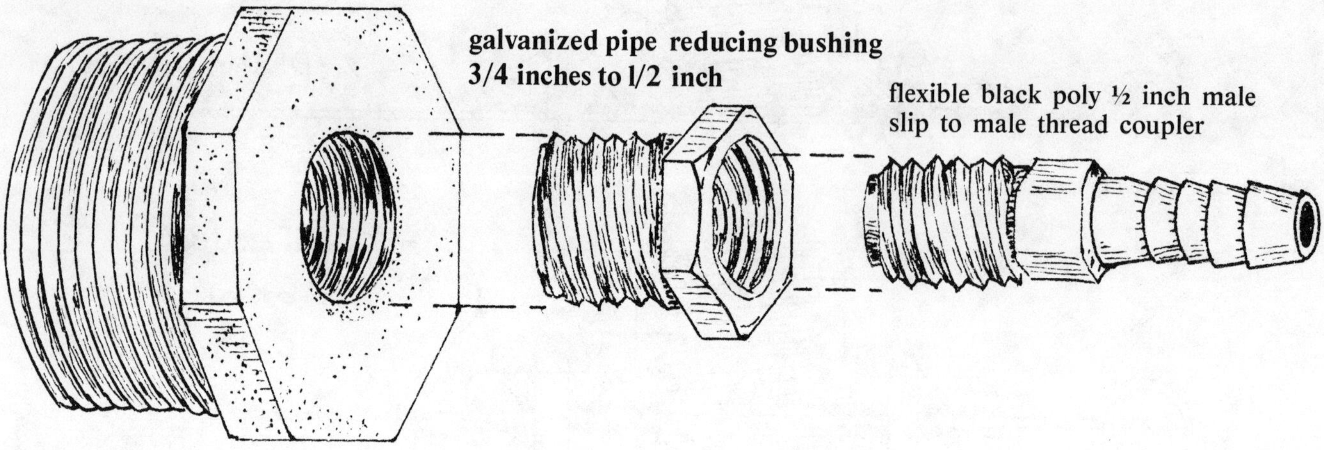

Fig. 46.22, Bung Coupler

be dissatisfied with that process, but you'll have to wash them anyway after burning, as that will leave a residue also.

Cleaning is a messy, frustrating, unrewarding, and difficult process, so it's good practice for real life. If you don't feel up to it, put your kids on it, or pay for clean barrels.

Gaskets

When putting on the lid, a gasket of some sort is needed to insure that the lid seals, airtight, onto the barrel lip. The cheapest gasket is produced when you fill the rim of the lid with plumber's putty. Both rim and barrel lip should be clean, dry, and oil free. Rubbing plumber's putty around the rim until a little of it sticks will provide a better seal.

After filling the rim with putty (and filling the barrel with substrate), push the lid down onto the lip, and tighten the ring while tapping the topside of the lid rim at different spots. When (if ever) you remove the lid, you'll find that much of the putty (which doesn't dry) has squeezed out from between the lip and the rim in flat ribbons which can be picked up, mashed together, and used again. Each time you take off a lid, fill the rim with putty before putting it back on for another period of use.

Reducers

In order to put a pipe, smaller than 2 inches, into a 2-inch bung, they must be fitted together with a plug reducer, a.k.a. a reducing bushing. The same principle applies, naturally, to a ¾-inch bung and a ½-inch pipe. Look at Figs. 46.10, 46.13, 46.15, and 46.16.

Coupler

For the designs herein, we'll generally hook pipes used to collect biogas or transfer cool liquids to a ½-inch male thread slip, flexible black poly coupling.

It's cheaper with one reducer, but it's harder to find a 2-inch to ½-inch reducer, and much easier to find both a 2-to ¾-inch and a ¾-to ½-inch reducer, which gets us to the same place. If you have time, compare the prices and consider ordering the odd size single reducer. Reducers are often sold by outside dimension, regardless of how small the inside hole is, so the cost of the single reducer is generally less than the cost of two reducers.

Valves

The purpose of these shenanigans is to avoid the use of a valve, since with the ½-inch coupler, we can attach a very flexible ½-inch diameter hose. Whenever we need a valve, all we have to do is bend it over and fasten it with a fence staple. It also makes the connection and disconnection of barrels to a gas collection system considerably easier.

Dip the end of the ½-inch hose in very hot water before putting it on the slip coupler. This will make it easier to stretch the hose to fit the coupler. Rubber hose is easier to use, but more expensive. Transparent plastic hose should be tried on the slip couplers before it is bought, as it may not fit easily, or at all. Thin wall plastic hose usually works best.

When we're collecting gas from several barrels with all the pipes feeding one common gas collection system, the low cost and simplicity of these valves will make a big difference in our final cost and interim hassle.

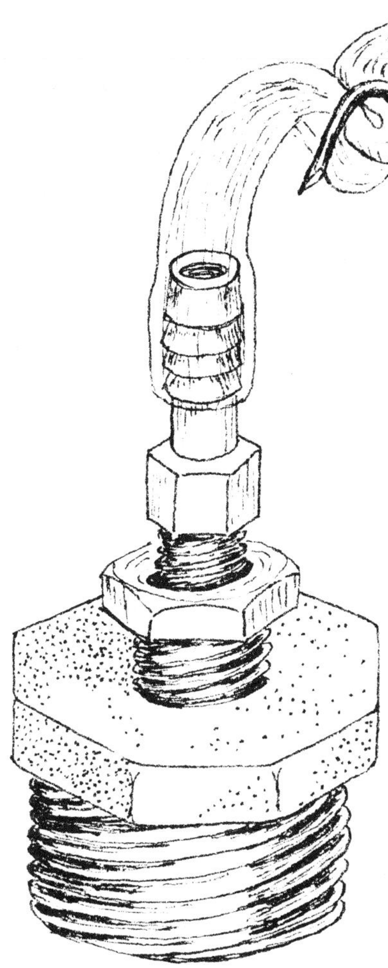

Fig. 46.23 Flexible Pipe Valve

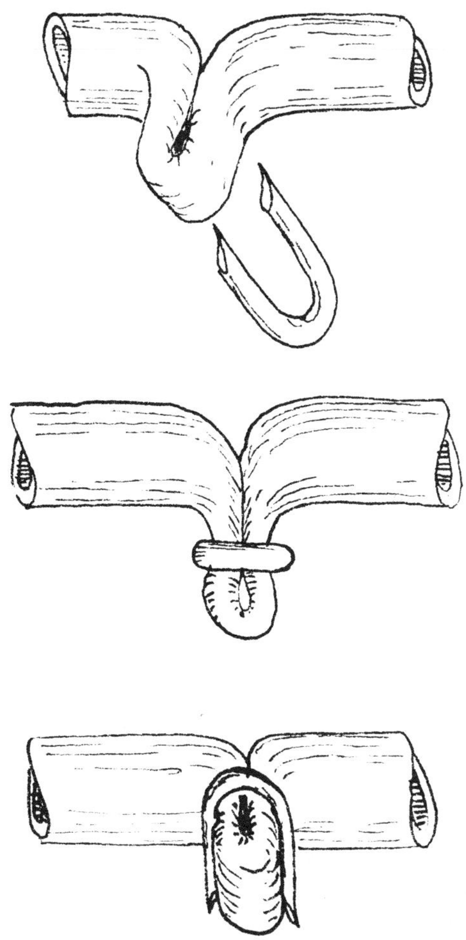

Fig. 46.24 Creating the Valve

To insure that the valves are not leaking gas, a little water poured in the open end of the pipe will show if bubbles are present. For more rigid pipe, rubber or cork stoppers may work as valves, but possibly not very well unless great care in taken to get the right size.

Dipping In

A 1-inch PVC slip to male thread coupler can be glued around a 3 foot piece of ¾-inch PVC, schedule 20 plastic pipe. The fit isn't very good, so a sliver or two cut out of a piece of the ¾-inch PVC will act as a shim to support the pipe while the glue dries. Stick it together as shown. Try the fit before you glue it. It should be snug, but not need forcing, because when you have glue all over it, it will suddenly become much harder to put together. Dry it well (for several days in a warm place) before you try to use it.

The device you end up with can be used in a number of ways, for liquid withdrawal or recirculation, or for gas recirculation or collection. We'll come to that. Meanwhile, we'll refer to this contraption as a plastic dip tube.

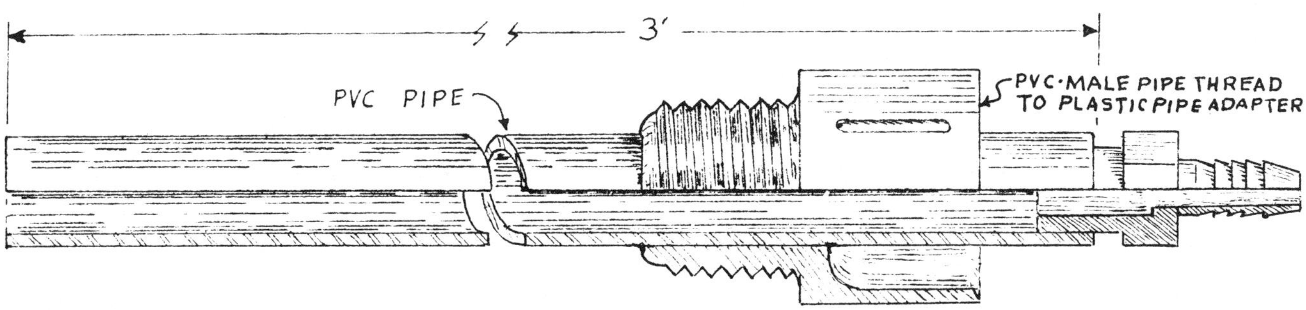

Fig. 46.25 Plastic Dip Tube

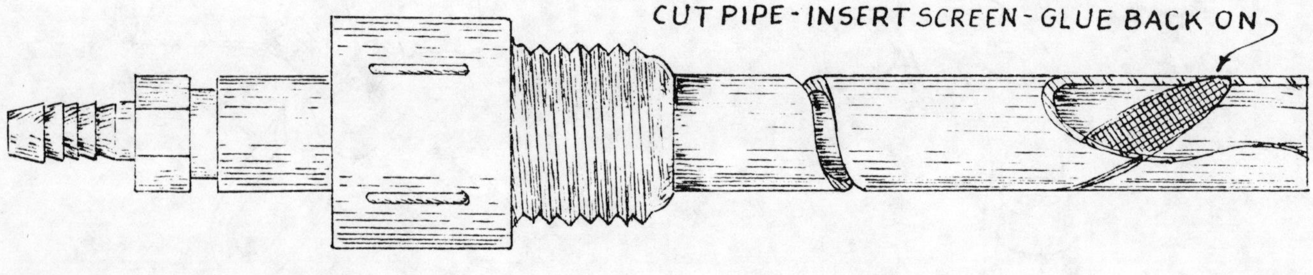

Fig. 46.26 Plastic Dip Tube with Screen

Handling and Draining

It is said that one picture is worth 10^3 words, and so this chapter will end with several figures which we hope will give you plenty of ideas about how to lift and transport drums. It should always be remembered that whenever we speak of transporting barrels full of substrate materials or effluent, we are speaking of barrels drained of whatever liquid will drain from them. The reason is simple. Two reasonably fit people can handle a drum weighing 70 to 100 kilograms (150 to 220 pounds). But when that drum, smooth and round, is full of water and substrate, it can weigh more than 200 kilograms, (440 pounds). Then it can no longer be handled easily by hand, and we need a mechanical advantage, (a winch, for example) to move it around. On the other hand, an empty drum-barrel, lid and ring weighs about 24 kilo (52 pounds). One person can carry one around fairly easily. A barrel which has been filled with leaves, then put through a batch digestion cycle, will weigh about 65 kilograms (140 pounds) after it has been drained of liquid, which will mean that it can be fairly easily handled by one or two people.

Liquid can be drained from the barrels by laying them on their side, or by using a pump. One variety to consider is made for 55-gallon drums full of gasoline. Ask at your hardware store. Clean the pump after using, as it is not designed for effluent and may rust.

A further way to drain liquid is by the siphon method. A hose, full of liquid which dips into the barrel and has its other end lower than the liquid level in the barrel, will continue to drain the liquid from a barrel until it is level with the liquid on whatever is being used as a receptacle, or the end of the hose, whichever is higher. To get into the barrel, use a plastic dip tube, modified with a fiberglass screen inserted in the bottom as a filter, as shown in Fig. 46.25.

(You may notice that the flexible poly slip to thread ½-inch fitting is simply stuck into the end of the PVC pipe on this dip tube. Sometimes, if the threads are carved off the flexible poly fitting, this can be done. However, for greater strength and certainty, the use of a PVC slip to female thread adaptor is generally advisable. Glue it on the protruding end of the PVC pipe and thread in the flexible poly fitting.)

Barrel Handling

Without a device such as the ones pictured, a barrel can be most easily moved by pulling back on one edge to tilt it at about 45 degrees, and rolling it on its edge to wherever it needs to go. However, this can put considerable pressure on the floor, resulting in a permanent groove in old linoleum or wooden floors, particularly when a heavy barrel is moved this way, or when a barrel is pivoted on the floor to swing it around to a new direction for rolling.

If you plan to keep the barrels in a structure, be it greenhouse, storage shed, or living room, handling should be given careful consideration—how will you get the barrels in and out, or up and down?

Consider the solutions offered in Figs. 46.27 and 46.28.

Terms

None

Question

None

Problems

None

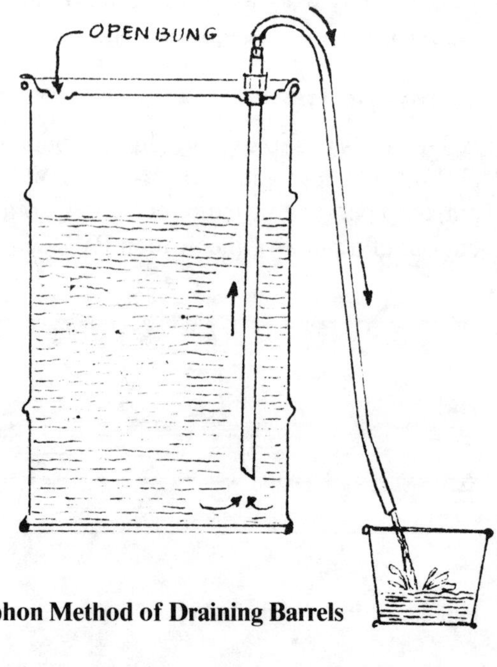

Fig. 46.28 Siphon Method of Draining Barrels

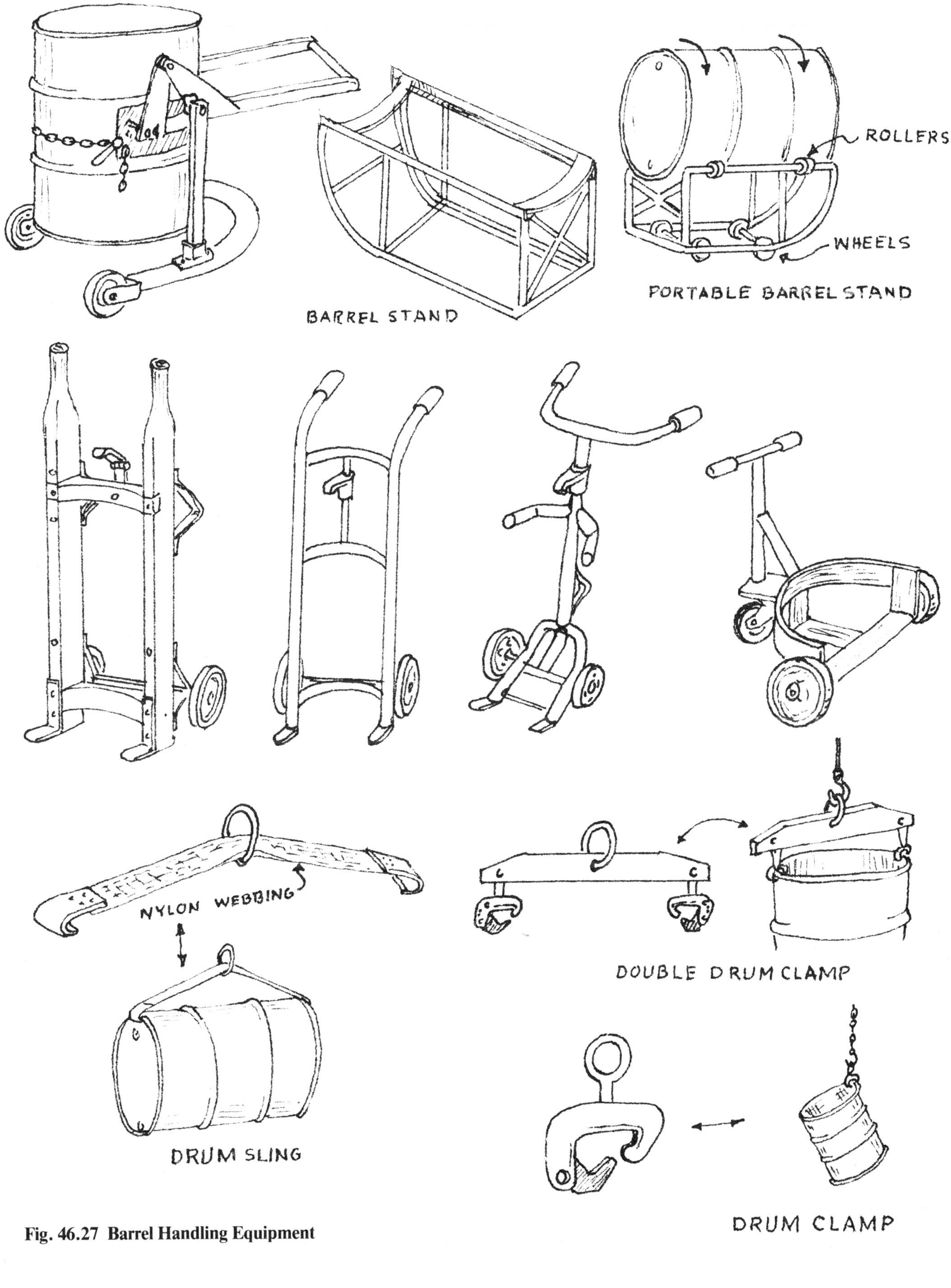

Fig. 46.27 Barrel Handling Equipment

47: Test Generator

A test generator is simply a 55-gallon drum acting as a batch generator to test some particular substrate material. The lid, which (in the simplest incarnation of this generator) needs only one hole, should be fitted with a plug reducer and a half-inch coupling, and some half-inch flexible hose should be attached thereon.

Heating

Any of the heating sources and methods previously described should work. When heating the skin with a biogas flame, place the burner underneath the generator, and place two or three inches (5 to 7.5 centimeters) of sand on its bottom before filling it. Steam heat will require the addition of a second pipe, either extending down through the top, up from the bottom (flange or welding), or, preferably, through a side bung.

Agitation

This can be accomplished by rolling the drum around on its side, by putting the drum on a frame which has rollers (possibly one which tilts the drum from an upright to a side lying position), or by simply sloshing the drum back and forth, although this will not be as effective as either of the first two.

Water

Depending on the substrate chosen, you may want to consider adding any necessary water after the generator is filled with substrate, the lid has been put on, and it is positioned where it will be. A 55-gallon drum, filled with dry leaves and the like, weighs about 35 kilograms (75 pounds), but once filled with water, it can weigh as much or more than 200 kilos (440 pounds).

A measured amount of water can be added through a funnel, or you can turn on a garden hose and measure the rate of flow (slow is better than fast). If you need 30 gallons of water, and the hose is flowing at the rate of 5 gallons per minute (as one example) then you can stick the hose in the barrel for 6 minutes (5 x 6 = 30) to get the amount of water needed. The amount of water added is not critical when using plant substrates, as long as there is enough to just barely cover the material in the barrel. When using liquid recirculation, less water needs to be used. Unless the substrate is kept very wet—whether or not it's underwater—it will not decompose.

Substrate Recipes

For those of you unfamiliar with the mysteries of C/N, percentage of solids, and the like, a few recipes for materials to fill a test generator can be found in Table 47.1.

If you want to get very far with biogas, you'll have to find out about C/N and all that, but if you're not going to use this form of alternative energy to any great degree, trying out one or two of these recipes will give you the basic idea.

Each recipe is given in both SI and American units, using units of weight for the solids and units of volume for the liquids. The recipes can be scaled up or down, depending on whether you want to fill a gallon jug or a 55-gallon drum.

Comments: The recipes in Table 47.1 are general approximations, based on the information presented in the C/N chapter and some casual assumptions. All of these combinations should have some buffering (see Chapter 7), before they were put to work in a generator, and none of them will do well without an initial culture (see Chapter 31). Unless a culture is added it may be months before gas production starts. Some other simple recipes, besides those in the Table, such as comfrey and water, may work well.

Some kinds of leaves will probably not work well, such as eucalyptus or walnut, but most deciduous leaves will do quite well. Newspaper exposed to outside sunlight for some time will be easier to shred. Urine is an excellent source of nitrogen, and available to everyone (but don't use too much, see Chapter 13).

American	Ingredients	Metric
1 pound	Air-dried Fall leaves	1 kilogram
5 pounds	Fresh cattle manure	5 kilograms
0.8 gallon	Water, (add not less than)	6.6 liters
1 pound	Air-dried Fall leaves	1 kilogram
2.75 pounds	Kitchen garbage	2.75 kilograms
2.25 gallons	Water, (not less than)	18.75 liters
1 pound	Very well shredded newspaper	1 kilogram
1.75 pounds	Kitchen garbage	1.75 kilograms
1.25 gallons	Water, (not less than)	10.4 liters
1 pound	Air-dried Fall leaves	1 kilogram
4 pounds	Lawn mowings	4 kilograms
1 gallon	Water, (not less than)	8.3 liters

Table 47.1 Substrate Recipes

If the effluent supernatent is recycled, much of this nitrogen will also be returned for re-use by the bacteria.

Gas Collection

Gas collection will probably be most convenient using one or more inner tubes or in a gasholder made from a 30-gallon drum inverted (open end down) into a 55-gallon drum partially filled with water. A 30-gallon plastic or galvanized trash can will also work. The gas holder types of collection will allow for easier calculation of gas production. See Appendix 1.

Liquid Recirculation

The recipes above are calculated at around 15% solids, if the minimum recommended amount of water is used. With sufficient buffering and liquid recirculation, less water might be used. Liquid recirculation can be done either by using a bubble pump (which will require a two-hole lid, unless the gas pump is inside the drum) or by directly recirculating the liquid. Direct liquid recirculation, when the pump is outside the drum, can be done using a two-hole lid if the gas is collected and the liquid returned, through one hole. One way of doing this is to have a one or two-inch T fitting come out of the lid.

Without the use of this, or a similar device, the drum will require three access holes—one for gas collection, one for liquid draw, and one for liquid return. With a side bung, a two-hole lid can be used, but any of these designs will be more flexible if only the lid is modified, since then any barrel may be attached to that lid and the generator will be complete. In this way, barrels can be used to gather, store, or transport substrate materials or drained effluent liquids or solids without the need to

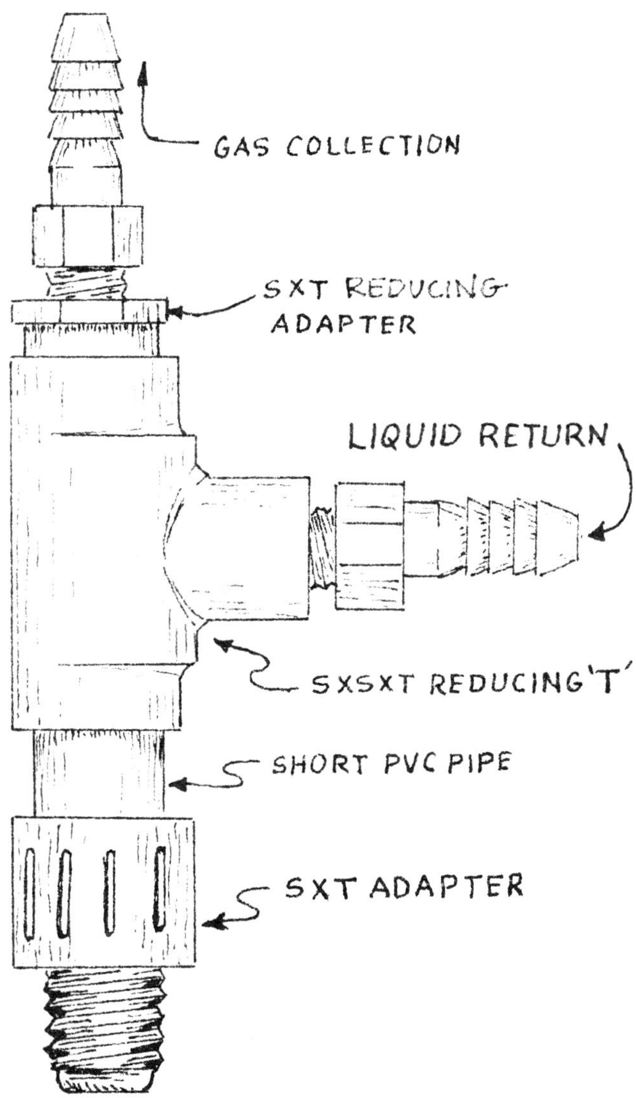

Fig. 47.1 I/O (Input/Output) Device

clean a particular barrel out before a certain generator can be put back on line, and producing gas.

When a barrel begins to produce biogas, large gas pockets will often form which will push the slurry up, unless the gas can escape. If this continues, the first sign of gas production will be a liquid overflow because the trapped bubbles are growing and squeezing the liquid out of the barrel.

To avoid this, when using the float-much substrates, several long sticks can be forced into the substrate before the lid is put on. This is an especially wise move when using unshredded leaves, or similar substrates, which become knitted together and readily trap gas.

Other kinds of substrates, including the mix-wells and shredded leaves or twigs, seem to more easily form gas passages, and though they may start with a liquid overflow, they only occasionally will continue to have this problem.

Your Tests

If you are considering using the test generator for its intended purpose—testing—then fill up a barrel with your substrate, arrange to keep it warm, and try to establish a culture of methane-producing bacteria. The procedure is covered generally in Chapter 31 and cultures are discussed in Chapter 30.

Probably the most frustrating time spent in biogas production is spent during the period of startup. Energy in the form of heat and hard work is poured into the generator, and, often no biogas is produced for weeks. Be patient, use a buffer, think. It really does work.

After the barrel is generating burnable gas, and has been for some time, (six weeks or more), agitate it vigorously (perhaps by rolling it around on its side for five or ten minutes) and drain the liquid into another test generator which has been filled with dry substrate. (Or drain half into each of two other waiting barrels; the effluent liquid is valuable, so don't waste it.)

A good practice whenever you start another barrel is to recycle the liquid portion from a previously generating barrel. Many plant waste substrates are nitrogen-poor, and this liquid is rich in nitrogen: it contains 50% – 80% of the total amount of nitrogen in the old effluent. The new barrel will also get a culture of bacteria and a certain amount of water. Thus, each time you empty a barrel, have another one standing by, filled with dry substrate and ready to be filled with effluent liquid.

Measure the gas output of this second test generator, and notice how it acts. How long does it take to get going? How much gas does it generate? When does it reach the peak of its productivity (there may be two peaks), and when does it fade away? The answers to these questions will give you hard information about your situation. The experience you gain—not to mention the bacterial culture you will have producd—will be valuable in other ways.

Peruse Appendix 4 for information on one way to make this research useful.

By the way, the actual volume of a 55-gallon drum is a bit more than its name would lead us to think—57 gallons, or 7.65 cubic feet, versus 55 gallons or 7.35 cubic feet.

Terms

None

Questions

None

Problems

None

48: Modular Batch Generator

Throughout the book we've referred to the modular-batch generator. Quite simply, the modular generator is a collection of test generators, as described in the last chapter, hooked together by a common gas collection system.

Advantages

The idea for this generator has grown out of attempts to overcome some of the disadvantages of conventional designs for use in most small scale situations. Among these:

1. *Money:* Conventional designs are non-modular: that is, they must be built as one complete unit, and so they require a considerable outlay of funds in one lump sum, or as a loan, which increases the cost of the generator by the amount of interest paid. A modular design, on the other hand, can be added to whenever extra money is available. Further, the cost of these modules, per unit volume, is less than the cost of a conventional generator. Depending mainly on the cost of the drum, the modules can be constructed for less than \$5.00. Considering their volume (215 liters, 7.65 cubic feet), this amounts to \$2.30 per 100 liters (65 cents per cubic foot), compared with the \$17.60 per 100 liters, (\$5.00 per cubic foot), commonly estimated for larger, non-modular, (and of course considerably more sophisticated) generators.
2. *Substrates:* Whereas conventional designs are almost entirely of the continuous-feed type, and cannot easily accept plant wastes, the modular-batch design will accept anything once alive. Therefore, using a modular-batch generator, the kinds of materials commonly found around the home can be made into biogas.
3. *Percentage of solids:* In order to be able to transport and agitate slurries, conventional designs generally call for 7% to 10% solids. This means that (turning the figures around) 93% to 90% of the weight of the slurry is water, and adds little or nothing to the biogas process. The modular batch design, using leaves or other plant wastes, can operate at 15% solids, and, with liquid recirculation, possibly more. This will mean that a greater portion of the volume of the generator will be filled with biogas producing substrate, and thus each unit volume of the generator will produce more biogas than a comparable volume of a conventional generator (although a 10% solids slurry will produce somewhat more than half the biogas of a 20% solids slurry because methane-forming bacteria face more difficulties in a slurry of higher percentage solids).
4. *Handling (and gas production):* In non-modular batch designs, handling the substrate materials and the effluent generally requires a considerable amount of hand labor. The generator must be filled, it generates, then it must be emptied. Not only does this require tremendous amounts of hand labor, but gas production is very irregular. The modular-batch design, while still requiring hand labor, breaks the workload down into manageable pieces, and because each module is at a different place in the cycle of anaerobic digestion, gas production is smoothed out.

Disadvantages

But the modular generator is not a panacea, a cure-all; nothing ever is. What are the disadvantages?

Heating is difficult with so many modules and this is the primary disadvantage. The modular-batch generator has a lot of surface area, which can either accept heat, or lose it. This means that the modules are most easily heated collectively, by either air or water.

In turn, this means that the modules generally have to be kept in a structure of some sort. If you're interested in using a modular generator as the heat

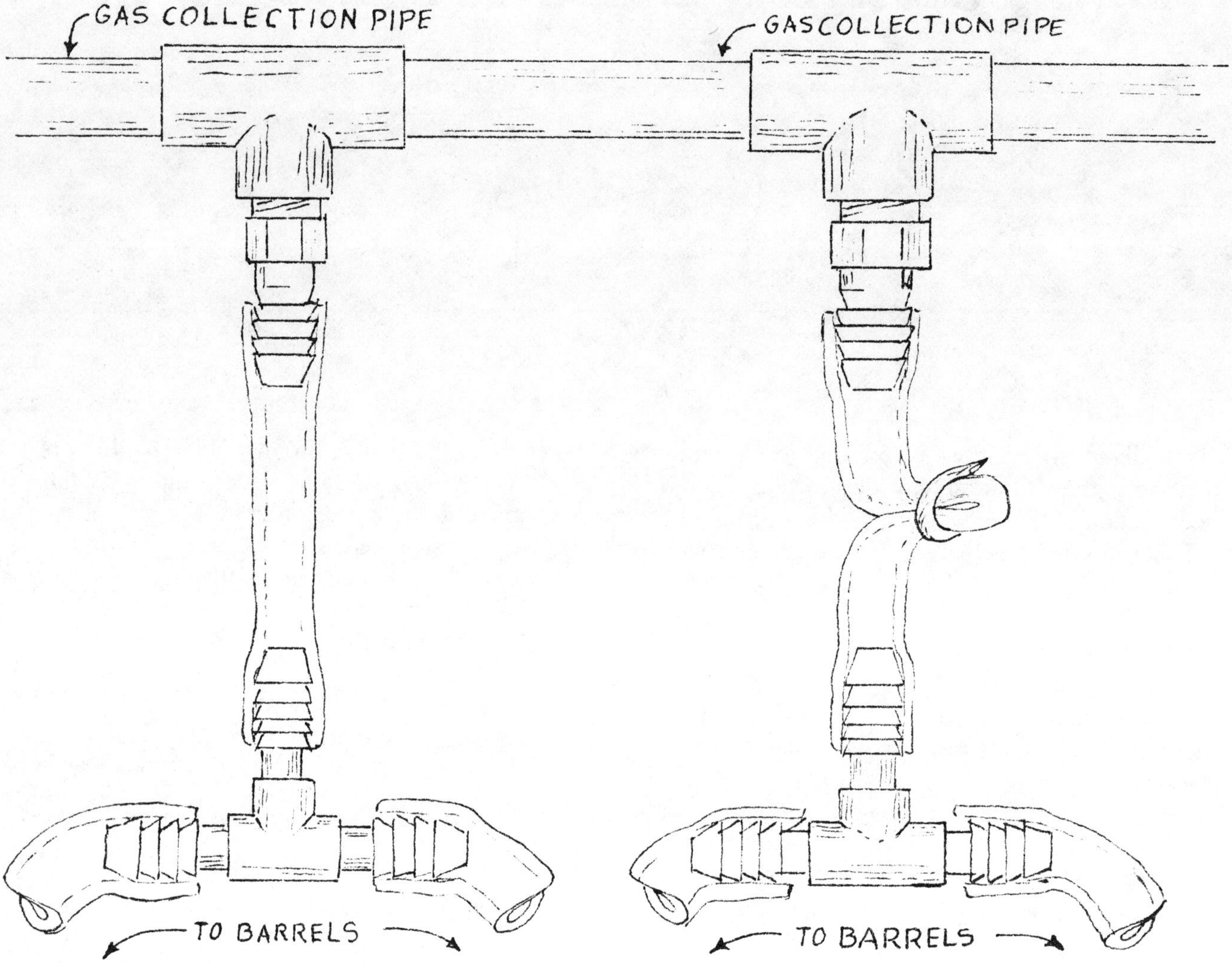

Fig. 48.1 Modular Collection Systems

storage unit on the back wall of a greenhouse, this is fine; but if you don't have a structure you can use, this is not very fine. For some, a portion of the garage can be insulated and heated. For others, a storage shed, or possibly even an extra bedroom (but protect the floors from spills) will do.

A small solar greenhouse is probably the best bet however, because the solar heat is free, and because the biogas can heat the greenhouse (at night) and provide CO_2 for the plants. Note, however, that H_2S and CH_4 may be toxic to the plants. When biogas is burned, the resulting chemicals are not toxic to plants in the amounts which could be expected. Thus, the only toxic gases we need to worry about are those that might come from leaks. The amount will probably be too small to worry about.

Handling is no cinch either, but draining the liquid out of the barrel will make the job much easier.

Using the Generator

For a 10-drum modular generator on a two month (60 day) rotation schedule, one barrel will need to be emptied and one started every 6 days. Since this will result in a peculiar schedule which changes from week to week, it is probably best in this situation to stick to a 70 day rotation. This may not increase the gas available from each drum by much, but it will make it easier to remember when the work should be done. For a seven barrel generator, a 49 day rotation will work if you double up every other week. If you are forgetful, write on the barrels with a grease pencil.

Gas Collection

The gas collection system is also centered around the slip coupler. Using 1-inch or ¾-inch PVC pipe, attach some threaded T connectors. Coming out of these are more slip connectors, and if two barrels need to be hooked up to the same part of the gas collection line, the cheapest way to do this is with a male flexible poly T. One version of this is shown in Fig. 48.1.

The main line will collect water from condensing water vapor, so be sure it has a slight slope, and the water can run down to a place where it can be collected, or the threaded T connectors can be put on the pipe pointing down, and the water will run back into the modules (assuming the modules are below the pipe, and the connecting hose has no loops in it).

Comments: The amount of gas you collect is pretty much directly related to the amount of work you do. The time allowed each module to continue the decomposition process, the percentage of solids, the type of substrate, and the temperature are other factors which act in concert to determine the amount of gas you can collect per day. But any of these parameters are largely affected by the amount of work you do.

The primary reason for this is quite simple. The biogas process, and the modular generator are not consumer items. You can't buy them in a store. Biogas belongs to those who can or will experiment. As time passes, this will change. All the factors will become better understood and biogas units will become an architectural cliche. But that all hasn't happened yet, so it's only the few who either want to, or must, that will enjoy the benefits.

Translated this means that you will have to find the factors which produce the best quality and greatest quantity of biogas for the least work in your particular situation. At first, then, it will be harder because of the many questions you must answer, and because you will not have a well established culture and tried and true procedures that you know will work. Keep after it, and remember that most people will not get beyond the first few steps. If you start, don't be among those who stop short. Do most of the work first, on paper and in your head, before you rush out and buy a lot of parts and pieces that you may not be able to use.

Helpful Hints

In working with this generator, the author has learned certain things which may be helpful.

1. When the modules start up, gas pockets may form in the slurry, forcing liquid out of the barrel. In turn, this will pool in the gas collection system or run into the water under the gasholder. Releasing the gas trapped in the slurry by rocking the barrel or other vigorous agitation will allow the liquid to drain back into the barrel. This problem can be largely avoided by putting three to five long wood sticks into the substrate prior to sealing the barrel, because rising gas can generally force its way up along the sticks to be collected.
2. Startup can be simplified by the use of buffers and cultures. Add buffer and culture, wait for two weeks, and keep the modules warm. If it doesn't work, it may be that the oxygen initially in the slurry killed the first culture. Seed the barrel again, to make sure.
3. Gas leaks, if they occur in a modular generator, most often occur at the threads of the barrel bungs. Some of the leaks will be intermittent, and they will hiss or bubble only when the pressure in the generator is high. Teflon plumbing tape will work better than pipe dope sealing compounds at eliminating these leaks, but keeping the whole system at low pressure may also solve the problems.

   Higher pressures—more than 5 or 10 centimeters of water—occur because the biogas, somewhere in the collection system, has to force its way through water, or because the gasholder is too heavy. Eliminate these problems if they occur and you may eliminate the leaks. You can test the gas pressure in the system with an open-tube manometer.
4. Agitation helps produce biogas, because it reduces internal pressures in the modules by releasing trapped gas bubbles (and thereby helps buffer a dissolved CO_2 induced, low pH), and it sloshes the liquid around, reducing localized volatile acid buildup. Putting a piece of wood under one edge (or both edges), of the bottom of the module— before you fill it up with water—will make it considerably easier to rock the module back and forth.

   If a lot of mixing is required (as when a buffer chemical, more water, or seed culture is added to a module) put the barrel down on its side and roll it around.

Terms

None

Questions

None

Problems

None

49: Hybrid Generator

The hybrid generator is a mix of batch and continuous, and is based on a separation of the acid-forming stage and the methane-forming stage of anaerobic digestion.

The acid-forming modules are batch-fed and unheated. They break down the complex organic substrate molecules into simple water soluble molecules. The liquid from these acid modules, rich in volatile acids (the primary food for the methane-forming bacteria) is then transferred to a heated methane-producing module.

It must be emphasized from the outset that the author built only one hybrid generator, which failed due to the toxic nature of the support material (see Comments on Both Designs). This is an idea which developed during the course of writing this book. However, further search in the literature has revealed that others have had this idea and published the results. What is found below draws heavily from this recent research.

The advantages to this generator are similar to those of the modular generator, except that the hybird generator:

1. needs less heat (since only part of the anaerobic cycle is receiving a heat input);
2. results in a gas higher in CH_4 content (as reported by researchers at the University of Pennsylvania);
3. is possibly easier to operate. (Since the methane module is more or less permanent you can have it in your house; and since the acid modules do not need to be heated, they can be more accessible and require less shelter, if any is needed.

The Nitty Gritty

Because, at this point, this design for hybrid generators has not been well tested, much of the information here is based on speculation. (That's the real nitty gritty.) However, the process itself is not mysterious, and much information exists on related questions which can guide our choices.

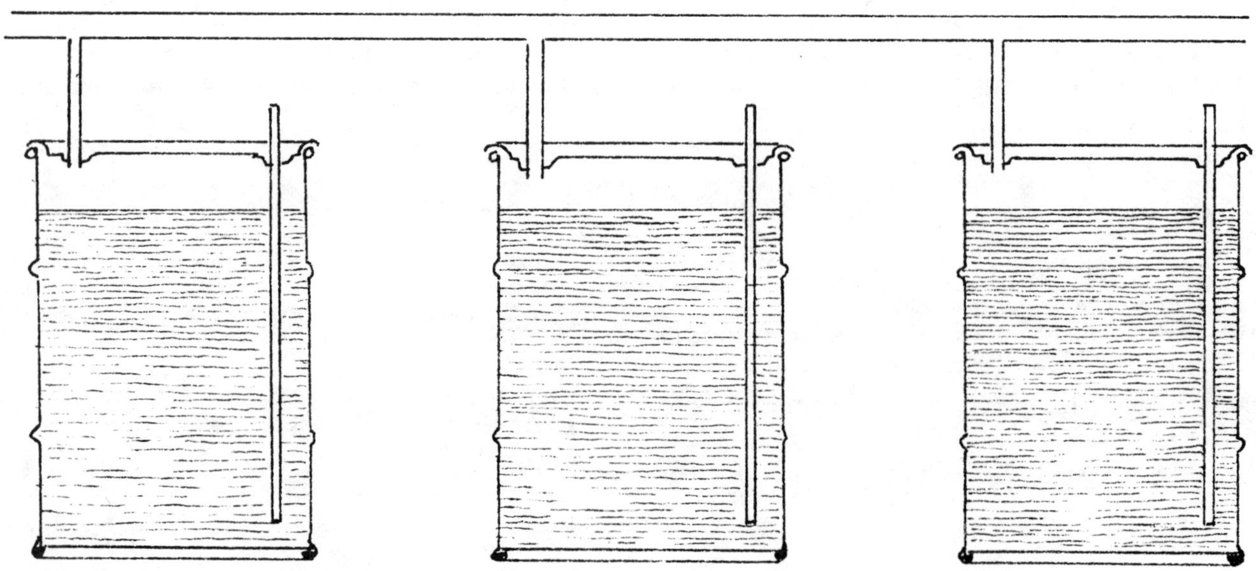

Fig. 49.1 Acid Production Modules

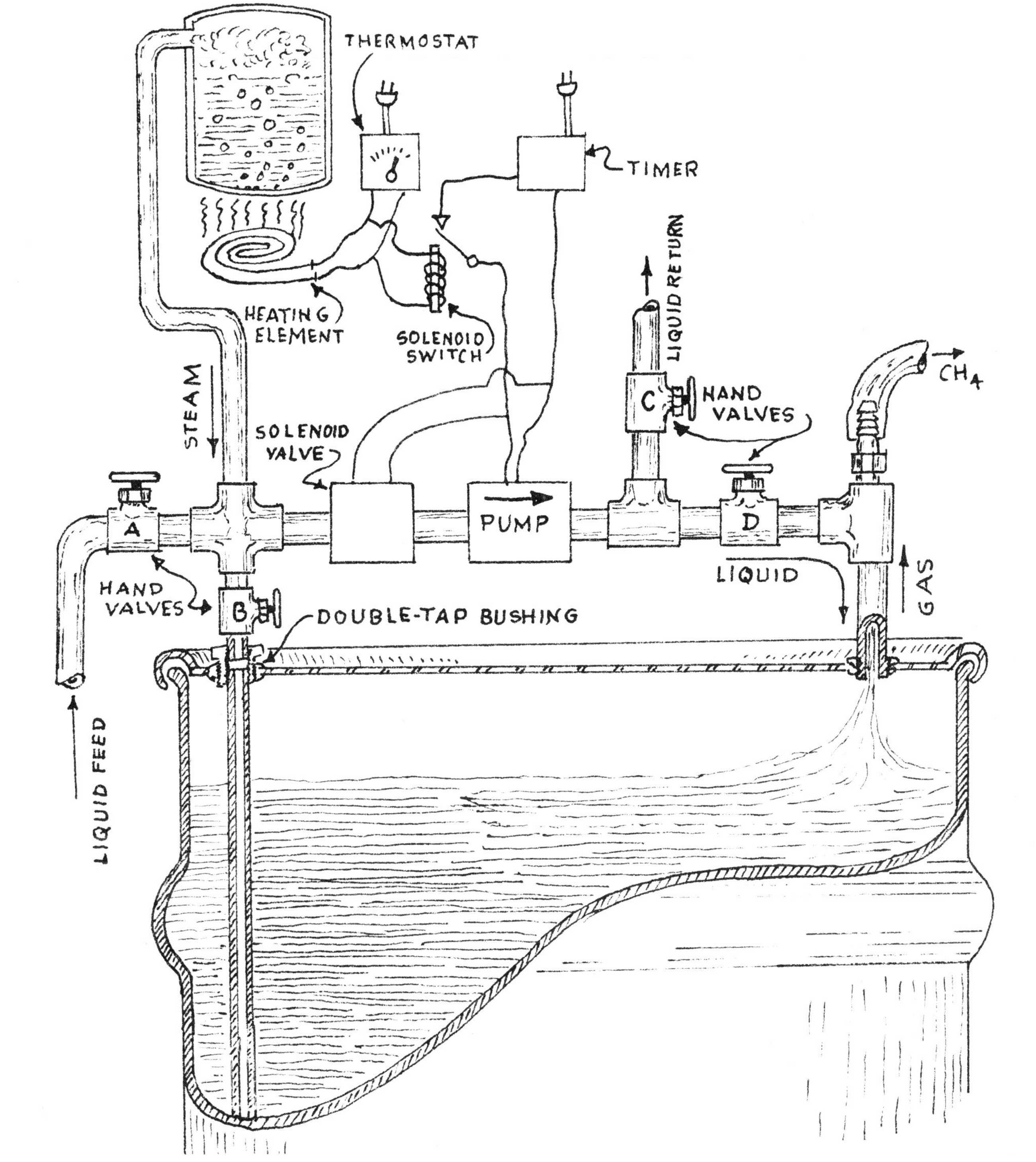

Fig. 49.2 Single Barrel Methane Module

System Description

The acid modules are simple to make. Only the lid needs modification. They are essentially a series of simple test generators, strung together. See Fig. 49.1.

The acid modules are hooked up to make one large acid-forming modular generator. Research has shown tat the gas formed in this stage is 90% – 98% CO_2, and acid-forming modular generator. Research has shown that the gas formed in this stage is 90% – 98% CO_2, and the remaining portion (10% – 2%) may be H_2. If the gas developed from your acid modules will burn, this is an indication that a significant amount of H_2 is being generated, and you may wish to scrub and store it. If the gas will not burn, don't collect it. The gas can be allowed to leak away, but don't release it indoors since it can suffocate you by replacing the oxygen. CO_2 does this because it is heavier than air and will, in still air, tend to settle in the lowest spots available, replacing the "good" air. Remember also—remember well—that the composition of the gas may depend on the outside temperature. As the ambient temperature increases, so will the percentage of methane.

The methane modules are more complex, and consequently, more expensive, but they are designed and built to stay in one place (such as your living room) for a long time. Shown below is a methane module using only one barrel.

Most pumps operate in only one direction. Based on this assumption, the four valves shown in Fig. 49.2 would be needed.

The main control in this design is the thermostat. As long as the module is warm enough, the thermostat will remain off, and the timer will cause the pump to cycle liquid intermittently. During this ordinary operation, valves A and C are closed, B and D are open.

When the thermostat indicates the module is too cold, it turns on the electricity to the heating element. The flow of current induces a magnetic field in the pump/solenoid switch, (called a relay switch), and the contact swings down, taking the pump out of action and closing the solenoid valve. This both protects the pump from steam, and forces the steam down into the barrel instead of letting it blow into the biogas through valve D and the right-hand pipe. (The drawing shows electric heating for the sake of schematic simplicity. Gas heating can certainly be used. All pipe in contact with steam must be galvanized.)

The steam goes down the pipe under valve B, and heats the module. The solenoid valve can be protected from most of the heat by placing it half a meter (2 feet) away from the four way T, using galvanized pipe, and not insulating the pipe. In the schematic, we show everything (pump, valves, etc.) in a straight line—it need not be. More room can be gained by using an L or U configuration of the pipes on top of the module.

When the module no longer produces sufficient biogas, valve D is closed, and C opened. (Valve B remains open; valve A remains closed.) The pipe shown under valve B, which extends to within about 15 centimeters (6 inches) of the bottom of the barrel will be used to pump the liquid out of the barrel. (The best place to put this supernatent effluent is in the waiting acid module, recently filled with dry substrate.)

The valves B and C are closed, A and D are opened, the module refilled, and normal operation is restarted.

Note carefully that Fig. 49.2 is schematic: the thermostat, for example, does not just hang in the air, rather it is placed underneath the insulation, in contact with the barrel. Similarly, the needed insulation is not shown.

Two Barrel Design

The two-barrel design, shown in Fig. 49.3 will probably be cheaper to make and use.

Here we have more valves, but no solenoids. Barrel one is lower than barrel two, and liquid transfer from two to one is by gravity feed. Because no steam is shunted through pipes that must also have other uses, all liquid transfer pipes can be PVC or thin wall plastic tube. If they are the latter, considerable money will be saved on valves.

Barrel one is filled and emptied, barrel two acts as a reservoir of methane bacteria. To operate the module, open valves A, E, and F. All others should be closed. The liquid will cycle according to how the timer is set, and a thermostat (not shown) will regulate temperature. Both barrels should be well insulated.

To drain the module, open valves A and C, and close all others. (Pump the supernatent effluent into a waiting acid module.) Should experience prove that the module can handle more than one barrel of acid effluent, open valves E, D, and A. Let the liquid flow from barrel two to barrel one, if the pump will allow it, and repeat the procedure of the first sentence.

To fill the module, if you've only drained barrel one, open valves B and D and close all others. If you've emptied both modules, open valves B, E, and F and close all others. Operate the pump.

Design Changes

You may wish to have a visible mechanical or electrical means of knowing what the liquid level in barrel one is; barrel two will not overflow as long as valve E is open. The simplest indicator of liquid level would be a piece of thin wall plastic tubing between the pipe above valve A and the gas collection line. The liquid level in this tubing will equal the liquid level in barrel one.

Alternatives to pumping should be explored. If the acid molecules are below the level of the methane module, siphon or gravity draining could be used. The design can be simplified if the pump will operate in both directions.

Comments on Both Designs

The methane modules accept the fatty acid enriched liquid, produced by the acid modules, and they, in turn, produce methane. Since numerous studies have shown that the methane-forming bacteria like to adhere to

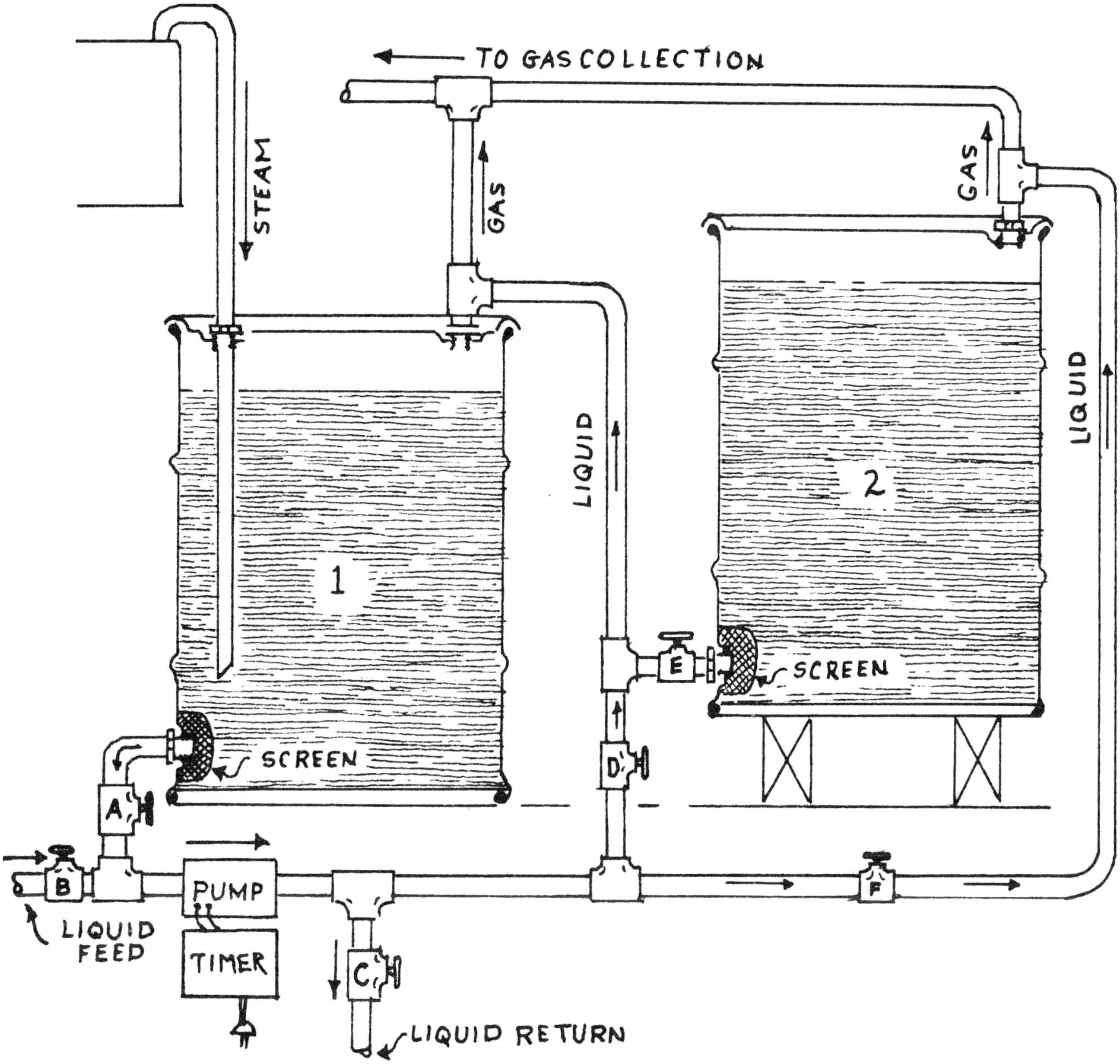

Fig. 49.3 Dual Barrel Methane Module

solids, the methane modules should be filled with some material that allows our friends to find a happy home—the support material. Also, the support material should be resistant to bacterial attack, so that it will last a long time; light, so that a methane module from which the liquid has been drained can be moved easily ; and it should have a large surface area, to give the bitty buddies more room. Materials which have one or more of those qualities should not be hard to find. Among them are wood chips from fir or pine trees, shredded branches, pottery shards, possibly even the little styrofoam "peanuts" that are used for packing material—if your tests indicate that they are not toxic to the methane bacteria. The author used a kind of pillow stuffing material in his hybrid, and the flame retardant with which it was treated proved to be toxic.

However, as yet, the questions of which materials will be suitable and which will not remains unanswered. It is probable that there are even some important questions which have not yet come to light, and further experimentation will reveal them. But exciting information is available from research done with similar generators.

For example, Ghosh, Conrad, and Klass (1975) report that in what they referred to as a "two phase digestion system," the second phase (or methane module in our terms) produced 8.9 volumes of biogas, per unit volume of generator per day. They further report that the gas was 70% methane. If we could achieve this kind of production from the two barrel methane module of the proposed hybrid generator, it would produce about 3,800 liters of biogas per day, or nearly 2,700 liters of methane per day (respectively, 135 and 95 cubic feet) or, in terms of the energy in that much methane, 2.16 x 10^4 Calories, or 8.5 x 10^4 Btu. That's no small potatoes from two little 55-gallon drums; however, in some cases heating "costs" would need to be subtracted.

However, such extrapolation can be misleading. Ghosh, et al., were working with sewage sludge, they heated the acid module, and they transferred soilds from the first (acid) to the second (methane) phase. How much difference these differences make, and whether the hybrid generator will produce results which will outshine or be overshadowed by these results remains to be seen.

Researchers at the University of Pennsylvania, with what they refer to as a two-stage generator, very cryptically report their results. From what we can decipher of it, it appears that they achieved about 3.1 volumes of biogas produced per unit volume of generator per day. (It is, at best, difficult to know if this is accurate. The results are reported in American and metric units, and the volume of the generator is never clearly stated.)

However, they do clearly state that the biogas was 80% methane, possible due to heavy buffering of their stage 1 (acid) module. As well, they point out that their experiments indicate "that the first (acid) stage can be operated at reduced temperatures without sacrifice of fatty acids productivity." The "reduced temperatures" to which they refer were around 22° C, or 72° F.

Using the Hybrid Generator

It is probable that the acid modules will be able to produce more than one charge of acid-enriched liquid, depending on the substrate used. Possibly encouraging the growth of fungus, gathered from samples growing on whatever materials are similar to the substrate in the barrel, will restimulate the production of fatty acids from a batch of substrate that seems unable to produce more, yet doesn't appear to be very well broken down.

The methane module should be started gradually, if the supernatent from a well-established test generator is not available. In any case, read Chapter 32 and do what seems appropriate. Don't add a lot of acid supernatent until the methane module is well established.

When using the two barrel module, barrel one should be drained into a waiting, filled with substrate (but dry) acid module. Then the acid-rich liquid from another acid module should be pumped into this lower barrel. Take care not to overfill it.

At this point, the lower barrel is full of a cold, aci rich liquid, poor in available methane-forming bacteria, and the higher barrel is full of a warm, acid-poor liquid absolutely swimming with methane bacteria. Don't mix them abruptly—the shock might be too much.

Rather, the first job is to warm up the cold barrel. If you've got a thermostat hooked up to it, this should begin to happen automatically. Otherwise, initiate the process. When the cold liquid has been warmed up to somewhere near the operating temperature of your unit, begin to transfer only a few liters an hour at first, then gradually more until the different batches of liquid have been fairly well mixed: then put it on "automatic pilot," so that the timer and pump take over.

Experience and experiments will show you how you need to operate the generator so that everything happens as you want it to. For example, you may not have a large storage facility, and therefore you may want to regulate production to be roughly equal to use. The hybrid generator will be flexible enough to allow a variation in production depending on the rate at which you feed the methane module. You may find that it is unnecessary, with well buffered acid effluent, to be so careful about mixing the acid-rich and methane-rich liquids, or you may find that they handle a two barrel loading satisfactorily.

The overall efficiency of this generator might be increased if the heat stored in the warm liquid that we pump out of the lower barrel could be retained. As it is, warm liquid is pumped into a waiting acid module and cold liquid is pumped out of another acid module with no attempt to retain the lost heat. If another insulated storage module were added near the methane module, the warm liquid could be pumped into this storage module and when the lower methane module was empty, this warm liquid could be pumped out of storage at the same time the cold liquid was being pumped into the methane module. This simultaneous transfer of liquids would allow them both to be pumped through a countercurrent heat exchanger. Experimentation will show whether the increased equipment costs and complexity can be offset by the savings of heat energy. A lot depends on how long one charge of acid-rich liquid continues to produce methane, and therefore the amount of heat needed for initial temperature rise versus the heat needed for temperature maintenance.

Another idea of possible merit is to establish a rotation among the acid modules. As you may remember, Buswell and Boruff (1927) showed that soaking cornstalks for four days in limewater (Calcium hydroxide and water) improved decomposition markedly. The idea then is to put each acid module through the following sequence. Fill with dry substrate, then either soak substrate in limewater or alternatively, encourage the growth of molds and fungi. Finally, fill with effluent liquid from the methane module for acid production.

Obviously, the limewater solution can be passed among the barrels, adding a bit to replace losses each time it is transferred. The solution remaining in the barrel will act as a buffer.

Terms

None

Question

None

Problems

None

50: Home Wastes Generator

The idea of generating methane from one's own excrement is a popular one, but as we have pointed out before, it is one of questionable practicality. As reported again and again by many researchers, the excrement of one person will only produce about 30 liters (1 cubic foot) of biogas per day.

In terms of biogas production, human excrement is, in many situations, better used as a source of nitrogen and added to some other source of carbon—leaves, kitchen wastes—in which case more abundant biogas production can take place.

However, if you don't have garden or yard wastes, or don't want to use them, but you do have home wastes, of the sort usually destined for the garbage disposal, then a home-wastes type generator may be useful to you. Kitchen wastes, which accumulate slowly, are not entirely suitable for use as the sole substrate for either a batch-modular or hybrid generator since it takes so long to gather a whole 55 gallons of such materials. And meanwhile, what do you do with them?

Thus, the home-wastes generator is conceived of as using a mixture of kitchen wastes and sewage in a situation where fairly low per-day biogas production is not seen as a handicap. The design is, again, fairly simple, but unlike the others presented so far, it requires welding.

To make this generator, cut the bottoms out of two open head barrels and weld them together, bottom to bottom. Neither lid will need modification, but unless you can find a two-sided or double-threaded reducing bushing, one of the bung stoppers should have a galvanized pipe coupler welded or brazed through it. The other welding or brazing operations needed are: pipe supports, slurry inlet, gas collection pipe, and surge chamber. (See Fig. 50.1.)

Not much room is allowed for gas collection in this design. The actual liquid level in the generator will be a bit lower than indicated if the gas is collected under pressure (*e.g.*, in a gasholder). This will cause some liquid to be displaced, to a depth equal to the pressure of collection in centimeters or inches of water.

Sometimes, however, a layer of scum is buoyed up with the gas production and enters the gas collection pipe. For this reason, the surge chamber is included, where, theoretically, scum materials may become separated from the biogas. It may prove more effective, if scum poses a problem, to make the surge chamber longer. Extend it down to the generator and replace the short section of gas collection pipe shown in Fig. 51.1 by welding it directly to the generator.

Pipe support brackets, easily made from any metal strips strong and handy enough (barrel bottoms?) should be large enough to allow the pipe cap to pass.

The gas or steam pipe is used alternately for steam heating or gas recirculation for agitation. Two possible problems with this arrangement are first, that sludge may cake on the pipe when steam is run through it, and second (less important), that the gas or steam may all come out of the first few holes unless they are further apart or smaller at the top than at the bottom. As long as you're welding anyway, you might consider having *two* pipes—one for steam and one for gas recirculation. If the steam pipe were open at the end, and the gas recirculation pipe was as described, this approach would solve these possible problems—and make the rest of the system (for steam generation and gas pumping) much less complex.

The inner tube should be secured with a hose clamp, of one of the types used on radiator hoses.

The inlet pipe should not exceed the diameter of the effluent pipe, since otherwise, slurry blockages will occur. The elbows necessary to give the proper height to the effluent pipe can be ABS sewer pipe (plastic) 22.5 degree elbows. For greater total volume, hook two of these generators together. The outlet of one will become the inlet of the other.

Operation

As mentioned, it is recommended that this generator be used for a mixture of home wastes, rather than on sewage alone. There will probably be more use for this design in a communal kitchen, or large household, than

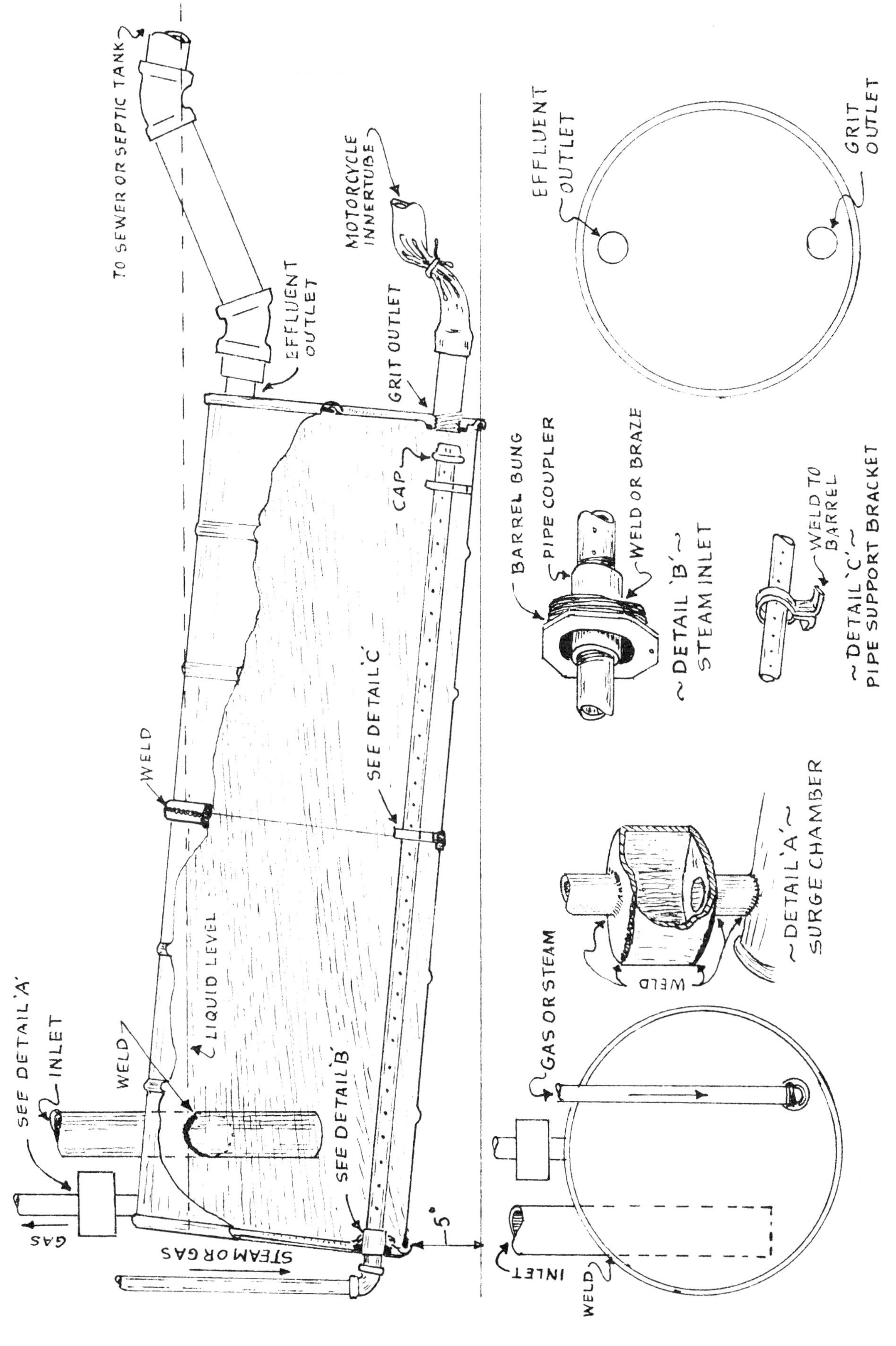

Fig. 50.1 Home Wastes Generator

there will be in some other, more usual situation. In any case, we will not be concerned with C/N or loading rate since, situationally, it will usually not be possible to calculate either of these. The loading rate will usually be low enough not to matter since the slurry will be dilute, and the C/N will be approximately correct (but probably low) for sewage and kitchen wastes.

We will, however, be concerned with the HRT, and thus, we don't want the slurry too dilute as this will cause the organic matter to be pushed through too fast.

For a single home wastes generator, assuming a minimum HRT of 19 days, the maximum per day volume of slurry allowable is about 35 liters (nine and a quarter gallons). Since one flush of an ordinary toilet requires 19 liters (five gallons), you can see that this is limited indeed. However, modifications for toilets are available which can reduce this to as little as 3.8 liters (one gallon) per flush.

It is entirely necessary to reduce the volume of wash water, as otherwise the dilution involved is extreme. Assuming the average individual uses the toilet twice a day, and produces 200 grams of feces, 1.3 liters of urine daily, ordinary toilet wastes are only three-tenths of one percent solids!

It may be however, that we are grasping at straws. Most people have the toilets they have, and they have the plumbing they have. In order to use any home sewage generator with ordinary toilets, given the above assumptions, each individual will require, as a bare minimum, 380 liters (100 gallons), the maximum possible useful volume of this design.

The American toilet gives us an end product which is neither suitable for use in a small-scale generator, nor safe to dispose of into the environment. Not only will it not give us biogas, but it kills fishes and spreads disease.

One way out of this dilemma is to ignore the toilet. Access to the waste water, from an existing sink (via modifications under the sink) or the addition of a special sink, used only for kitchen garbage, may be a better answer. (A garbage disposal is a definite plus.) Besides, for one or two people, the added biogas from toilet wastes will probably not amount to a great deal.

"But," you may protest, "I want to make biogas from my food (after I have eaten it)." So be it. But not with an average toilet hooked into the system, or only with a great deal of well-shredded dry substrate in the bargain, if an average toilet is used.

Assuming you use your toilet, you will need a minimum of 200 liters of generator volume per flush per day, (assumes a 10-day HRT), and you should add between 1 and 1.8 kilograms (2.2 to 4 pounds: TS weight) of well-shredded substrate materials (such as leaves or kitchen wastes) per flush—if you can. (In other words, minimum generator volume of ten times the toilet waste volume per day, plus added materials to bring the total solids up to between 5% and 9%.)

If you load kitchen wastes, this will greatly increase your gas production. Again, aim at 5% and 9% solids, minimum 10-day HRT. For the design shown, as mentioned before, this means that total slurry volume should not exceed 35 liters, total TS weight added at 5% to 9% of that, or 1.75 to 3.15 kilograms TS (3.9 to 7 pounds) per day.

Terms

None

Questions

None

Problems

None

51: Continuous-Fed Fibrous Substrates Generator

That seems like a cumbersome name for a generator design. It should have a more elegant name—but what should it be? Oh, well. This design (modified slightly) is by Buswell and Boruff (1933), and was patented (U.S. Patent 1880 772, October 4, 1932) by them. It grew out of their very small-scale batch experiments, in which they discovered that scum is, as we have been saying all along, a problem with float-much substrates. So, they hit on a simple idea for breaking the scum up. They "invented" a bottle from which gas could be collected whether it was upright or inverted. Fig. 51.1 should give you the idea.

Using this "generator" in the lab, they found that gas production from plant substrates could be greatly increased when the scum blanket was broken up. Based on this simple idea, they designed a continuous-fed plant wastes generator, similar to that shown in Fig. 51.2.

Inside the cylinder of the generator tank, a smaller cylinder constructed mainly of wire screen was placed. This was able to rotate freely when a handle connected the main shaft of the wire screen cylinder was turned. The scum solids were prevented from escaping into the generator at large by interlocking circular lips on the wall of the generator and the end wheels of the wire screen cylinder.

Except for the difference in substrates, this generator can be used very much like any other continuous-fed generator. The inventors report that the maximum loading rate tried was 3.8 grams TS per liter of generator per day for sewage screenings; 1.7 grams TS per liter for cornstalks. For the individual with a good volume of float-much substrate materials which can be shredded or chopped, the generator may prove a valuable idea. It could be used as the acid module in a hybrid generator.

The basic concept of a continuous-fed, plant wastes generator is valid and valuable. This is one excellent answer to the general problems involved, but it does not exhaust the possibilities—can you think of other designs?

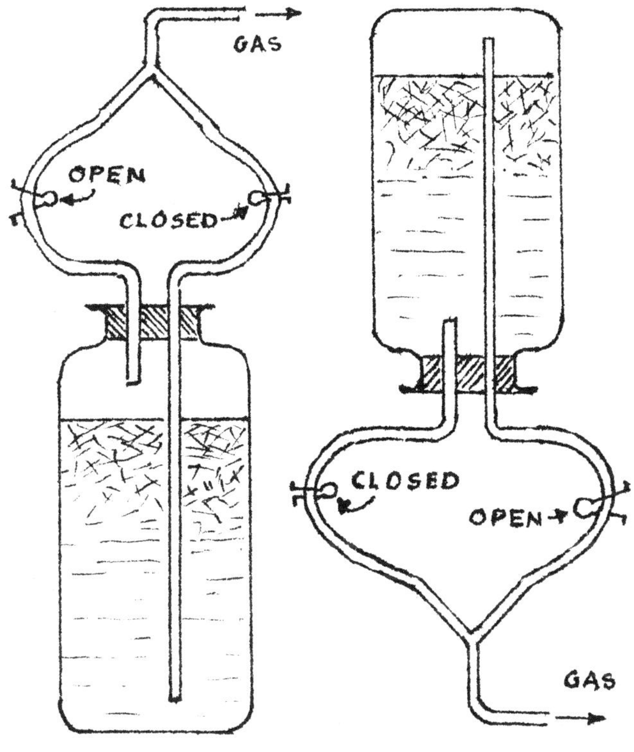

Fig. 51.1 Experimental Generator

Terms

None

Questions

1. Now that you had some time to think, can you think of a better design?

Problems

None

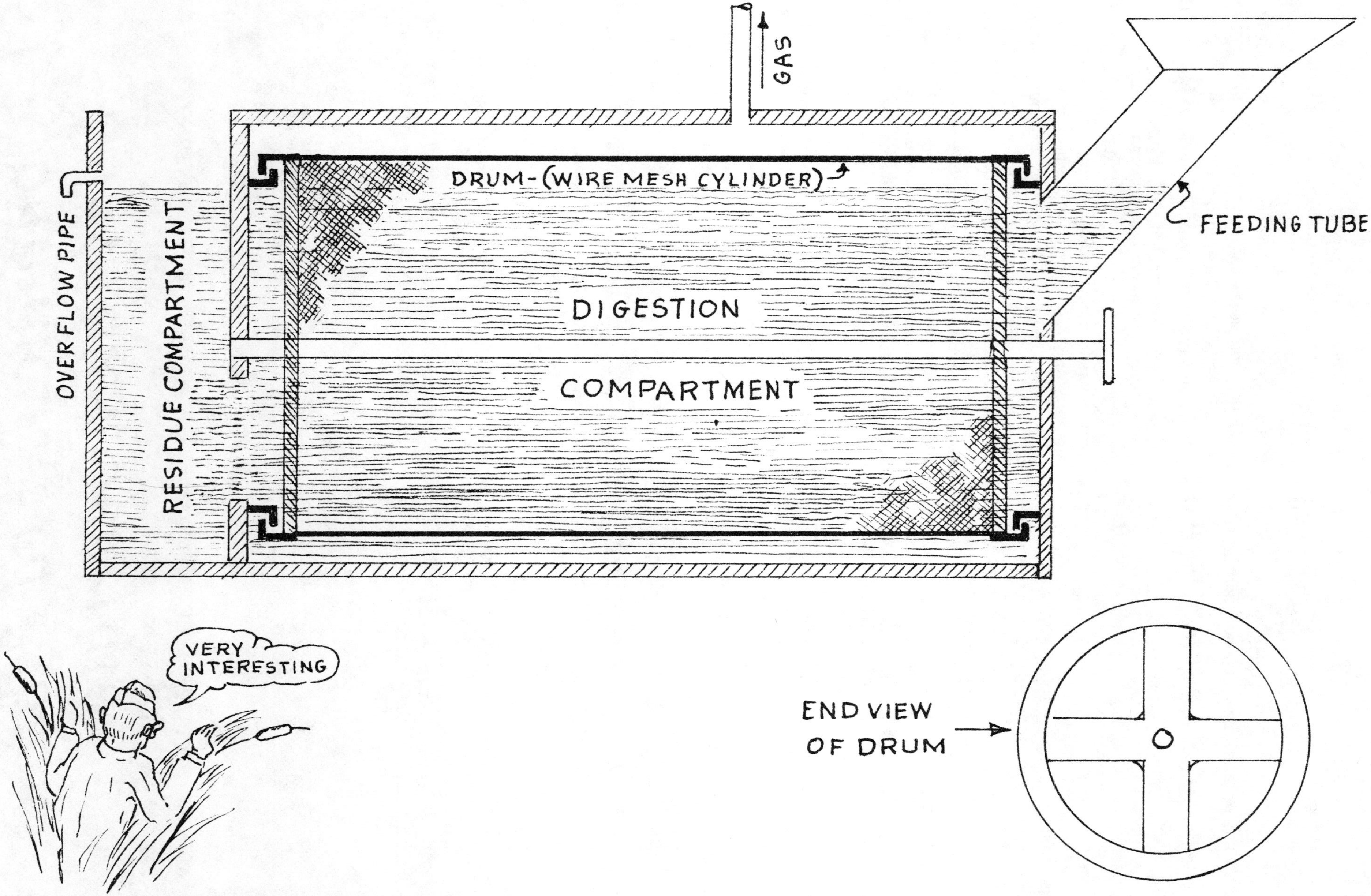

Fig. 51.2 Drum Digester for Fibrous Materials

Appendix 1: Analysis

As you must suspect, this chapter is not about talking to psychiatrists. It is about finding out the various simple physical and chemical truths we might want to know when dealing with biogas. It stresses analyses which can be simply undertaken, since, for the most part, complex analysis requires extensive background information and expensive equipment. Further, information on complex analysis would be redundant, since it exists in sufficiently clear form elsewhere. So, this chapter will talk about analyses which can be undertaken with simple tools and inexpensive chemicals and equipment.

One indispensible piece of equipment is a scale, for weighing small (under a kilogram or two pounds) amounts. A postal scale will work.

Gas Analysis

Biogas consists of only two important kinds of gases most of the time and it is these with which we should be concerned. There are the flammable and the non-flammable gases. Coincidentally, the major constituents of almost any biogas—CH_4 and CO_2—fall neatly into these two categories.

Since CH_4 and CO_2 generally comprise 98% or 99% of most biogas, we can get a reasonable estimate of the amount of CH_4 in biogas by simply extracting the CO_2. Now, you may react negatively to the idea of determining how much CH_4 you have by assuming that everything that isn't CO_2 is CH_4. "What about H_2S, H_2O, N_2, or NH_3?" you may cry.

Here's the straight information: at this level of science, care and awareness are much more important than equipment. We can only expect a certain level of accuracy. With much better equipment, we might be able to better that by 5%. To get still greater accuracy, we'd have to spend quite a bit more money on equipment, and be quite a bit more competent.

In any case, nothing else we measure (volume or the heat requirement estimates for example) is any more accurate, so why worry? We could go through all kinds of gymnastics, figuring out the partial pressure of water vapor and the amount of CH_4 which will dissolve in the solution we are using, but we won't. For those who wish to, some information is presented in the charts and tables which follow, but the rest of us will be happy with the simple, good-enough-for-government-work approach outlined below.

CO_2, CH_4

First, fix up a jar or bottle so that it has a definite volume. Any glass jar will do as long as it has a lid that fits well and seals tightly. It can be calibrated for volume by weighing it partly filled with water, and then weighing it while empty. The volume of water which was in it can then be calculated by using the formula that follows.

$$Vj = \frac{Mt - Mj}{Dw}$$

where:

Vj = volume of the jar (to whatever point it's filled with water)

Dw = density of water (one kilogram per liter, 8.3 pounds per gallon)

Mt = mass (or weight) of water and jar

Mj = mass (or weight) of the dry empty jar

Subsequent calculations will be easier if you aim for a specific volume, such as a liter, or a quart, rather than 0.658 gallons, or some other such absurdity. Obviously, it may be beneficial to start with a jar which already rates as being approximately some particular unit volume.

After you determine the right volume of water, mark the jar in 3 places (around its circumference) at the bottom of what is called the meniscus. Use a grease

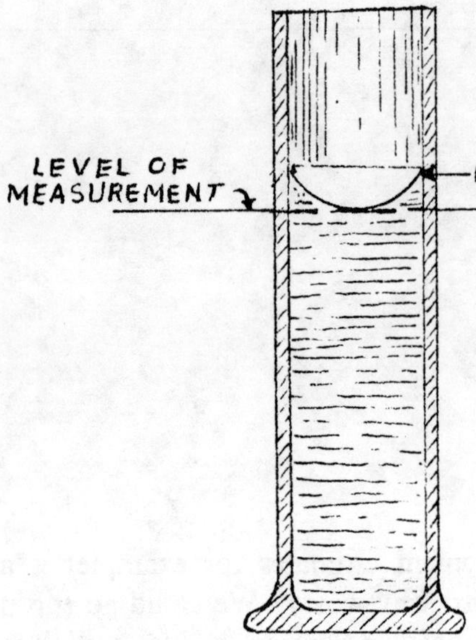

Fig. A1.1 Meniscus (cutaway view)

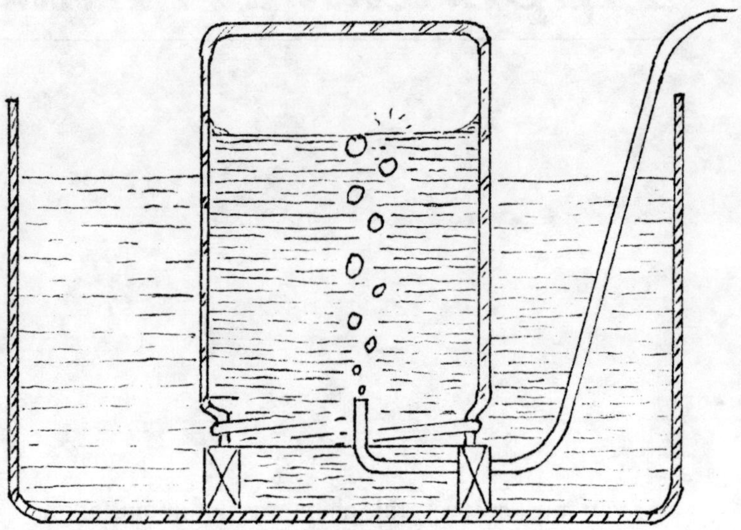

Fig. A1.2 Gas Measuring Apparatus

pencil, or some other kind of marker that will not wash off the glass. Shown in Fig. A1.1 is a very narrow jar. In narrow jars, the meniscus is more visible than in wide jars.

Then fill the jar with clear cold water, invert it into a larger vessel filled with water, and bubble the biogas up into it to the particular measured volume you calibrated. (Remember that the bottom of the meniscus should just hit the line, like before.)

Now comes the tricky part: put the lid on the jar and transfer its contents to another large container. This container should be partly filled with an alkaline solution—sodium hydroxide (NaOH) or calcium hydroxide (CaOH) and water. *Be careful.* Such solutions can avidly eat through your flesh, and a small splash can result in a myriad of holes in your clothes. Wear rubber gloves, *eye shields,* and a piece of plastic sheet for an apron; move slowly and have a large bucket full of vinegar and water ready to wash off any drops of the alkali that get on you. Make sure your gloves don't leak.

The solution in the container should be shallow enough so you can set the inverted jar on the bottom of the container and still have enough room to comfortably grab the bottom of the jar, and pull it out. Use a plastic bucket, if possible. Plastic generally won't react with the alkaline solution.

Have tongs ready to remove the lid or cap as soon as you get it under the surface of the solution. Practice this a couple of times with air in the jar and water in the larger container until you're sure you can do it safely and with skill. When going through the calibration process, swirl the jar gently to mix the water in the jar with the solution in the large container, being careful not to get air into the jar, or alkali solution out into the world at large.

Then just leave it alone. Come back occasionally, put on your protective equipment, slow yourself down, and gently swirl the jar a bit more. After a day or so, come back (put on your equipment) and lift the jar a bit, being careful not to get air into it. Notice that the biogas has less volume. This is because the CO_2 (hopefully, all of it) has reacted with the solution and has been pulled out of the biogas.

Using your tongs, find the lid on the bottom of the large container, set the jar in it, and twist the jar to put it back on, tightly enough so it won't fall off.

Bring over a container filled with water (don't take the jar over to it) and put the jar in it *carefully* to wash it off.

Using the rubber gloves, reach under the water and tighten the lid. Lift the jar out, set it, bottom down, lid up, on a level surface. Mark the bottom of the meniscus in three places, as before. With all the respect you have developed for the alkali solution, gently empty the jar's contents into the alkali container, and again wash the jar in the water container. You no longer need to be concerned with the gas in the jar, as you already have the marks you need. Don't wash off the marks.

Fill the jar with water up to the new marks, weigh it, and using the formula for Vj that appeared earlier in the chapter, calculate the volume of gas left in the jar. The approximate percentage of CO_2 is:

$$\%CO_2 = \frac{(Vb - Ve)\,100}{Vb}$$

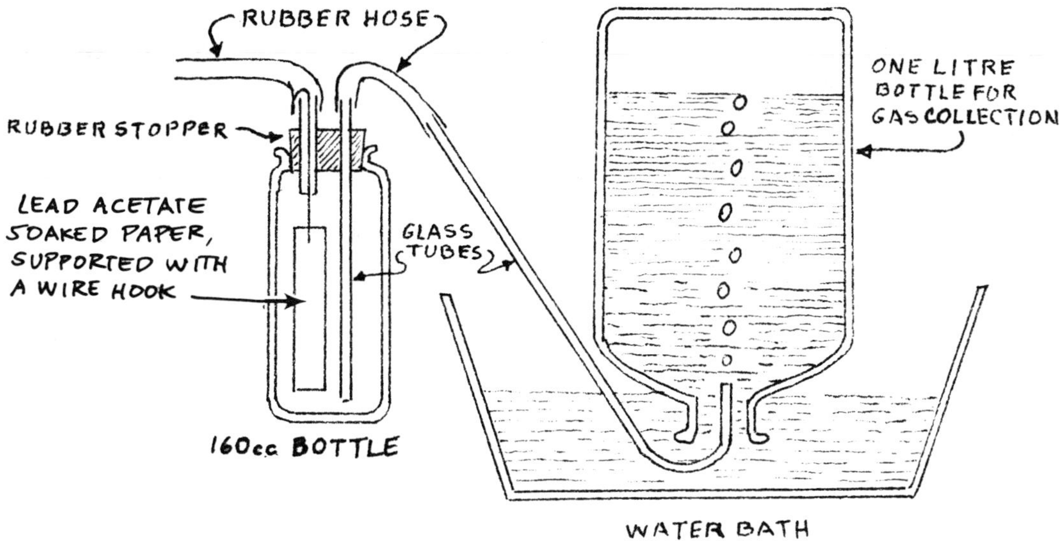

Fig. A1.3 H_2S Detection

where:
Vb = beginning gas volume

Ve = ending or final gas volume

And, of course, the assumption is that:

$$100 - \%CO_2 = \%CH_4$$

See Chapter 29 for a means of making this information useful.

H_2S

It just so happens that there is an inexpensive, moderately accurate method of determining the percentage of H_2S in biogas.

It was developed by McBride and Edwards and reported by Hazeltine (1933). Unfortunately, the test as developed will only tell if H_2S is present in amounts ranging from a trace, to 15% of the maximum amount in which we are interested. In other words, the upper limit of sensitivity of the test as described and developed is 0.015% H_2S by volume, rather than the critical concentration of 0.100% H_2S by volume which we have taken as the upper limit of safe use in an engine.

The basis of the test is that lead acetate—Pb $(COOH)_2$—reacts with H_2S to form brown-colored lead and sulfur compounds. Thus, 22 by 77 millimeter (seven-eighths of an inch by three inches) strips of paper, soaked in a 10% solution of lead acetate (*e.g.*, 10% lead acetate, 90% water by weight), dried, and then exposed to one liter of biogas, using the aparatus shown, would develop a characteristic brown coloration, as follows in Fig. A1.3.

It seems clear that, since these colors are the result of a certain amount of H_2S reacting with lead acetate,

Color	Percentage H_2S
No color	0.0004 or less
Trace of color	0.0005 to 0.0008
Light color	0.0015 to 0.0025
Moderate color	0.005 to 0.008
Dark color	0.015 or more

Table A1-1 Estimation of Percentage of H_2S

the test could be modified slightly so that, if a dark brown were present, the test could be run again with half the former volume of biogas. Twice the above percentages of H_2S would then produce the color densities listed; this would give us a better idea about whether scrubbing was indicated or not. The test can also be used to indicate the effectiveness of scrubbing.

Substrate Analysis

Most of us will not be able to do very much substrate analysis, as most substrate analysis requires rather elaborate equipment and technical knowledge.

However, one quite important analysis which is also very simple is the percentage of moisture, or dry weight. All it requires is to weigh some amount of substrate, then dry it in an ordinary oven (at around 220°F, or 105°C) for several hours. The dried weight is, of course, TS, total solids.

To figure the percentage of water, simply divide the weight of water by the total weight of the *wet* substrate, and multiply by 100:

$$\% H_2O = \frac{Ww - Wd}{Ww} \times 100$$

where:
Ww = weight of wet substrate
Wd = dry weight of substrate

In Chapter 9, we were working with what we called the "H_2O number," designated here as H*n* and simply the reciprocal of the % H_2O, times 100:

$$H_2On = \frac{100}{\%H_2O}$$

If you're starting with wet and dry weights, an easier way to directly derive the H*n* is to divide the wet weight of the substrate by the weight of water:

$$Hn = \frac{Ww}{Ww - Wd}$$

(Likewise, the "CN number," designated here as CN*n*, is the reciprocal of that percentage proportion of the substrate which is C + N times 100.)

Appendix 2: Biohydrogen

Hydrogen gas appears to be routinely evolved whenever organic materials undergo anaerobic decomposition, but before it can escape from the anaerobic environment, it is snatched up by the methane-forming bacteria and used to make CH_4, methane.

The factors which lead to the rapid formation of hydrogen are, therefore, little understood, since the hydrogen so rarely shows up in the final biogas. Further, hydrogen, as a gas, is a more dilute energy source than methane. Where 12.5 liters of methane has 100 Calories (net) energy available, the same volume of hydrogen, has only 30 Calories (net) energy available. So, why shoot for hydrogen when you can have methane?

The answer may be that we can have our cake and eat it too (in other words, maybe we can get both gases). Hydrogen is also desirable for its higher flame velocity; mixed with biogas, hydrogen will make that biogas harder to blow off of pilot lights, and easier to keep burning on stoves. But very little research has been directed toward the production of hydrogen by anaerobic decomposition.

However, several things are known:

1. Hydrogen is produced during the first (acid) stage of anaerobic breakdown, but not, as far as is known, during the second stage.
2. Low pH seems to favor H_2 production, or possibly just its release. Whether the mechanism is release or production is not known because the low pH data is based on studies of ordinary generators, where both stages of digestion are active simultaneously. Thus while low pH is related to an increase in H_2 in biogas, it may be that it merely inhibits the methane formers, which could then allow H_2 to get past them, rather than stimulating the production of hydrogen.
3. A high C/N seems to stimulate the production of H_2.
4. Substrates high in carbohydrates, such as cellulose, seem to stimulate H_2 production.

Omelianski (1902), an early pioneer in anaerobic studies, found that anaerobic cultures obtained from horse manure or soil, when heated to 75°C and held there for about 15 minutes, produced hydrogen when the culture was then kept at 35°C. He was apparently using cellulose as a substrate.

A. W. Shorger in his book, *Chemistry of Cellulose and Wood* (1926), recounts research from the *Journal of the Society of Chemistry and Industry,* v. 42, 1923, page 169. This research found that methane-rich biogas was produced more than 12 times as rapidly as hydrogen-rich biogas.

While H_2 may be evolved slowly, hybrid generation, which can separate the acid-forming stage from the methane-forming stage, may show some promise for H_2 production. In their cryptic, poorly written, tantalizing report, researchers at the University of Pennsylvania (1974) report that a gas of up to 10% H_2 was produced from the first (acid) stage reactor.

No lower methane production should come from a slurry used first to produce hydrogen and then methane. Chemically and biochemically, iron can replace hydrogen (as an electron donor), as reported by Thimann (1955), and thus both hydrogen and methane production should be compatible in the proposed hybrid generator.

The key seems to be in finding the right culture, keeping the pH slightly acid, and keeping the acid modules warm. This last requirement throws us back into a hydrogen and fatty acids producing modular batch generator situation, but if hydrogen is what you want, this, apparently, is what you'll have to do.

Recently, other researchers have proposed and

tested methods of producing hydrogen photosynthetically, and although it looks to be high tech, you're welcome to wade through it. The references, found in full in the bibliography, are Mitsui, 1974 and 1975, and Newton, 1976.

One final note. Travin and Buswell (1934) (see Bibliography), reported a biogas that was about 30% hydrogen by volume. By weight (and assuming the other 70% to have been CO_2) that is only about 2%. Not startling, but still, another reference for you. Hydrogen filled balloons from a compost pile? Another interesting fantasy, perhaps.

Appendix 3: Agricultural Use of Effluent

Behold this compost! Behold it well!
Perhaps every mite has once form'd part of a sick
person—yet behold!
The grass of spring covers the prairies,
The bean bursts noiselessly through the mould in the
garden,
The delicate spear of the onion pierces upward,
The apple-buds cluster together on the apple-branches,
The resurrection of the wheat appears with pale visage
out of its graves.
What chemistry!
That the winds are not really infectious.
That all is clean forever and forever,
That the cool drink from the well tastes so good,
That blackberries are so flavorous and juicy.
That the fruits of the apple-orchard and the orange-
orchard, that melons, grapes, peaches, plums, will
none of them poison me,
That when I recline on the grass I do not catch any
disease.
Now I am terrified at the Earth, it is that calm and
patient,
It grows such sweet things out of such corruptions,
It turns harmless and stainless on its axis, with such
endless succession of diseased corpses,
It distills such exquisite winds out of such infused
fetor,
It gives such divine materials to men, and accepts such
leavings from them at last.

"This Compost," Walt Whitman

Some background on fertilizer and plant growth seems in order for this discussion. Of necessity, it will be brief, but this is quite obviously a subject of great concern to anyone striving for greater self-sufficiency (or rational interdependence—a more reasonable goal), and it is of considerable interest to anyone concerned with the other major useful by-product of the biogas process—the effluent, and its potential fertilizing value.

The modern school of agriculture is based in a narrow view of the soil/plant ecosystem which perceives it almost exclusively in chemical and mechanical terms. For example, one book on plant nutrition describes plants as ". . . those fixed, silent, chemical machines . . ." (Epstein 1972). Such a view is not surprising, considering the tools that have been used to explore the life of plants. For example, the most common method of comparing one plant with another is to burn it and analyze the ashes. Nearly all of the tools developed for basic agricultural research similarly involve the death of the plant or the dissection and destruction of its environment to gain knowledge about the nature and function of the plant and its surrounding ecosystem.

The elements which comprise a human body are worth about $2.00 on the open market—but how much does this tell us about human beings and their nature and function? Modern agriculture is very much like modern medicine, which knows a great deal about disease, but has yet to come to significant scientific conclusions about health.

This is not to say that the facts discovered through science are incorrect—agricultural science has made tremendous progress in the last one hundred years. However, no true scientist would believe that humanity's progress in knowledge has ended, or that any of today's cherished theories will not be regarded by future scientists very much as today's scientists look upon yesterday's theories.

NPK

The modern method of agriculture generally compares all materials used for soil amendments (fertilizers and composts) based on their chemical analysis, and particularly on their relative amounts of the nutrient elements N, P, and K (nitrogen, phosphorus, and potassium). This is because the quantity of these three nutrients in most plants, as compared with the other elements which comprise most plants, is great. Also present in great abundance are C, H, and O, but these elements are easily gained from air and water, rather than from the soil. Often chemical fertilizers are spoken of as "5-2-2," or "10-5-2." These numbers refer to the percentage of N, P, and K found in the fertilizer.

Strictly in these chemical terms, dried sludge (the settled solids portion of effluent) is a poor fertilizer. However, the *whole* effluent is a good-to-excellent fertilizer in terms of its chemical analysis. The difference lies in the fact that a good portion of the N in the effluent is in the liquid (supernatent) and in the form of ammonia (and related compounds), which rapidly evaporate (or are washed away) when the solids are drained and dried.

Local use of effluent, however, should mean that it will not have to be dried—indeed, it may have to be diluted for ease of pumping. This, and careful handling, will mean that a much higher portion of the N in effluent will reach the soil than would be the case were only dried sludge used.

Economics

It has been demonstrated over and again that the N in the original substrate remains in the anaerobic slurry to a greater degree than in aerobic composting. Therefore in NPK terms, the biogas process produces a "superior fertilizer." Essentially, the whole NPK value of the original substrate remains and is available, and one method of evaluating the fertilizer value of the effluent is to value the amount of N, P, and K in the effluent according to what those amounts would cost if they were purchased as chemicals.

Remember, however, that these chemicals are not created by the biogas process: they are available in any case, although sometimes to a lesser degree in the original substrate or an aerobic compost made from that substrate. Further, we are not interested in mere chemicals. What we really want to know is: how well does effluent make plants grow?

The true fertilizer value of the effluent, then, becomes a matter of weighing factors other than a simple chemical analysis. Probably the most accurate comparison is with aerobic compost, since this is often the destiny of substrates if they are not used in making biogas.

Aerobic composting will result in a 25% loss of N as compared with anaerobic effluent, but the aerobic compost will have more of that N in a form which is not so easily lost—e.g., not as ammonia, but rather tied up in some kind of slow release form. For application on pasture and grass-related crops (corn, grains), ammonia N is often preferable. For application on tree crops, legumes, and most vegetables, slow release N is generally preferable.

The large-scale production and use of aerobic compost will involve time and equipment comparable to the time and equipment necessary for the biogas process. It is probable that the time involved in running a well-conceived and constructed biogas generator and in spreading the pumpable liquid effluent will be less than the time involved in an aerobic compost venture, but the equipment (generator plus systems, pumps, etc.) necessary for the biogas process will, on the other hand, generally be more expensive than the equipment (tractor with skip loader, manure spreader) necessary for making aerobic compost.

For aerobic compost, substrate handling may be more difficult, since it is not feasible to simply add water—for example, to a manure substrate—and pump it away to the compost bin as could be done were it to be used in biogas production. Aerobic composting, unless well done, can also be a source of flies, and manure cannot be stored in the open without rain water washing out nutrients, so a storage structure may be necessary.

Generally, the economic benefits of using effluent as fertilizer are the lowered costs of handling. If we are not considering the value of the biogas produced, then anaerobic composting will have an economic benefit over that of aerobic composting where some factor or factors make handling expensive. Feedlots, or other intensive animal production situations, are often of this nature.

But these factors are situational, and the relative economic benefits of aerobic versus anaerobic change according to a person's needs. Often a more expensive option in a particular situation will still be so much more beneficial that it will prove to provide a greater dollar return than an option initially less expensive. The economics question involves factors particular to your own situation.

Biology

Biologically speaking, it is nearly impossible to compare aerobic compost and anaerobic effluent in terms of their effect on crop quality. The reason lies near the heart of our comments about modern agriculture:

it has not developed inexpensive tools to give such comparisons. The literature often refers to such things as "percent nitrogen uptake" or "dry weight increase"—we can roughly compare compost with effluent, or either one with chemical fertilizers—but that doesn't really tell us anything good or bad—or even very useful—if we want to know about the *qualities* of the plants grown with these different nutrient sources, rather than their *quantities*.

Food, after all, is more than "dry weight." It is discouragingly difficult to answer a seemingly simple question such as: Which process produces more healthful food? We cannot blame the scientist for wanting to answer questions which are more answerable: Which nutrient source provides a greater increase in dry weight?—but we might blame a science which seems to feel that only the presently answerable questions are important.

All we can do then, to answer our presently unanswerable question, is to extrapolate. Since plants have evolved in a certain biotic situation, it is likely that they will respond best to attempts to enhance rather than radically alter that situation. Mother Nature makes aerobic compost, for the most part. This doesn't mean that she can't be improved upon, but it does tend to indicate that compost will produce a better, more healthful plant under most circumstances than will effluent.

Agricultural Use

Dr. H. H. Koepf, an authority on soil biology (and Biodynamics) has suggested that effluent be treated with straw and stinging nettle (1974) to help balance its effect upon the soil. Biodynamics has interesting, useful, and subtle answers to some of the questions raised earlier.

Nettle and straw could be used either in lagoon storage of the effluent, or in conjunction with the composting technique suggested by Ransome (1944). Using 45 centimeters (18 inches) of straw, effluent of 6% solids was applied at the rate of 6 liters effluent per kilogram of straw (9.63 cubic feet per 100 pounds, 1400 gallons per ton). The pile is built up in layers, and treated like ordinary compost. For air-dried sludge, 5 centimeters (2 inches) of sludge is used for every 45 centimeters of straw.

More often, however, effluent is used directly, or the sludge is settled and dried. For information on the special problems of the agricultural use of effluent from human excrement, see the relevant subsection in Chapter 15.

Since it is unlikely that, in the circumstances in which most of us find ourselves, we will be drying the effluent, we will discuss only liquid effluent. (Briefly, use dried sludge like compost.)

Spreading

Liquid sludge can be spread in many ways. Commonly, it is spread either by a truck with a tank on it, or by irrigation. There are five kinds of irrigation to be considered: (1) sub-soil; (2) furrow and ditch; (3) flood; (4) open pipe; and (5) spray. Because of the high solids content, and (often) the large particles of biogas effluent, spray irrigation is not always possible. Also, the higher pressures required necessitate higher energy and equipment costs. For soils without any appreciable slope, furrow and ditch irrigation is not always possible. If there is not enough water or effluent to make flooding practical, then the use of a tank truck or a movable open-ended pipe to spread the effluent may be required. Sub-soil irrigation requires buried pipe, and may suffer in orchard situations from root invasion of the pipe. The higher initial cost of buried pipe may be outweighed in some situations by greater safety (when an effluent with possible parasite or pathogen contamination is used) and/or less evaporation. Porous clay pipe can be used. In a situation where home sewage is used, a leach line system may fulfill sanitation code requirements, yet still allow biogas production before the sewage effluent is put in the leach lines.

Soils

Some soils respond better to effluent than others. Open porous soils—sandy or loamy—will in general be more apt to remain friable (loose, tillable) than silt or clay soils, when effluent is used as a soil amendment.

The nutrients in effluent encourage the growth of soil bacteria, an occurrence which can have many benefits for the soil structure and humus content. However, if excessive use, or excessive soil saturation, causes the soil to become "clogged" with the products of this growth, slime organisms begin to grow. Water percolation is then seriously reduced, and the CO_2 released by decomposition processes and plant roots cannot leave the soil environment, causing it to become more acidic. A close check on the soil pH will provide indications that this is happening. The use of effluent could probably be increased if the soil is tilled (harrowed or plowed) a few days after application. More compact clay soils respond to effluent by clogging more rapidly than sandy, open, or porous soils.

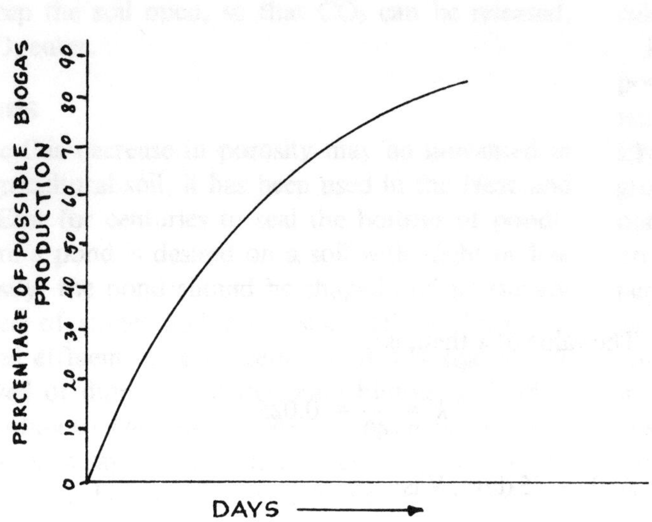

Fig. A4.1 K Value Equation Graphed

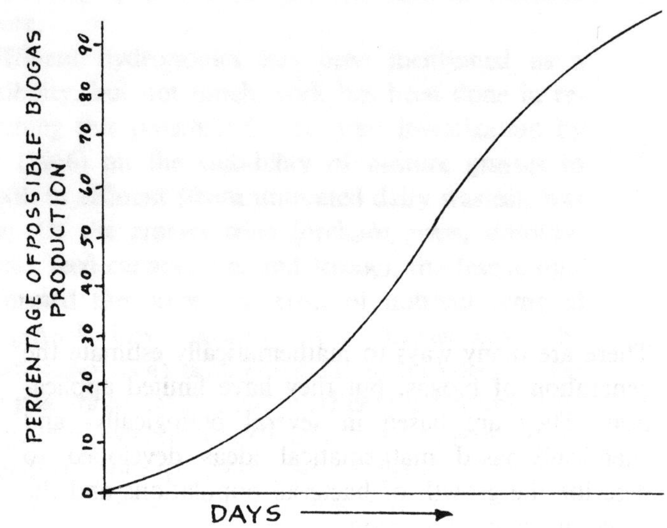

Fig. A4.2 Real Life Situation Graphed

The time (A) required for 50% of the potential biogas production is:

$$A = \frac{0.639\ \text{(days)}}{k}$$

For a k value of 0.05, A = 12.8 days. For a k value of 0.15, A = 4.3 days.

We can easily see that tropical vegetable matter digests less rapidly than sewage sludge, but there is more biogas (per unit dry weight) available in vegetable matter than in sewage solids, and thus vegetable matter will generate biogas longer than sewage sludge.

If you graph the k value equation, you may notice that it starts out with a bang (as it were) rather than moving into things gradually, as in Fig. A4.1.

Whereas, in many actual real life situations, it takes a while for things to get going, as in Fig. A4.2.

The difference between the instant takeoff and the gradual increase types of biogas generation is *seeding*. The k value equation assumes a heavy seeding (for example, a 1:1 ratio between old active slurry as a seed, and fresh new slurry as a food source). When enough bacteria are added to the new material, indeed, there is instant gas production.

This k value equation, then, describes a batch reaction (where one-half the material added is older slurry), or it describes the gas production from a certain portion of material put into a continuous-feed generator, where there is good agitation and therefore excellent inoculation. The k value must be derived by batch studies, however, and careful monitoring of gas production, so that the exact time it requires for 63% of the total gas production to occur is known. As mentioned before, k is the reciprocal of that time period. So, if 63% of the total gas is produced in 12.8 days, then k is:

$$k = \frac{1}{12.8}$$

$$k = 0.078$$

If you wish to use the k value equation, then, you should follow the procedure outlined in Chapter 47. Keep a careful record of gas production.

Compound Interest Equation

Other equations, however, give us readouts for situations in which there is a time lag before digestion starts, so we can describe the case where a minimal seed, or no seed, is used. One of these is a two-part equation from Fair and Moore (1932, No. 2):

$$\log V = (\log 0.5\ Vt) - K1\ (Ti - T) \quad \text{first stage}$$
$$\log (Vt - V) = (\log 0.5\ Vt) - K2\ (T - Ti) \quad \text{second stage}$$

where:

V = volume of gas produced by the time "T"
Vt = total volume of possible gas production
$K1$, $K2$ = rate constants, derived as shown below
T = time at which gas production is to be predicted
Ti = time of inflection, or transition between first and second stage

Appendix 5

The density of something is, of co
mass per unit volume. Ordinary earth
and 76 centimeters of mercury air
density of 1.215 grams per liter. (Just
a nickel weighs about 5 grams.)

The specific gravity of a gas is its
to the density of air (when both ar
standard temperature and pressure).
gravity is misleading, really. It migh
relative density, but we have what
through spelled that way? Same reaso

If we assume that biogas is just
then we can find its specific gravity b
percentage of each by a certain nun
as shown

Sg = (%CO_2) 0.01529 + (%CH

where:

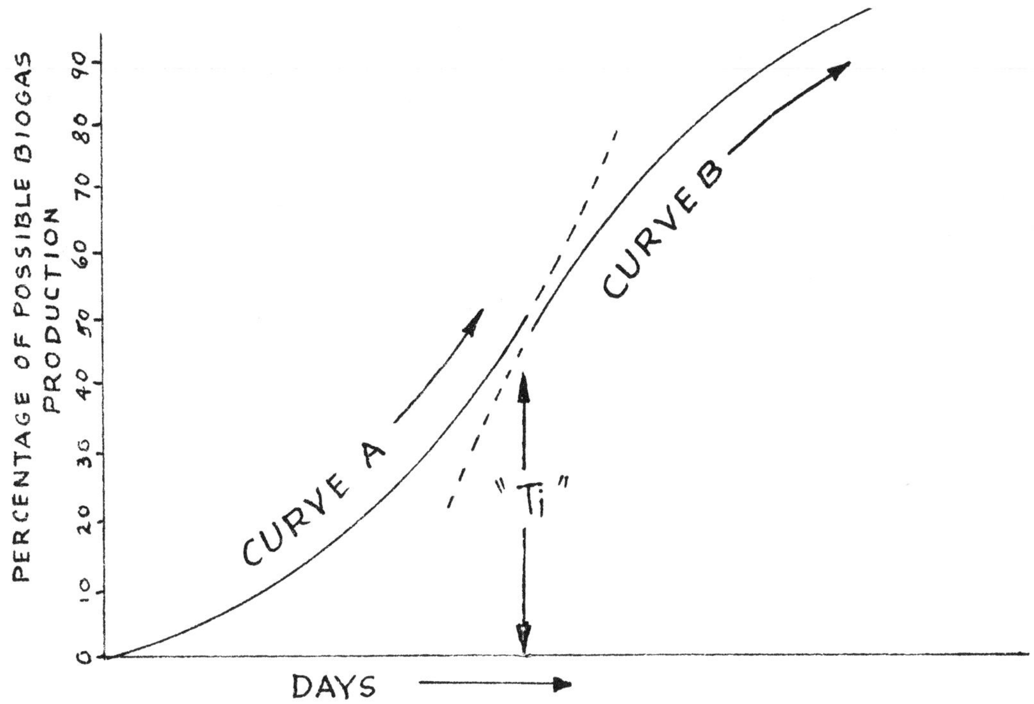

Fig. A4.3 Compound Interest Curve

These equations, and all the equations using logs in this chapter, are log base 10.

Essentially, both of these equations describe a kind of "compound interest" situation. The top equation shows gas production increasing daily over a period of time, by a certain percentage of yesterday's production, and the bottom equation shows the opposite—gas production decreasing daily by a certain percentage of yesterday's production. Together, they graph as in Fig. A4.3.

Thus, while the top equation, represented by curve A, is increasing, the bottom equation, represented by curve B, is decreasing. At some point, Ti, they meet, and a continuous, S-shaped (ogee) curve is produced by adding the proper portions of each one together. Point Ti is the time of inflection, meaning the time when the trend toward the vertical begins to be diverted into the trend toward the horizontal.

The constants K1 and K2 are found by the equations:

$$K1 = \log\left(1 + \frac{R1}{100}\right)$$

$$K2 = \log\left(1 - \frac{R2}{100}\right)$$

where:
R1 = the daily percentage increase in biogas production
R2 = the daily percentage decrease in biogas production

The k values reported earlier can be transposed to $K2$ values via a peculiar equation:

$$K2 = \log(e^{k})$$

Fair and Moore report values of $K1 = 0.0989$, $K2 = 0.0732$, Vt = 672 liters per kilogram (or cubic centimeters per gram), Ti = 14.37 days, for one sample of sewage.

For most purposes, Ti is taken as being the time at which V = 0.5 Vt, e.g. the time when exactly half the gas has been produced. Or, alternatively and more exactly:

$$T = Ti,$$

whenever

$$V = 0.5\left(Vt - \frac{K1}{K2}\right)$$

But the difference between the two ways of figuring Ti is generally rather minor.

Continuous-Fed

Here's a lovely one. The amount of gas available from a given amount of substrate in a continuous-fed generator, can be approximated by the use of the equation:

$$V = \frac{Ge\ Lr\ Vgen\ (1 - e^{-kT})}{T}$$

where:

V = volume of gas produced per day
Ge = maximum gas production expected per unit of VS or TS of substrate used

Lr = loading rate, expressed in eit
as long as both Ge and Lr ar
same way
Vgen = volume of the generator
e = 2.71828 etc.
k = reaction velocitiy constant
T = HRT of the generator

Appendix 6: Drill Statistics

Tables A6.1 and A6.2 show drill sizes for use in enlarging orifices on burners using natural gas, propane or butane. Biogas requires a larger orifice, as will be explained below.

The numbers under the column labeled *Designation* are steel-wire gauge numbers (U.S.); the letters are standard twist drill letters; the fractions are fractions of an inch.

Designation	Diameter inches	Area sq in.
1/2	.5000	.1963
31/64	.4844	.1843
15/32	.4688	.1726
29/64	.4531	.1613
7/16	.4375	.1503
27/64	.4219	.1398
Z	.413	.1340
13/32	.4063	.1296
Y	.404	.1282
X	.397	.1238
25/64	.3906	.1198
W	.386	.1170
V	.377	.1116
3/8	.375	.1104
U	.368	.1064
23/64	.3594	.1014
T	.358	.1006
S	.348	.09511
11/32	.3438	.09281
R	.339	.09026
Q	.332	.08657
21/64	.3281	.08456
P	.323	.08194
O	.316	.07843
5/16	.3125	.07670
N	.302	.07163
19/64	.2969	.06922
M	.295	.06835
L	.29	.06605
9/32	.2813	.06213
K	.281	.06202
J	.277	.06026
I	.272	.05811
H	.266	.05557
17/64	.2656	.05542
G	.261	.05350
F	.257	.05187
E-1/4	.2500	.04909
D	.246	.04753
C	.242	.04600
B	.238	.04449
15/64	.2344	.04314
A	.234	.04301
1	.288	.04083
2	.221	.03836
7/32	.2188	.03758
3	.213	.03563
4	.209	.03431
5	.2055	.03317
6	.204	.03269
13/64	.2031	.03241
7	.201	.03173
8	.199	.03110
9	.196	.03017
10	.1935	.02940
11	.191	.02865

Designation	Diameter inches	Area sq in.	Designation	Diameter inches	Area sq in.
12	.189	.02806	46	.0810	.00515
3/16	.1875	.02761	47	.0785	.00484
13	.185	.02688	5/64	.0781	.00479
14	.182	.02602	48	.0760	.00454
15	.1800	.02545	49	.0730	.00419
16	.1770	.02461	50	.0700	.00385
17	.1730	.02351	51	.0670	.00353
11/64	.1719	.02320	52	.0635	.00317
18	.1695	.02256	1/16	.0625	.00307
19	.1660	.02164	53	.0595	.00278
20	.1610	.02036	54	.0550	.00238
21	.1590	.01986	55	.0520	.00212
22	.1570	.01936	3/64	.0473	.00173
5/32	.1563	.01917	56	.0465	.001698
23	.1540	.01863	57	.0430	.001452
24	.1520	.01815	58	.0420	.001385
25	.1495	.01755	59	.0410	.001320
26	.1470	.01697	60	.0400	.001257
27	.1440	.01629	61	.039	.001195
9/64	.1406	.01553	62	.038	.001134
28	.1405	.01549	63	.037	.001075
29	.1360	.01453	64	.036	.001018
30	.1285	.01296	65	.035	.000962
1/8	.1250	.01227	66	.033	.000855
31	.1200	.01131	67	.032	.000804
32	.1160	.01057	1/32	.0313	.000765
33	.1130	.01003	68	.031	.000755
34	.1110	.00968	69	.0292	.000670
35	.1100	.00950	70	.028	.000616
7/64	.1094	.00940	71	.026	.000531
36	.1065	.00891	72	.025	.000491
37	.1040	.00849	73	.024	.000452
38	.1015	.00809	74	.0225	.000398
39	.0995	.00778	75	.021	.000346
40	.0980	.00754	76	.020	.000314
41	.0960	.00724	77	.018	.000254
3/32	.0938	.00690	78	.016	.000201
42	.0935	.00687	1/64	.0156	.000191
43	.0890	.00622	79	.0145	.000165
44	.0860	.00581	80	.0135	.000143
45	.0820	.00528			

Table A6.1 Drill Sizes

Calculations based on natural gas: 1,050 Btu per cubic foot, specific gravity of 0.65. Liquid propane has 2,500 Btu per cubic foot, and a specific gravity of 1.55.

Gas (by % CH_4)	Orifice area multiplier
Propane	**1.00**
100%	1.51
95%	1.66
90%	1.82
85%	1.99
80%	2.19
75%	2.41
70%	2.66
65%	2.93
60%	3.27
55%	3.66
50%	4.15

Table A7.2 Orifice Conversion based on Propane Gas

Gas (by % CH_4)	Orifice area multiplier
Butane	**1.00**
100%	1.71
95%	1.88
90%	2.06
85%	2.26
80%	2.48
75%	2.73
70%	3.01
65%	3.32
60%	3.71
55%	4.15
50%	4.70

Table A7.3 Orifice Conversion based on Butane Gas

of area 0.00238 square inches, we find that for biogas of 70% methane, the area of the original orifice must be increased by a factor of 1.73. Doing the math (1.73 x 0.00238), we find that the needed area is 0.00412 square inches. Table A6.1 does not list any area of 0.00412 square inches exactly, but it does show that a number 49 twist drill makes a hole with an area of 0.00419 square inches, fractionally larger than required. A number 49 drill then will do the trick.

For propane or butane, the changes are a bit more radical, since both of these gases are much more concentrated sources of energy per unit volume than is biogas. This means that a much greater volume of biogas (and hence a greatly enlarged orifice) will be required when attempting to run butane or propane appliances with biogas.

Table A7.2 is calculated (using the previous formula) for propane conversions. Propane has a specific gravity of 1.562, and it generally has 2,572 Btu per cubic foot.

The final simple conversion chart is for conversions of butane appliances and burners. The specific gravity of butane is 2.067; it generally has 3,353 Btu per cubic foot.

In Appendix 6, Table A6.2 relates various burner Btu outputs to gas pressure and is calculated for natural gas. To get the same Btu output for biogas, use Table A7.1 to figure the area coefficient. For further information on the safe use of burnable gas, *The Natural Fuel Gas Code* is recommended and should be available at your library.

Appendix 8: Vital Statistics of Gases

Herein we list some interesting numbers, all having to do with different gases.

Solubility in Water

When scrubbing biogas with water, it is pleasant to be able to predict the amounts of various gases which will dissolve in a given amount of water. In Table A8.1 below, the numbers in columns labeled a, a-2, or a-3 tell how many grams of various gases will dissolve in 1,000 grams of pure water at 76.2 centimeters (30 inches) of mercury pressure. In columns labeled b or b-1 the numbers tell approximately how many liters of the listed gas will dissolve in a liter of pure water under a pressure of 76.2 cm. Four of the columns are labeled either a-2, b-1, or a-3. This kind of designation means that the listed number of kilograms or liters should be multiplied by that power of ten to get the actual solubility. So, for example, the actual solubility of CH_4 at 5°C is 3.41×10^{-2}, or 0.0341.

Because Table A8.1 is calculated on a weight per weight, volume per volume basis, "A" can be considered as pounds per 1,000 pounds, and B as cubic feet per cubic foot.

temperature		CO_2		H_2S		CH_4		H_2	
°C	°F	a	b	a	b	a-2	b-1	a-3	b-1
0	32	3.35	1.79	7.10	4.84	3.97	5.83	1.93	2.26
5	41	2.77	1.48	6.00	4.11	3.41	5.02	1.83	2.14
10	50	2.33	1.24	5.13	3.52	2.98	4.36	1.74	2.04
15	59	1.97	1.05	4.42	3.03	2.60	3.83	1.67	1.96
20	68	1.69	0.90	3.85	2.63	2.32	3.41	1.60	1.88
25	77	1.45	0.77	3.38	2.31	2.10	3.08	1.54	1.80
30	86	1.26	0.67	2.99	2.05	1.91	2.80	1.47	1.73

Table A8.1 Solubility in Water

Name	Formula	Molecular Weight	Specific Gravity
Hydrogen	H_2	2.02	6.95–2
Methane	CH_4	16.03	5.54–1
Ethane	C_2H_6	30.05	1.05
Propane	C_3H_8	44.06	1.56
Butane	C_4H_{10}	58.08	2.07
Hydrogen sulfide	H_2S	34.08	1.19
Carbon dioxide	CO_2	44.00	1.53
Air	—	28.97	1.00

Table A8.2 Molecular Weight and Specific Gravity

Appendix 10: Steam Temperature, Pressure, Water Depth

Consider, for an example, a steam temperature of 110°C, and a vapor pressure of steam at 107.5 centimeters of mercury, or 31.5 centimeters of mercury pressure over and above ordinary (76 centimeters mercury) air pressure. At this vapor pressure, steam will be able to overcome the pressure found at a depth of 4.28 meters under water. (When steam is at 100°C, its vapor pressure equals sea level atmospheric pressure.) This table is useful for determining the temperature of steam which is injected into a generator using its own vapor pressure, at a given depth.

Steam Temperature		Pressure		Water Depth	
°C	°F	cmHg	psi	meters	feet
100	212	76	14.70	0	0
101	213.8	78.8	15.23	0.37	1.23
102	215.6	81.6	15.78	0.76	2.49
103	217.4	84.5	16.34	1.16	3.80
104	219.2	87.5	19.92	1.56	5.13
105	221	90.6	17.52	1.99	6.52
106	222.8	93.8	18.14	2.42	7.93
107	224.6	97.1	18.77	2.86	9.39
108	226.4	100.4	19.42	3.32	10.90
109	228.2	103.9	20.09	3.79	12.44
*110	230	*107.5	20.78	*4.28	14.03
111	231.8	111.1	21.49	4.77	15.66
112	233.6	114.9	22.21	5.28	17.34
113	235.4	118.7	22.96	5.81	19.06
114	237.2	122.7	23.73	6.35	20.84
115	239	126.8	24.52	6.91	22.66
116	240.8	131.0	25.33	7.48	24.53
117	242.6	135.3	26.16	8.06	26.45
118	244.4	139.7	27.02	8.66	28.42
119	246.2	144.3	27.90	9.28	30.44
120	248	148.9	28.80	9.91	32.52
121	249.8	153.7	29.72	10.56	34.65
122	251.6	158.6	30.67	11.23	36.84
123	253.84	163.6	31.64	11.91	39.09

Table A10.1 Steam Injection

Appendix 11: Heat Transfer in Pipes

To design a conventional hot water heating system for a generator, you must know how much heat can be forced out of a specified length of a given diameter pipe, with the water circulating inside at a given temperature and velocity. You probably never thought that being a designer was so difficult, did you? For the circulation of hot water in pipes, the C values of a particular pipe, in Btu ft^{-2} hr°F^{-1}, can be found with this formula:

$$C = 0.0014\,(t + 100)\,\frac{V^{0.8}}{D^{0.2}}$$

where:

C = heat transfer coefficient of the pipe in question, under the stated conditions
t = temperature of circulating water, in °F
V = velocity of that water, in feet per hour
D = inside diameter of the pipe, in feet

(In this Appendix, we will emphasize American values, since that is what most of us will be working with, for pipe sizes and such. Approximations to the S.I. system can be fairly easily made.)

In using this equation, Tables A11.1 and A11.2 should simplify matters.

Standard Weight Galvanized Pipe			
D, nominal inches	D, actual inches	D, actual feet x 10^{-1}	$D^{0.2}$ feet
3/8	0.493	0.411	0.528
1/2	0.622	0.518	0.553
3/4	0.824	0.687	0.585
1	1.049	0.874	0.614
1-1/4	1.380	1.15	0.649
1-1/2	1.610	1.34	0.669
2	2.067	1.72	0.703
2-1/2	2.469	2.06	0.729
3	3.068	2.56	0.761
4	4.026	3.36	0.804
6	6.065	5.05	0.872

Table A11.1 D Values for Pipe

	Tubes	
D, actual inches	D, actual feet x 10^{-2}	$D^{0.2}$ feet
0.25	2.1	0.461
0.375	3.1	0.500
0.50	4.2	0.530
0.75	6.3	0.574
1.00	8.3	0.608
1.25	10.4	0.636
1.50	12.5	0.660
2.00	16.7	0.699
2.50	20.8	0.731

Table A11.2 D Values for Tubes

Velocity feet per hour	$V^{0.8}$
50	22.9
100	39.8
250	82.9
500	144.3
750	199.5
1000	251.2
1500	374.4
2000	437.3
2500	522.8
3000	604.9
3500	684.3
4000	761.5

Table A11.3 Velocity

Now plug into the formula above, assuming values for three of the variables—t, V, and D. For example, at a water temperature of 130°F, and a velocity of 100 feet per hour in standard-weight, half-inch pipe, the expression becomes:

$$C = 0.0014\,(130 + 100)\,\frac{39.8}{0.553}$$

$$C = \frac{23.17 \text{ Btu}}{\text{hr ft}^{2}\,{}^{\circ}\text{F}}$$

Since we want to insure that the generator will remain warm under the worst case conditions—e.g., the middle of winter during a blizzard just after having loaded some cold slurry—then we can calculate our design heat load for these conditions. Referring to Chapter 28 gives a good idea of how to go about this. The answer we are looking for will be in terms of Cal or Btu per hour, since this is the best way to describe a heat loss. That heat loss must be balanced by a heat gain, measured in the same units. The present equation gives us those kind of answers.

Let's assume that we've run the heat loss equations and come up with a figure of 6,000 Btu per hour. That means that we will have to replace that lost heat to keep the generator warm.

Remembering that:

$$Ht = C\,A\,\Delta t\,T$$

where (in this case):

Ht = design heat load (worst case conditions)
C = C value of pipe
A = area of pipe's whole surface
Δt = temperature difference between the desired generator temperature and the circulated water temperature
T = time required to replace the lost heat (here taken as one hour)

Assume a heat load of 6,000 Btu per hour, and a desired temperature of 90°F. The remaining variable is A, area, as shown below:

$$6{,}000 = 23.17\,A\,(130 - 90)\,1$$

This follows the format of the equation above. If you know your algebra, you will realize the need to isolate A on one side of the equals sign. The equation below is identical to the one above, except that A has been isolated:

$$A = \frac{6{,}000}{(23.17)\,(130 - 90)\,1}$$

$$A = 6.47\ \text{f}^{2}$$

So, we know that we need a surface area of a bit more than six-and-a-half square feet, given our assumptions. Now we have to translate this into a certain length of a certain diameter pipe:

$$A = \pi D L \qquad D = \frac{\pi L}{A}$$

$$D = \pi L / A$$

$$L = \pi D / A \qquad L = \frac{\pi D}{A}$$

where:

A = area
π = 3.1416, pi
D = diameter of the pipe
L = length of the pipe

Pipe comes in certain fixed diameters, so there is no point in trying to find out how to get the surface area needed using a fixed length, since then the answer would be in terms of a variable diameter. It makes more sense to use the third equation above, plug in the area we have, assume we are working with a common pipe size, and find out what length will satisfy us. In the example, we are using half-inch pipe:

$$L = \frac{6.47}{(3.1416)\,(0.0518)}$$

L = 39.76 feet of half-inch pipe, with the above conditions.

Therefore, if we need 6,000 Btu per hour to replace heat lost to the Great Outdoors, we can do this by constructing a heat exchange unit made of at least forty feet of half-inch pipe (this will end up as a rough spiral on the bottom or up the sides of the generator), through which we will need to circulate water of 130°F at the rate of one hundred feet per hour (that's slow; about a foot-and-a-half each minute). Of course, our heater will need to be sized properly to heat that much water at that rate. Larger pipe sizes may allow the use of thermosiphon as a circulating means. Otherwise, a pump will have to be used.

Appendix 12: Decimals and Fractions

The tables in this appendix can help solve a number of problems, for those of you without recourse to a calculator. (It is recommended that you get one, however, if possible. Money can no longer be said to be an obstacle, since there are adequate calculators on the market now that run eight dollars or less.)

For example, what part of a pound is three ounces? Remembering that there are (and no one knows why) sixteen ounces in a pound, we realize that three ounces is 3/16 of a pound. The tables show us that this equals 0.188 parts of a pound. Or if you need to know how many pounds and ounces there are in 1.68 pounds, the table will get you close—0.68 pounds is just a bit less than eleven ounces.

The numbers are given to three-place accuracy.

Fractions	Decimals	Fractions	Decimals
1/16	0.063	9/16	0.563
1/8	0.125	5/8	0.625
3/16	0.188	11/16	0.688
1/4	0.250	3/4	0.750
5/16	0.313	13/16	0.813
3/8	0.375	7/8	0.875
7/16	0.438	15/16	0.938
1/2	0.500	1	1.000

For those who are interested in greater accuracy, the following values are exact:

Fraction	Decimal
1/100	0.01
1/64	0.015625
1/50	0.02
1/32	0.03125
1/25	0.04
1/16	0.0625
1/10	0.1
1/8	0.125

Table A12.1 Fraction into Decimal Conversions

Likewise, if decimal portions of feet need to be converted to inches or vice versa, the following chart will help. Seven inches is 0.583 feet, 0.425 feet is a bit more than 5 inches.

Feet and Inches, decimals

Foot decimals	Inches and decimal inches	Foot decimals	Inches and decimal inches
0.083	1.00	0.583	7.00
0.100	1.20	0.600	7.20
0.167	2.00	0.667	8.00
0.200	2.40	0.700	8.40
0.250	3.00	0.750	9.00
0.300	3.60	0.800	9.60
0.333	4.00	0.833	10.00
0.400	4.80	0.900	10.80
0.417	5.00	0.917	11.00
0.500	6.00	1.00	12.00

Table A12.2 Decimal Feet into Decimal Inches

Appendix 15: Temperature Conversion

The time honored formulas are:

$$°F = ((1.8)(°C)) + 32$$

$$°C = 0.56(°F - 32)$$

There are, however, simpler ways of making the conversion. For each degree Fahrenheit rise, there is five-ninths (0.56) of a degree rise in the Centigrade scale. For each degree Centigrade rise, there is nine-fifths (1.8) of a degree rise in the Fahrenheit scale. Add the commonly known fact that 0°C is 32°F., and you have the makings of a simple version of temperature conversion. For example, 41°F. is nine degrees above 32°F. Clearly, 41°F. is 5°C, since a nine degree Fahrenheit rise is a five degree Centigrade rise. Likewise, −5 °C. is 23 °F.

Below, then, we have a conversion table and interpolation table. The conversion table gives direct conversion information: 10°C. equals 50°F. The interpolation table shows that a certain number of degrees change in one system equals a certain number of degrees change in the other.

To find out what 233°C is on the °F scale:

30°C	equals	86°F change	(from the conversion chart)
+		+	
200°C	change is	360°F change	(from the interpolation chart)
+		+	
3°C	change is	5.4°F	(from the interpolation chart)
233°C	equals	451.4°F	

Again, only the conversion table gives actual conversion figures. The interpolation table simply shows that a change of a certain number of degrees in one system equals a change of a certain number of degrees in the other.

Conversion

°C	= °F	°F	= °C
0	32		
10	50	10	-12.2
20	68	20	-6.7
30	86	30	-1.1
40	104	40	4.4
50	122	50	10.0
60	140	60	15.6
70	148	70	21.1
80	176	80	26.7
90	194	90	32.2
100	212	100	37.8

Table A15.1 Centigrade and Fahrenheit Conversion

Interpolation

°C	= °F	°F	= °C
1	1.8	1	0.6
2	3.6	2	1.1
3	5.4	3	1.7
4	7.2	4	2.2
5	9.0	5	2.8
6	10.8	6	3.3
7	12.6	7	3.9
8	14.4	8	4.4
9	16.2	9	5.0
10	18.0	10	5.6
100	180	100	56
200	360	200	111
300	540	300	167
400	720	400	222
500	900	500	278
1000	1800	1000	556

Table A15.2 Interpolation

Appendix 16: Where to Find It

This is a short list of sources for products, information, and such. If you live within ICBM range of a fair-sized city you can, with an hour's work in the yellow pages, come up with a better list. Nevertheless, if all comes to naught, here are some addresses.

The list has two parts. Part one is alphabetical by product, and contains names of companies (and in some cases individuals) which carry this product or service. Part two is alphabetical by company (or individual) name, and for simplicity, is alphabetized by its first letter, whether of first or last name. Thus, J.D. Gould Co. is under "J."

ACR: Dometic, Norcold

Air compressor: Airborne, Aircraft, Brookstone, Brunner, Fesco, Nash, Palley, Sears, Watsco

Alternators: Airborne, Fesco, Palley, Sears

Anti-corrosion materials: American, Tapecoat

Carburetors: Dual Fuel, Jerry Friedburg, J & S, L. P. G.

Centrifugal pumps: see Manure pumps.

Chemicals: MC/B

Chemical analysis: La Motte

CO_2 indicators: Bacharach, Dwyer, Lee, National Appliance

Combustion bases test equipment: Bacharach, Dwyer, Lee, S. G. A.

Compressors: see Air compressors.

Diesel engines: Palley

Drum-handling equipment: MECO

Drum lids: Clarks Products Co.

Electric generators (powered by ICE): Empire, Katolight, Kohler, Onan, W.W.

Electric generators (without engine): Airborne, Aircraft, Fesco, Palley, W.W.

Flow meters: Dwyer, Manostat, National Sonics, Singer

Flue gas analysis: see Combustion.

Furnaces, lab (for VS determinations): Grieve, Nurnberg Scientific, Thermolyne

Gas analysis and detection equipment: Bacharach, Burrell, Lee, Nurmberg Scientific, S. G. A.

Gas/electric refrigerators: Dometic, Norcold

Gas production meters: Singer

Generator main tanks: A.O. Smith, Independent

Heat exchangers: Edwards

Heaters, Instantaneous water: Paloma, John Condon, Pressure, Lauren

Heat transfer cement: Thermon

H_2S hazard indicators: Bacharach

Information access: Alternative Sources, Energy Primer, Harrowsmith, Mother, RAIN

Lab equipment: Nurnberg Scientific

Linen tester: Edmund

Litmus (pH) paper: MC/B, Nurnberg Scientific

Liquid level sensors: Lee, National Sonics

Manometers: Dwyer

pH test equipment: Chemtrix, Hatch

Pressure switches: Dwyer, Mercoid, Palley

Pressure tanks: Airborne, Palley

Propane appliances (possible conversion): Century, Teeco

Publications access: Alternate, CTT, Ears, Henry, Mother, RAIN

Pumps, manure, slurry (low pressure centrifugal): A.O. Smith, Badger, Beatty, Clay, Holz, J.F. Farm, Mitchel, Parma, Sahlstrom, Starline, Vaughn, Wader

Pumps, (medium to high pressure centrifugal): Barnes, Construction, Enpo, Gorman-Rupp, Hale, Mitchel

Pumps, (positive displacement): Challenge, Holz, Monyo, Pearson, Robbins

Pumps, water: Airborne, Aircraft, Brookstone, KJ (cheapest small), Marine, W.W.

Refrigerators: see ACR or gas/electric.

Sludge density indicators: Lee, National Sonics

Solenoid valves: J. D. Gould, Johnson, W.W.

Steam engines, generators: Automotive, GCOE, Lancer, Semple

Thermometers, electric: Bacharach, United

Thermometers, plain or bimetalic: Nurnberg Scientific, Palmer, P. T. C.

Thermometer, recording: Bacharach, Lee, Mercoid, P. T. C. (cheapest), United

Toilets: Marine

Tubing: Chicago

Valves: see Solenoid valves.

Water heaters: Paloma

Water pumps: see Pumps.

Now a list, giving the addresses of those listed above.

A. O. Smith Harveststore, 550 W. Algonquin Rd., Arlington Heights, IL 60007

Airborne Sales Co., Inc., 8501 Steller Drive, Culver City, CA 90230

Aircraft Components, Inc., P.O. Box 1188, Benton Harbor, MI 49002

Alternative Sources of Energy, Rt. 2, Milaca, MN 56353 (4 issues/$5.00)

American Gilsonite Co., 1150 Kennecott Bldg., Salt Lake City, UT 84133

Automotive Steam Systems, 8591 Pyle Way, Midway City, CA 92655

Bacharach Instrument Co., 625 Alpha Dr., Pittsburgh, PA 15238

Badger Northland, Inc., Kaukauna, WI 54130

Barnes Pumps of Canada Ltd., 1711 Mattan Ave., Mississauga, Ontario, Canada

Beatty Farm Equipment Co., Fergus, Ontario, Canada

Brookstone Co., 120 Vose Farm Rd., Peterborough, NH 03458

Brunner Mfg. Corp., 100 Locke Road Edge Moor, Wilmington, DE 19809

Burrell Corp., 2223 Fifth Ave., Pittsburgh, PA 15219

CTT (Conservation Tools and Technology), 143 Maple Rd., Surbiton, Surrey, KT6 4BH Great Britain

Century Mfg., 1462 U.S. Route #20, P.O. Box 188, Cherry Valley, IL 61016

Challenge Mfg. Co., 1308 67th St., Oakland, CA 94608

Chemtrix, Inc., 163 S.W. Freeman Ave., Hillsboro, OR 97123

Chicago Speciality Co., 7500 Linder Ave., Skokie, IL 60076

Clarks Products Co., 415 N. Poplar, P.O. Box 400, South Hutchinson, KS 67501

Clay Equipment Co., Cedar Falls, IA 50613

Construction Machinery Co. of Canada, Box 231, Waterloo, Ontario, Canada

Dometic, 2320 Industrial Parkway, P.O. Box 490, Elkhart, IN 46515 (line discontinued)

Dual Fuel Systems, Inc., 720 West Eight Street, Los Angeles, CA 90017

Dwyer Instruments, P.O. Box 373, Michigan City, IN 46360

Ears, 2239 East Colfax, Denver, CO 80206

Edmond Scientific, Barrington, NJ 08007

Edwards Engineering Co., 101 Alexander Ave., Pompton Plains, NJ 07444

Energy Primer, $4.50, Portola Institute, 558 Santa Cruz Ave., Menlo Park, CA 94025

Enpo-Cornell Pump Co., Pique, OH 45356

Fesco, P.O. Box 3426, El Monte, CA 91733

GCOE Corp., 7701 N. Stemmons Freeway, Suite 245, Dallas, TX 75247

Gorman-Rupp of Canada Ltd., 70 Buswell Rd., St. Thomas, Ontario, Canada

Grieve Corp., 500 Hart Rd., Round Lake, IL 60073

Hale Fire Pump Co., 708 Spring Mill Ave., Conshohocken, PA 19428

Harrowsmith, P.O. Box 1600, Ithaca, NY 14850 (6 issues, $6.00)

Hatch Chemical Co., Box 907, Ames IA 50010

Henry Doubleday Research Assn., Convent Lane, Bocking, Braintree, Essex, Great Britain

Holz Mfg. Co., P.O. Box 1359, Newburgh, NY 12550

Independent Power Developers, Inc., Box 1467, Noxon, MT 59853 (catalogue $1)

J & S Carburetor, P.O. Box 10391, 2634 N. Beckley St., Dallas, TX 75207

J. D. Gould Co., 4707 Massachusetts Ave., Indianapolis, IN 46218

J. F. Farm Machinery Ltd., P.O. Box 760, Exeter, Ontario, Canada

Jerry Friedburg, Rt. 2 Box 96C, Leslie, AR 72645

John Condon Co., 1103 N. 36th Street, Seattle, WA 98103

Johnson Corporation, 805 Wood Street, Three Rivers, MI 49093

K. J. Miller Corp., 2401 Gardner Rd., Broadview, IL 60153

Kohler Co., Kohler, WI 53044

L. P. G. Automotive, Box 2253, Eugene, OR 97402

LaMotte Hydroculture, Chestertown, MD 21620

Lancer Engineering, Box 544, Pekin, IL 61554

Lauren Nichols, 8285 S.W. Cashmur Lane, Portland, OR

Lee Co., Box 66099, Seattle, WA 98166

MC/B Mfg. Chemists, 2909 Highland Ave., Norwood, OH 45212

MECO, 2011 Cuming St., Omaha, NE

Manostat, 519 Eighth Ave., New York, NY 10018

Marine Center, 2130 Westlake Ave. N., P.O. Box G19030, Seattle, WA 98109

Mercoid Corp., 4201 Belmont Ave., Chicago, IL 60641

Mitchell-Lewis-Staver, 801 S.E. Alder, Portland, OR

Monyo: see Robbins & Myers

Mother Earth News, P.O. Box 70, Hendersonville, NC 28739

Nash Engineering, 310 Wilson Ave., Norwalk, CT 06856

National Appliance Co., 10855 S.W. Greenburg Rd., Portland, OR 97223

National Sonics, 250 Marcus Blvd., Hauppauge, NY 11787

Norcold, 1620 West Bristol St., Elkhart, IN 46514

Nurnberg Scientific, 127 Merrick Rd., Rockville Center, NY 11201

Onan, 1400 73rd Ave. N.E., Minneapolis, MN 55432

Compression ratio: The greatest volume of a cylinder (with the piston at the bottom) divided by the least volume (with the piston at the top).

Conduction: Heat transfer by contact.

Continuous feed: Generators into which slurries are daily or more constantly put.

Convection: Heat transfer by the motion of fluids; air currents, for example.

Conversion efficiency: The percentage of one kind of energy, transformed into another useful kind; mechanical motion into electricity, as one example.

Cubic centimeters per gram VS added: A measure of biogas production, where the amount of VS decomposed is not considered.

Cubic centimeters per gram VS decomposed: A measure of biogas production, where the amount of VS decomposed is important.

Culture: A population of bacteria, used to establish a population in a biogas generator.

Ecosystem: The interactive web of life which covers the whole planet, or the web of life in one small portion of it—such as a pond or a biogas generator.

Effective volume: That portion of the generator which is used, and/or how well it is used to produce biogas. For example, if only eight cubic meters of a ten cubic meter generator is full of slurry, the generator cannot have more than 80% effective volume.

Effluent: The used slurry, sludge, supernatent, or scum.

Energy: Whatever has energy has the ability to do work. That definition includes us.

Entraining air: Mixing air or some other gas into a primary gas stream.

Entropy: An annoying tendency of energy to become less useful and less concentrated.

Enzymes: Chemicals which help form and break down molecules.

Extracellular enzymes: Enzymes which operate outside the bodies of small biogas buddies.

Design heat load: The amount of heat required, per unit time, to keep a structure of generator warm under the worst weather conditions ordinarily experienced in that location.

Diesel: An engine which has no spark plugs; air alone is compressed into the cylinder, and the fuel is injected when combustion is desired.

Digestion: The biological breakdown of organic materials.

Digestor: Here, an anaerobic composter designed primarily with the digestion of organic matter in mind, rather than the efficient production of biogas.

Draft tube: A tube which assists the development of a convection-type current in a generator when gas is recirculated into the tube.

Dry digestion: Digestion at above 10% solids.

Dry slurry: Slurry with more than 10% solids.

Dual-fuel engine: An engine using both a liquid and a gaseous fuel simultaneously.

Dung: Manure, usually without urine.

Dynamic head: The total force a pump must overcome to pump a fluid through a system.

Dynes per square centimeter: A means of measuring pressure.

Facultative: Switch-hitters, able to be either anaerobic or aerobic.

Fatty acids: The major general food source of the methane bacteria.

Film: When speaking of heat transfer, this term refers to the surface of gas or liquid which touches an object, that resists the flow of heat from or to that object, into or out of the main body of the gas or liquid.

Float much: Those substrates which refuse to make easily pumpable slurries.

Flocculation: The tendency of algae to coagulate under certain conditions. See Chapter 18.

Fluffing/percolation: The process described by one researcher which indicated healthy dry digestion. See Chapter 10.

Food pyramid: The food chain, another reason to believe that all flesh is grass.

Force: The cause of motion.

FS: Fixed solids; ashes.

Gas-diesel: An engine which is designed to operate on both diesel fuel and biogas, where both are injected into the cylinder for firing. Sometimes used incorrectly to describe engines which have been adapted to run on both fuels, rather than an engine which is designed to do so.

Gas digestion: The second stage of biogas generation, during which the CH_4 of the biogas is produced.

Gasholder: Usually, a container, open at the bottom, floating on water, for the collection of biogas.

Generation: Here, the production of biogas.

Generator: Here, the word is used to describe an anaerobic compost unit that is designed primarily to efficiently produce biogas.

Grass roots research: Homegrown, relevant information.

Head: The vertical distance to which a liquid must be pumped to get it from source to point of use.

Heat: Thermal force.

Heat load: The amount of heat required per defined period of time, to keep a structure of generator warm, under the prevailing weather conditions.

Heat of vaporization: The amount of heat required to evaporate a unit volume of a liquid with no change in temperature or pressure.

High-rate generators: Biogas generators which operate at optimum parameters—optimum temperature, low HRT, high rate agitation, etc.

HRT: Hydraulic Retention Time; the number of days an average unit volume of slurry stays in the generator.

Hybrid generators: A two-stage biogas generator design consisting of a larger batch-fed, cold, acid-producing phase, and a smaller continuous-fed, methane-producing, heated phase. Alternatively, any generator which separates acid and methane production.

ICE: Internal Combustion Engine.

Incandescent lamp: A gas-powered light with a mantle.

Inches of water: A way of expressing pressure measurements.

Injector: That part of a Bunsen or Fisher burner which injects gas into a tube where it is mixed with air prior to burning.

Inoculation: Adding a culture of anaerobic bacteria to a generator.

Intensive animal production: Flesh-factory or feed-lot animal production.

Ions: Loosely speaking, atoms or molecules with a positive or negative charge.

Kinetic: Energy in motion.

Latent heat of vaporization: The amount of heat required to evaporate a unit volume of a liquid with no change in temperature or pressure.

Lignin: A common compound in plants which resists digestion.

Lime: Here, $CaCO_3$—calcium carbonate.

Lime water: Here, $Ca(OH)_2$—calcium hydroxide.

Litmus paper: A pH test paper.

Loading rate: The total weight of VS fed into the generator each day divided by the volume of the generator.

Low technology: A technology available to mom, pop, and the kids. AKA low tech, kitchen sink technology.

Mantle: A net of material which incandesces and gives light when it is heated.

Median: The middle value, midway between highest and lowest. This is not the average in the common sense of the word.

Mesophilic: Middle-lovers (referring to anserobic bacteria that enjoy temperatures from 20° to 40°C.—70°-105°F.).

Metabolism: The cellular mechanics of life—the process of using biological energy.

Methane: CH_4, the main combustible gas in biogas.

MF: Methane forming.

Millimeters of mercury: A way of expressing pressure measurements.

Mix well: Those substrates which make pumpable slurries—chiefly the different manures.

Moderate-rate generators: Biogas generators which operate near optimum temperature, but with longer HRTs and less agitation than might be used.

Molecules: CH_4, CO_2, H_2O, and their brethren and sistren.

Non-lignin C/N: The C/N based on the available rather than the total carbon. Because lignin is so common, a significant part of the total carbon is tied up in the lignin, and not available to the biogas process.

7: pH

Answers to Questions

Add a buffer, remove CO_2 from the biogas and recirculate, or add old effluent.

9: C/N

Solutions to Problems

1. 203 − 38 = 165; 38 − 8 = 30

Using the combined number technique:

$$(165)(1) = (30)(Wb)$$
$$5.50 = Wb$$

$$Wwa = (165)(4.9) = 808.5$$
$$Wwb = (5.5)(5.4) = 29.7$$

In cases like this, we can divide both numbers by ten if we wish:

$$Wwa = 80.85, Wwb = 2.97$$

leaves purslane

2. Combining the substrates as suggested, for beets and whole turnips, we have:

$$38 - 27 = 11; 44 - 38 = 6$$
$$(11)(1) = (6)(Wb)$$
$$1.83 = Wb$$

$$(11)(22.2) = 244.20$$
$$(1.38)(32) = 58.56$$

Then, for timothy and clover hay, it's

$$58 - 38 = 20; 38 - 24 = 14$$

At 50% moisture, the combined number of timothy would be (2 x 2) 4:

$$(20)(1) = (14)(Wb)$$
$$1.43 = Wb$$

$$(20)(4) = 80$$
$$(1.43)(8.4) = 12$$

3. Impossible. The C/Ns of both of these are above 38. How will you get to a target C/N of 38 from substrate C/Ns of 58 and 44?

11: HRT

Solutions to Problems

1. $\frac{100 \text{ liters}}{15 \text{ days}} = 6.67$ liters/day, loading rate

2. (200 grams TS per liter) (0.80 grams VS per gram TS) = 16 grams VS per liter.

(2.5 grams VS per liter of generator per day) (100 liters of generator) = 250 grams VS per day.

$$\frac{250 \text{ grams VS per day}}{16 \text{ grams VS per liter}} = 15.63 \text{ liters per day,}$$

. . . or an HRT of 6.4 days

If you got this wrong, don't worry. You were on your own and we didn't give you any formulas for it. Chapter 42 will help.

12: Agitation

Answers to Questions

Agitate to increase gas production, control scum, maintain a uniform temperature throughout the generator, promote more rapid digestion, prevent "caking," increase substrate particle surface area (possibly), and regulate pH through the release of CO_2.

16: Pure Substances

Answers to Questions

Carbohydrates, cellulose, sugars, starch, alcohol, fats —all of them, really, except perhaps protein.

25: Steam

Solutions to Problems

The energy required to bring the water to its boiling point is:

(5) (100 − 20) = 400 Cal

The energy required to entirely convert the hot water to steam is:

(5) (539) = 2,695 Cal

At 100% efficiency, then, (400 + 2,695) 3,095 Cal is required to convert 5 liters of 20°C water to steam.

27: Engines

Solutions to Problems

At 100% conversion efficiency, the engine would require 641 Cal per horsepower hour (from the metric conversion charts). At 20% conversion efficiency, it would need (100 ÷ 20) 5 times that amount, or 3,205 Cal per horsepower hour.

(3,205) (3) (20) = 1.92×10^5

28: Space Heating

Solutions to Problems

First, we need to find the C numbers:

C of brick = 101 ÷ 10 = 10.1
C of polyurethane = 1.86 ÷ 5 = 0.37
C of air space = 5.37

$$\frac{1}{Ct} \text{ of the wall} = 0.12 + \frac{1}{10.1} + \frac{1}{0.37} + \frac{1}{5.37}$$

Ct of wall = 0.32

Ct of window = 3.03 (from text)

wall area = (5) (5) (4) = 100 m^2

Ht = (3.03) (2) 10 °C.) (10 hr) + (0.37) (100) (10°C) (10 hr)

Ht = 606 + 3700

Ht = 4306 Cal

Since both Cts are in terms of Cal m^{-1} °C^{-1} hr^{-1}, or:

$$\frac{\text{Calories}}{\text{meters °C. hours}}$$

and since we are multiplying by units of meters, °C, and hours, then meters, °C. and hours all drop out of the final expression, leaving only Cal. To put it another way, a number in terms of Cal °C^{-1} times some other number in terms of °C. comes out in terms of Cal.

29: Gas Requirements

Solutions to Problems

1. (0.08) (72) = 5.76 Cal 1^{-1}
2. An ACR requires 20 Cal 1^{-1} da^{-1} (see Table 29.1, so 60 liters of ACR requires ((20) (60)) 1,200 Cal da^{-1}. The light requires 350 Cal hr^{-1}, or ((3) (350)) 1,050 Cal for 3 hours. This is 2,250 Cal da^{-1} total.

Hv = (65) (0.08) = 5.2 Cal 1^{-1}

$$Vb = \frac{(2250)\,(1)}{5.2}$$

Vb = 432.7 liters

$$Vb = \frac{(2250)\,(1)}{(78)\,(0.08)}$$

Vb = 360.6 liters

31: Startup

Answers to Questions

There are four general methods: heavy seed, compound interest, load-and-hope, and percent of solids (gradual increase). Always use a buffer.

36: Heating

Answers to Questions

Either steam or hot water would do.

40: Pumping

Answers to Questions

1. Any of the mentioned pumps would work.
2. The best, however, would probably be a centrifugal type (since it might be cheapest), if there was a low enough dynamic head.

42: Sizing Generators

Solution to Problems

1. Vgen = HRT (Vau) (Nau) (Df)

$$Vgen = \frac{15\text{ da }(37.8\text{ 1 }au^{-1}\,da^{-1})\,(550\text{ kg }cow^{-1})\,(3\text{ cows})\,(2)}{450\text{ kg }au^{-1}}$$

Vgen = 15 da (37.8 1 $au^{-1}da^{-1}$) (3.67 au) (2)

Vgen = 4162 liters

The loading rate is immaterial in this equation.

43: Safety

Answers to Questions

Any electric appliance that has a motor, or produces heat, or has metal-to-metal contact switches that produce sparks. Don't forget static electricity either.

45: Economics

Solutions to Problems

1. If we assume 55% CH_4 in the biogas (and for generators where O_2 is strictly excluded, this is conservative), then we obtain [(300) (0.55)] 165 cubic meters of methane per day. A cubic meter (from Appendix 14) is 1,000 liters. In Gas Requirements, we learned that one liter of methane has a saturated net heat value of 8 Cal. (Actually, it's 8.12 Cal, a difference of 1.5 percent.)

   This means that we have approximately [(8.00 Cal 1^{-1} (165 m^3) (1000 1 m^{-3})] 1.32 million Calories of methane (saturated net heat value) produced per day. In the operation under question, a propane-fired grain dryer, biogas would probably yield a similar conversion efficiency. Grain drying is a seasonal operation, but let's assume for simplicity that the farmer can use the methane energy to replace propane year round. Further, he has calculated his generator heat load at an average of 40% of the per day heat value of the biogas. (This also changes day-to-day and summer-to-winter.) Since 60% of the CH_4 is left after the generator heat requirements have been met, this means that the net annual available heat is:

$$(0.60)\ (3.2 \times 10^6 \text{ Cal da}^{-1})\ (365 \text{ da yr}^{-1})$$
$$= 7.01 \times 10^8 \text{ Cal per year.}$$

Since propane runs at \$21.64 per million Calories, this is:

$$(7.01 \times 10^8 \text{ Cal})\ (10^6 \text{ Cal})\ (\$21.64) = \$15{,}165 \text{ per year.}$$

Thus, our equation becomes:

$$\text{Anb} = (15{,}165 + 4{,}200) - 4{,}800 - (35{,}000)\ (0.156)$$
$$\text{Anb} = \$9{,}105$$

The conclusion seems to be that propane is a very expensive fuel. It is well to remember that a generator does not generate dollars, it generates biogas and saves labor to an extent that can be roughly correlated with a certain number of dollars. If the above farmer had other energy sources available, or if he simply could not find a use for all that energy, the factors would change accordingly.

2. From the above, we have 3.2 million Calories per day. Since a biogas-powered electric generator will give us waste heat, the heat load of the generator will not need to be subtracted from this figure—unless the engine does not produce sufficient waste heat to keep the generator warm.

   Assuming, again, year round replacement of the present energy source and 100% use of biogas, we then need to know how much energy we can obtain as electricity. Perusing again Chapter 27, we find that scrubbing the biogas can make a remarkable difference in the efficiency with which it is used. We've assumed 55% CH_4, and using Neyloff and Gunkel's figures, we could expect each horsepower hour to cost us 4,395 Cal. Ordinary conversion efficiency of mechanical to electrical energy runs about 50% (total conversion efficiency from biogas to electrical is about 7.3%). This means that our 3.2 million Cal gives:

$$\frac{(3.2 \times 10^6)\ (0.50)}{(4395 \text{ Cal hp}^{-1} \text{ hr}^{-1})} = 364 \text{ hp}$$

$$(364 \text{ hp})\ (\text{hr})\ (0.746) = 272 \text{ kwh per day}$$

In a year, this is:

$$(272 \text{ kwh da}^{-1})\ (365 \text{ da yr}^{-1}) =$$
$$9.9 \times 10^4 \text{ kwh per year}$$

In dollars, assuming 25¢ per kwh, that's:

$$(\$0.025)\ (\text{kwh}^{-1})\ (0.2 \times 10^4 \text{kwh hr}^{-1}) = \$2{,}480$$

If the biogas were scrubbed, this would improve conversion efficiency and therefore the annual dollar benefit would rise. Of course, the cost of the scrubber would then become part of the cost of the facility.

Our equation becomes:

$$\text{Anb} = (2480 + 4200) - 4800 - (35000)\ (0.156)$$
$$\text{Anb} = -\$3{,}660$$

This is, of course, a per year loss in benefits.

If we could add a scrubber and improve CH_4 percentage to 80%–90%, thereby increasing the efficiency of biogas to mechanical power efficiency from 14.6% to 22.5%, and if we could increase our mechanical to electrical conversion efficiency to 60%, this would be 13.5% total conversion efficiency, versus the 7.3% we had before.

This would increase the daily or yearly energy output to 85%, or the per year benefit in biogas produced kwh would rise from \$2,480 to \$4,586. Thus, even though in this case the change won't pull us through, we can see the importance of insuring that our energy transfers and transformations take place with maximum efficiency.

Appendix 19: Bibliography

In the bibliography, we have used some abbreviations due to the fact that the titles of some journals recur again and again. As a matter of fact, one particular journal has had three names over a period of years *(Sewage Works Journal, Sewage and Industrial Wastes,* and *Journal of the Water Pollution Control Federation)* and its value is reflected in the fact that, time and again, useful and applicable research has been published in its pages.

The abbreviations are:

Ind. Eng. Chem. = Industrial Engineering and Chemistry
Sewage Ind. = Sewage and Industrial Wastes
Sewage Works J. = Sewage Works Journal
J. Water Pollu. Control Fed. = Journal of the Water Pollution Control Federation

Abeles. 1977. *Design Considerations for Anaerobic Farm Digesters.* Minneapolis: i/e Associates.

______. 1980. *A Farm Integrated Utility System (FIUS): A Study of Farm Energy Use.* Minneapolis: i/e Associates.

Abeles and Ellsworth. 1978. *Task VI-Biological Production of Gas.* Minneapolis: i/e Associates.

Abeles, Freedman, DeBaere, and Ellsworth. 1978. *Bibliography of Anaerobic Digestion.* Minneapolis: i/e Associates.

Abeles, Freedman, Ellsworth, and DeBaere. 1978. *Energy & Economic Assessment of Anaerobic Digesters for Rural Waste Management.* Minneapolis: i/e Associates.

Acharya. 1935. Studies on the Anaerobic Decomposition of Plant Materials. IV: The Decomposition of Plant Substrates of Varying Compositions. *Biochemical Journal* 29: 1459.

______. 1958. *Preparation of Fuel Gas and Manure by Anaerobic Fermentation of Organic Materials.* Indian Council of Agricultural Research. Research Publication 15.

Alexander. 1956. Chemical Control of Digester Scum. *Sewage Ind.* 28:931.

Alich and Inman. 1976. Utilization of Plant Biomass as an Energy Feedstock. In *Energy, Agriculture and Waste Management,* p. 453. Ann Arbor: Ann Arbor Science Publishers.

Althausen, Boruf, Buswell, and Symons. 1933*a*. Comments on Fair and Moore's Articles on "Heat and Energy Relations in the Digestion of Sewage Solids." *Sewage Works J.* 5:34.

______. 1933*b*. Additional Comments on Fair and Moore's Articles on "Heat and Energy Relations in the Digestion of Sewage Solids." *Sewage Works J.* 5:233.

Ameen. 1974. *Community Water Systems Source Book.* n.p., Technical Proceedings.

American Society of Heating, Refrigeration, and Air-Conditioning Engineers. n.d. *American Society of Heating, Refrigeration, and Air-Conditioning Engineers Handbook of Fundamentals.* New York.

Andersen. 1943. Overhauling and Repairs to Methane Gasholder. *Sewage Works J.* 15:535.

______. 1963. *Audels Home Appliance Service Guide.* Audels.

Anderson, n.d. *The Production of Methane From Manure.* Microbiology Dept., Bio Science Bldg.) Oregon State Univ. Corvallis, OR 97330.

Anonymous. n.d. *Natural Gas Vehicle Project.* Los Angeles: Dual Fuel Systems, Inc.

______. 1923. Hydraulics. In *International Library of*

Technology, v. 190. Scranton, PA: Scranton International Textbook Co.

______. 1943 Tips and Quips: Instructions for Using Sludge Fertilizer. *Sewage Works J.* 15:754.

______. 1954. Dual Fuel Engines Show Economy at Lansing. *Sewage Ind.* 26:800.

______. 1955. Check Chart for Centrifugal Pump Troubles. *Sewage Ind.* 27:867.

______. 1973. Biomass Offers Renewable Energy Sources. *Modern Power and Engineering* (Canada). July.

______. 1974*a*. *Methane Gas Generation from Agricultural Wastes.* Folder 160, April, 1974. Cooperative Extension Service, Montana University, Bozeman, MT 59715.

______. 1974*b*. Sewage Plant Changes Carburetion Systems to Lower Maintenance and Operating Costs. *Diesel and Gas Turbine Progress,* April.

Aroian, et al, eds. 1958. *Van Nostrand's Scientific Encyclopedia.* New York: Van Nostrand and Co.

Auerbach. 1973. *A Homesite Power Unit: Methane Generator.* Madison, CT: Alternative Energy Systems.

Babbit and Bauman. 1958. Sludge Digestion. In *Sewerage and Sewage Treatment,* p. 581. New York: Wiley.

Bacon. 1944. Digester Heating and Mixing Equipment. *Sewage Works J.* 16:534.

Bakewell, et al. 1977. No Tanks. *Rain: Journal of Appropriate Technology* 3:12.

Banta and Pomeroy. 1934. Hydrogen Ion Concentration and Bicarbonate Equilibrium in Digesting Sludge. *Sewage Works J.* 2:234.

Bardoch, Ryther, and McLarney. 1972. *Aquaculture: The Farming and Husbandry of Freshwater and Marine Organisms.* New York: Wiley Interscience.

Barker. 1913. *Barker on Heating: The Theory and Practice of Heating and Ventilation.* New York: Van Nostrand.

Baumgartner. 1934. Effect of Temperature and Seeding on Hydrogen Sulfide Formation in Sewage. *Sewage Works J.* 6:399.

Baxter and Horlacher. 1954. Submerged Sludge Heating at Philadelphia's Northeast Works. *Sewage Ind.* 26:961.

Beinhauer. 1933. Use of Gas from Municipal Treatment Plants as Motive Power for Trucks. *Sewage Works J.* 5:728.

Bell, Boulter, Dunlop, and Keller. 1973. *Methane: Fuel of the Future.* Andrew Singer.

Bernard, Denit, and Anderson. 1971. Effluent Discharge Guidelines and Animal Waste Management Technology. In *Animal Waste Management.* Virginia, Arlie House.

Bigger. 1974. 350 Valley Homes Get Electricity from Landfill. *Southern Sierran* 29:5.

Bio-Energy Council. 1979. *Bio-Energy Directory.* 1625 Eye St. N.W., Suite 825A, Washington, D.C. 20006.

Black. n.d. *Pumps.* n.p., Audel and Co.

Boruff and Buswell. 1929. Fermentation Products of Cellulose. *Ind. Eng. Chem.* 21:1181.

______. 1930. Fermentation Products from Cornstalks. *Ind. Eng. Chem.* 22:931.

______. 1934. The Anaerobic Fermentation of Lignin. *J. of the American Chemical Society* 56:886.

Boshoff. 1963. Methane Gas Production by Batch and Continuous Fermentation Methods. *Tropical Science* 5:155.

______. 1967. Reaction Velocity Constants for Batch Methane Fermentation on Farms, Notably in the Tropics. *J. of Agricultural Science* (Gt. Brit.) 68:347.

Braithwaite and Capener, n.d. *The Feasibility of Methane Production From Dairy Animal Wastes: The Farmer's Perspective.* Department of Rural Sociology, State Col. of Agri. & Life Sci., Cornell Univ., Ithaca, N.Y. 14853.

Braker and Mossman. 1971. *Matheson Gas Data Book.* n.p., Matheson Gas Products.

Bunsen. 1857. *Physical and Chemical Properties of Gases.* London: Walton and Moberly.

Burford and Varani. n.d. *Methane on the Move, Energy Potential Through Bio-Conversion of Agricultural Wastes.* Loveland, CO: Bio-Gas of Colorado.

Burlew, ed. 1953. *Algal Culture from Laboratory to Pilot Plant.* Publication 600. Washington, D.C.: Carnegie Institute of Washington.

Buswell. 1929. Control of Scum in Sewage Tanks. *Ind. Eng. Chem.* 21:322.

______. 1930. Production of Fuel Gas by Anaerobic Fermentations. *Ind. Eng. Chem.* 22:1168.

Buswell and Boruff. 1932*a*. The Relation Between the Chemical Compositions of Organic Matter and the Quality and Quantity of Gas Produced During Sludge Digestion. *Sewage Works J.* 4:454.

______. 1932*b*. Patent for Fibrous Waste Generator. U.S. Patent 1880 772, October 4, 1932.

______. 1933. Mechanical Equipment for Continuous Fermentation of Fibrous Materials. *Ind. Eng. Chem.* 25:147.

Buswell, Pagano, and Sollo. 1949. Effect of Sodium Sulfate and Chloride on Methane Fermentation. *Ind. Eng. Chem.* 41:596.

Buswell, Boring, and Milam. 1960. A Paper Chromographic Method for Volatile Acids. *J. Water Pollu. Control Fed.* 32:721.

Buswell and Sawyer. 1963. Biochemical v.s. Physical Factors in Digester Failure. *J. Water Pollu. Control Fed.* 35:205.

Bylinsky and Schuyten. 1976. Green Plants Might Provide the Cheapest Energy of All. *Fortune* Sept., 1976, p. 152.

Canada Animal Waste Management Guide Committee. 1972. *Canada Animal Waste Management Guide.* n.p., Canada Committee of Agriculture.

Cassell and Sawyer. 1959. A Method of Starting High-rate Digesters. *Sewage Ind.* 31:123.

Chan. 1970. *Hydrolisis Rate of Cellulose in Anaerobic Fermentation.* Report 70-3. Berkeley, CA: Sanitary Engineering Research Laboratory, University of California.

Charpentier. 1902. *Timber.* n.p., Scott Greenwood and Co.

Chung Po. n.d. *Production and Use of Methane from Animal Wastes in Taiwan.* Taipei, Taiwan: Joint Commission on Rural Restoration.

______. n.d. *Swine Wastes and Methane.* Taipei, Taiwan: Joint Commission on Rural Reconstruction.

Cobb. 1963. *Islamic Contributions to Civilization.* Washington, D.C.: Avalon.

Coffey. 1973. "Briton Runs Car on High-Octane and Pollution-Free Pig Manure." *Los Angeles Times,* May 2, 1973.

Coker. 1966. The Value of Liquid Digested Sewage Sludge. *J. Agricultural Science* Jan.: 91-107.

Corin. 1965. *Professional Engineers Examination: Chemical Engineering.* Fall River, MA: Aero.

Cravath and Lansing. 1907. *Practical Illumination.* New York: McGraw.

Crooks. 1949. The Dual Fuel Engine and its Application to Sewage Treatment Plants. *Sewage Works J.* 21:957.

Cross and Bevan. 1902. "Researches on Cellulose." n.p.

Crouch. 1972. *The Autonomous Servicing of Dwellings—Design Proposals.* Cambridge, England: Department of Architecture, Technical Research Div., University of Cambridge.

Dale. 1971. Status of Dairy Waste Treatment and Management Research. *Animal Waste Management* Sept. :85.

Dalrymple and Proctor. 1967. Feasibility of Dairy Manure Stabilization by Anaerobic Digestion. *Water Sewage Works* Dec.: 361

Davis. 1966. Cattle-Manure Handling and Disposal Systems on the West Coast. In *Management of Farm Animal Wastes.* ASAE Publication SP-0366. American Society of Agricultural Engineers.

DiLallo and Albertson. 1961. Volatile Acids by Direct Titration. *J. Water Pollu. Control Fed.* 33:356.

Dobbins. 1962. Mechanism of Gas Absorption by Turbulent Liquids. *J. Water Pollu. Control Fed.* 34:245.

Duggan. 1950. "Pilot Plant Production of Methane by Anaerobic Fermentation of Pear Waste." Master's thesis, Oregon State University.

Dunkley. 1921. *Some Conditions Affecting the Usefulness of Iron Oxide for City Gas Purification.* Bulletin 119. University of Illinois Engineering Experiment Station, Urbana, IL.

Dunlop. 1968. Survival of Pathogens and Related Disease Hazards in Municipal Sewage Effluent for Irrigation. Louisiana Tech. Alumni Foundation, Box 1, Tech. State, Ruston, LA 71270.

Dye. 1973. "Garbage Dump Has Lighted Christmas Tree—and It's a Gas." *Los Angeles Times,* Dec. 2, 1973.

Eby. 1966. Evaluating Adaptability of Pasture Grasses to Hydroponic Culture and Their Ability to Act as Chemical Filters. In *Farm Animal Wastes Symposium,* May 5-7, 1966, p. 117.

Ecotope Group. 1976. *Report on the Design and First Year Operation of a 50,000 Gallon Anaerobic Digester at the State Honor Farm Dairy, Monroe, Washington.* 2332 E. Madison, Seattle, WA 98112.

Elder and Buswell. 1929. Changes of Sulfur Compounds During Sewage Treatment. *Ind. Eng. Chem.* 21:560.

Eliassen. 1951. 1951 Industrial Wastes Forum (Application of Submerged Combustion). *Sewage Ind.* 24:867.

Emerich. n.d. *Heating Handbook.* New York: McGraw-Hill.

Epstein. 1972. *Mineral Nutrition of Plants: Principles and Perspectives.* New York: Wiley.

Fair. 1934. Time and Rate of Sludge Digestion, and Their Variations with Temperature. *Sewage Works J.* 6:3.

______. 1934*b*. Digestion of Garbage. *Sewage Works J.* 6:259.

Fair and Moore. 1930. Observations on the Reliability of Different Methods for Determining the pH of Sewage Sludge. *Sewage Works J.* 2:3.

______. 1932*a*. Heat and Energy Relations in the Digestion of Sewage Solids I. *Sewage Works J.* 4:242.

______. 1932*b*. Heat and Energy Relations in the Digestion of Sewage Sludge II. *Sewage Works J.* 4:428.

______. 1932*c*. Heat and Energy Relations in the Digestion of Sewage Solids III. *Sewage Works J.* 4:589.

______. 1933*A*. Author's Reply. *Sewage Works J.* 5:40.

______. 1933*b*. Author's Closure. *Sewage Works J.* 5:236.

______. 1937. Observations on the Digestion of Sewage Sludge Over a Wide Range of Temperatures. *Sewage Works J.* 9:3.

Fiat Auto Group. 1977. *TOTEM, Total Energy Module.* Advertising brochure.

Finney, Evans, and Finney. 1976. *Fast Production of Methane by Anaerobic Digestion: Quarterly Report, May 24, 1976–August 23, 1976.* Natural Dynamics Publication #COO-2900-5. Des Moines, IA: Natural Dynamics.

Fischer. 1929. Separate Sludge Digestion Studies. *Sewage Works J.* 1:236.

Fischer, Sievers, and Fulhage. 1976. Anaerobic Digestion in Swine Wastes. In *Energy, Agriculture and Waste Management,* p. 307. Ann Arbor, MI: Ann Arbor Science Publishers.

Fong. 1973. *Methane Production from Animal Wastes by Anaerobic Decomposition.* Master's thesis, University of Manitoba.

Foot, et al. 1976. *Studies on Livestock Wastes.* n.p., Agricultural Research Council.

Fowler and Joshi. 1920. Studies on Vegetable Wastes. *J. of the Indian Institute of Science* 3:39.

Friedberg, 1972. *How to Convert Your Auto to Propane—A Manual.* Arrakis Volkswagen. (No current address. If anyone knows Jerry's address, please let us know. DH)

Frook. 1952. Removal of Carbon Dioxide from Digester Gas. *Sewage Ind.* 24:1473.

______. 1957. Volatile Acids by Direct Phosphoric Acid Distillation. *J. Water Pollu. Control Fed.* 29:18.

Fry. 1974. *Practical Building of Methane Power Plants for Rural Energy Independence.* Standard Printing, 1223 No. Nopal, Santa Barbara, CA 93103.

Fry and Merrill. 1973. Methane Digesters for Fuel Gas and Fertilizer. Woods Hole, MA: New Alchemy Institute.

Fuhrman. 1954. Scum Control in Sludge Digestion. *Sewage Ind.* 26:453.

Fulweiler. 1930. Use of Sewage Gas as City Gas. *Sewage Works J.* 2:424.

Gaddy, Park, and Rapp. 1974. Kinetics and Economics of Anaerobic Digestion of Animal Waste. *Water, Air and Soil Pollution* 3:161.

Gallese. 1977. "Garbage Power." *Wall Street Journal,* Aug. 4, 1977.

Geigy Pharmaceuticals, eds. 1956. *Documenta Geigy—Scientific Tables.* n.p.: Geigy Pharmaceuticals.

Gerdel. 1946. Digestion Tank Scum. *Sewage Works J.* 18:543.

Gerhard. 1908. *The American Practice of Gas Piping and Gas Lighting in Buildings.* New York: McGraw-Hill.

Gessel. 1960. Composting Wood Wastes with Other Organic Materials. *Compost Science* Spring: 26.

Ghosh, Conrad, and Klass. 1975. Anaerobic Acidogenesis of Wastewater Sludge. *J. Water Pollu. Control Fed.* 47:30.

Gill. 1917. *Gas and Fuel Analysis for Engineers.* New York: Wiley.

Golueke. 1960. Composting Manure by Anaerobic Methods. *Compost Science* Spring: 44.

______. 1963. Power from Solar Energy Via Algae-Produced Methane. *Solar Energy* 7:86.

______. 1971. Closed Waste Treatment System for High Intensity Animal Production. *Waste Age* Mar./Apr.:10.

______. 1972. *Composting: A Study of the Process and Its Principles.* Emmaus, PA: Rodale.

Golueke and Oswald. 1959. Biological Conversion of Light Energy to the Chemical Energy of Methane. *Applied Microbiology* 7:219.

______. 1965. Harvesting and Processing Sewage-Grown Planktonic Algae. *J. Water Pollu. Control Fed.* 37:471.

Goodman. 1974. *Methane Gas Generators.* n.p.: Fallbrook Farms.

Gotaas. 1956. *Composting: Sanitary Disposal and Reclamation of Organic Wastes.* Geneva: World Health Organization.

Green. 1961. *International and Metric Units of Measurement.* New York: Chemical Publishing Co.

Griswold. 1946. *Fuels, Combustion and Furnaces.* New York: McGraw-Hill.

Grune and Sload. 1954. Biocatalysts in Sludge Digestion. *Sewage Ind.* 26:1425.

Gunson. n.d. Sludge, Heating and Circulation by Direct Steam Injection. *Sewage Works J.* 19:661.

Hagglund. 1951. *Chemistry of Wood.* New York: Academic.

Hammer and Borchardt. 1969. Dialysis Separation of Sewage Sludge Digestion. *J. of Sanitary Engineering Division.* Proceedings of the American Society of Civil Engineers SA5:907.

Hammerton. 1945. The Corrosion of Cement and Concrete. *Sewage Works J.* 17:403.

Harnik. 1948. A Preliminary Investigation on Methane Gas Production from Pear Waste. Master's thesis, Oregon State University.

Harris. 1912. *Talks on Manures.* n.p., Orange Judd Co.

Hart. 1963. Digestion Tests of Livestock Wastes. *J. Water Pollu. Control Fed.* 35:748.

Hart, Moore, and Hale. 1966. Pumping Manure Slurries. In *Management of Farm Animal Wastes.* ASAE Publication SP. 0366. American Society of Agricultural Engineers.

Haskins. 1917. *Compilation of Analysis of Agricultural Chemicals, Refuse Salts, Phosphates, Guanos, Ashes, Lime Compounds, Marls, By-Products, Refuse Substances, Animal Excrements, Insecticides and Fungicides.* Rhode Island Experiment Station, Bulletin #21.

Hass and Hill. 1928. *Chemistry of Plant Products: Volume I.* New York: Longmans Green.

Hatfield, Symons, and Mills. 1928. Gases from Sewage Sludge Digestion. *Ind. Eng. Chem.* 20:174.

Hatfield. 1938. Automatic Shutoff for Gas Still. *Sewage Works J.* 10:891.

______. 1940. Safe Handling of Gas at Sewage Plants. *Sewage Works J.* 12:128.

Hauschka. 1950. *The Nature of Substance.* n.p.: Vincent Stuart.

______. 1951. *Nutrition.* n.p.: Stuart and Watkins.

Haseltine. 1933*a*. The Use of Sludge Digestion Gas for Power Production. Cause and Effect of Variations in Composition of Gas. *Sewage Works J.* 5:482.

______. 1933*b*. Comments on Rudolf's and Miles' Articles on "Heat Losses from Sludge Digestion Tanks." *Sewage Works J.* 5:844.

Hecking. 1940. Utilization of Sludge Gas for Power Production. *Sewage Works J.* 12:132.

Henry. 1961. Hydrogen Sulfide in Sludge Gas. *J. Water Pollu. Control Fed.* 33:140.

Heukelekian. 1929. Effect of Certain Pure Substances on Sludge Digestion. *Sewage Works J.* 1:545.

______. 1942. Effect of Yeast on the Stabilization of Sewage and Sludge Digestion. *Public Works.* Sept.:17.

______. 1943. Effect of the Addition of Sodium Nitrate to Sewage on Hydrogen Sulfide Production and B.O.D. Reduction. *Sewage Works J.* 15:255.

Heukelekian and Berger. 1953. Value of Culture and Enzymes Additions in Promoting Digestion. *Sewage Ind.* 25:1259.

Heukelekian and Heinemann. 1939*a*. Studies of the Methane-Producing Bacteria: I. Development of a Method for Enumeration. *Sewage Works J.* 11:426.

______. 1939*b*. Studies on the Methane Producing Bacteria: III. Digestion of Sewage Solids By the Addition of Enriched Cultures of Methane Producing Organisms. *Sewage Works J.* 11:571.

Heukelekian and Kaplovsky. 1948. Effect of Change of Temperature on Thermophilic Digestion. *Sewage Works J.* 20:806.

Heukelekian and Rudolfs. 1928. Carbon and Nitrogen Transformation in Fresh Sewage Digestion. *Ind. Eng. Chem.* 20:177.

Hill. 1939. Digestion Studies on Paper Pulp. *Sewage Works J.* 11:864.

Hill and Barth. 1977. A Dynamic Model for Simulation of Animal Waste Digestion. *J. Water Pollu. Control Fed.* 49:2129.

Hills. n.d. *Fuel Plus Fertility.* Henry Doubleday Research Association, Convent Lane, Bocking, Braintree, England.

Hinden and Dunstan. 1963. Anaerobic Digestion of Potato Processing Wastes. *J. Water Pollu. Control Fed.* 35:486.

Hommon. 1931. Imhoff Tank Sludge as Farm Fertilizer—Canton, Ohio. *Sewage Works J.* 3:232.

House. 1973. *Methane Systems.* Earthmind, 4844 Hirsch Rd, Mariposa, CA 95338.

Hundley. 1949. Use of Internal Combustion Engines in Sewage Treatment Plants. *Sewage Works J.* 21:2.

Hussey, Row, and Allison. 1934. *Fuel from Sewage and Cellulosic Waste.* Bulletin 18. Virginia Polytechnic Institute Experiment Station.

Imhoff. 1927. Operating Results of the Essen Activated Sludge Plant. *Engineering News Record* 99:790.

______. 1946. Digester Gas for Automobiles. *Sewage Works J.* 18:17.

______. 1952. Sludge Gas as Fuel for Motor Vehicles. *Water and Sewage Works* July:284.

Inglett, ed. 1973. *Symposium: Processing Agricultural and Municipal Wastes.* Westport, CT: AVI Publishing.

Ingram and Bartsch. 1960. Graphic Expression of Biological Data in Water Pollution Reports. *J. Water Pollu. Control Fed.* 32:297.

Jewell. n.d. *Bioconversion of Agricultural Wastes for Pollution Control and Energy Conservation.* Department of Rural Sociology, Cornell University, Ithaca, NY 14853.

______. ed., 1976. *Energy, Agriculture and Waste Management.* Ann Arbor: Ann Arbor Science Publishers.

Johnson. n.d. Methane Power & Auto Information Sheet. Oregon Museum of Science and Industry Energy Center, 4015 S.W. Canyon Rd., Portland, OR 97221.

Jones. 1932. What Are Offensive Odors About a Sewage Works? *Sewage Works J.* 4:60.

Keefer and Kratz. 1929. Digesting Sludge at Its Optimum pH and Temperature. *Engineering News Record* Jan. 17 :103.

______. 1932. The Effect of Gases on Sewage Sludge Digestion. *Sewage Works J.* 4:24.

______. 1933. The Interchange of Heat During Sludge Digestion. *Sewage Works J.* 5:3.

______. 1934. The Digestion of Garbage with Sewage Sludge. *Sewage Works J.* 6:14.

Keenan. n.d. *Two-Stage Methane Production from Solid Wastes.* Philadelphia: University of Pennsylvania.

______. 1977. Fuels Via Bioconversion. *Energy Conversion* 16:95.

Kern. 1974. *The Owner Built Homestead.* New York: Scribners.

Keyes. 1929. Renaissance of the Absorption Cycle Refrigerator. *Ind. Eng. Chem.* 21:497.

King. 1911. *Farmers of 40 Centuries.* London: Cape.

Kinsel. 1935. Gas Collection and Sludge Heating. *Sewage Works J.* 7:895.

Klock. 1972. Sequential Processing in Wastewater Lagoons. *J. Water Pollu. Control Fed.* 44:241.

Koepf. 1974. Personal communication.

Krishnamoorthi, Abdulappa, Sarkar, and Siddiqi. 1973. Productivity of Sewage Fertilized Fish Ponds. *Water Research.* 9:269.

Kristal and Annett. 1940. *Pumps: Types, Selection, Installation, Operation and Maintenance.* New York: McGraw-Hill.

Kunsch. 1948. Sludge Digestion Experiences with the External Heat Exchanger. *Sewage Works J.* 20:999.

Lambrecht. 1948. "Methane Production by Anaerobic Fermentation of Pear Waste." Master's thesis, Oregon State University.

Latham. 1873. *Sanitary Engineering*. n.p.: E. & F.N. Span.

Laura and Idnani. 1971. Increased Production of Biogas from Cowdung by Adding Other Agricultural Waste Materials. *J. of the Sciences of Food and Agriculture* 22:164.

Law. 1968. *Agricultural Utilization of Sewage Effluent and Sludge—An Annotated Bibliography*. Federal Water Pollution Control Administration, Dept. of the Interior, Washington, D.C.

Lea. 1946. Sewage Pumping Stations. *Sewage Works J.* 18:238.

Leach. 1972. "British 'Ecological House' Aims at Halting Pollution & Waste." *Los Angeles Times,* Oct. 18, 1972.

Levine, Nelson, Anderson, and Jacobs. 1935. Utilization of Agricultural Wastes: I. Lignin and Microbial Decomposition. *Ind. Eng. Chem.* 27:195.

Lindsey, Haskins, Smith, and Beals. 1919. *Compilations of Analyses.* Special Bulletin of the Massachusetts Agricultural Experiment Station.

Lunge and Ambler. 1934. *Technical Gas Analysis.* n.p.: Gurney and Jackson.

Maly and Fadrus. 1971. Influence of Temperature on Anaerobic Digestion. *J. Water Pollu. Control Fed.* 43:641.

Maramba. 1978. *Biogas and Waste Recycling: The Philippine Experience.* Maya Farm, Antipolo, Rizla, Philippines.

Matthias. 1942. *How to Design and Install Plumbing.* Chicago: American Technical Society.

Mau. 1956. Applying Recent Research to the Design of Separate Sludge Digesters. *Sewage Ind.* 28:1200.

McCarty. 1964*a*. Anaerobic Waste Treatment Fundamentals: II. Environmental Requirements and Control. *Public Works* Oct.:123.

______. 1946*b*. Anaerobic Waste Treatment Fundamentals: III. Toxic Materials and Their Control. *Public Works* Nov. :91.

McClung and Lindberg. 1957. The Study of Obligately Anaerobic Bacteria. In *Manual of Microbiological Methods,* p. 120.

McGee. 1955. Gasoline Engine Conversion. *Sewage Ind.* 27:107.

McKinney. 1953. Biocatalysts and Waste Disposal: I. Fundamental Biochemistry of Waste Disposal. *Sewage Ind.* 25:1129.

McKinnie. 1960. Operation and Maintenance of Gas Engines at Santa Rosa. *J. Water Pollu. Control Fed.* 32:311.

Merz and Stone. 1968. Quantitative Study of Gas Produced by Decomposing Refuse. *Public Works* Nov. :86.

Meyers and Brown. 1961. *Gas Regeneration and Food Production in a Closed Ecological System.* National Academy of Sciences—National Research Council, Publication 893. Washington, D.C.

Meynell. 1979. *Methane: Planning a Digester.* New York: Schoken Books.

Miles. 1947. Evaluation of Methods of Heating Sludge. *Sewage Works J.* 19:161.

Mitsui. 1974. The Utilization of Solar Energy for Hydrogen Production by Cell-Free System of Photosynthetic Organism. In *Hydrogen Energy (Part A).* New York: Plenum Press.

______. 1975. Photoproduction of Hydrogen via Microbial and Biochemical Processes. In *Hydrogen Energy Fundamentals, Symposium Proceedings.* Miami: University of Miami.

Moore. 1935. Thermal Considerations in the Design of Heated Sludge Digestion Tanks. *Sewage Works J.* 7:618.

Morgan. 1954*a*. Studies of Accelerated Digestion of Sewage Sludge. *Sewage Ind.* 26:462.

______. 1954*b*. Efficient Utilization of Digester Capacity. *Sewage Ind.* 26:1340.

Morris. 1976. Methane for Home Power Use. *Compost Science.* Nov./Dec.

Morris, Jewell, and Casler. 1976. Alternative Animal Waste Anaerobic Fermentation Designs and Their Costs. In *Energy Agriculture and Waste Management,* p. 317.

Mother Earth News. n.d. *A Drum Methane Generator.* Mother Earth News Reprint #160. Box 70, Hendersonville, NC 28739.

______. 1971. Gobar Gas: Methane Experiments in India. 12:28.

______. 1972. The Plowboy Interview: Ram Bux Sing. 18:6.

______. 1973. The Plowboy Interview: L. John Fry. 23:6.

Muehling. 1971. The Handling and Treatment of Swine Wastes. In *Animal Waste Management,* p. 111. Virginia: Arlie House.

Mueller. 1972. Nutrient Removal by Natural Gas Fermentation. *J. Water Pollu. Control Fed.* 44:25.

Mullen. 1955. *Modern Gas Analysis.* New York: Interscience.

Murray. 1976. Weed Holds Promise for Pollution Cleanup. *Chemical and Engineering News* March 22:23.

Myers. 1955. Algae as an Energy Converter. *Proceedings of the World Symposium on Applied Solar Energy,* Phoenix, Arizona, Nov. 1–5, 1955. Palo Alto, CA: Stanford Research Institute.

National Academy of Sciences. n.d. *Methane Generation from Human, Animal & Agricultural Wastes.* Natl. Acad. of Sci., 2011 Constitution Ave., Washington, D.C. 20418.

Nelson, Straka, and Levine. 1939. Effect of Temperature of Digestion, Chemical Composition, and Size of Particles on Production of Fuel Gas from Farm Wastes. *J. of Agricultural Research* 58:273.

Newton. 1976. Photoproduction of Molecular Hydrogen by a Plant-Algal Symbiotic System. *Science* 191: 559.

Neyeloff and Gunkel. 1976. Methane-Carbon Dioxide Mixtures in an Internal Combustion Engine. In *Energy, Agriculture and Waste Management,* p. 397.

Niles. 1956. 1955 Operators Forum (Digester Scum Control). *Sewage Ind.* 28:579.

Nishihara. 1935. Digestion of Human Fecal Matter, with pH Adjustments by Air Control. *Sewage Works J.* 7:798.

Novak and Ramesh. 1975. Stimulation in Anaerobic Degradation. *Water Research* 9:963.

Nussberger. 1948. Determination of Hydrogen Sulfide in Digester Gas. *Sewage Works J.* 20:499.

Oswald and Golueke. 1960. Biological Transformation of Solar Energy. *Advances in Applied Microbiology.* 2:223.

______. 1964. Solar Power via a Botanical Process. *Mechanical Engineering* Feb. :40.

Pacific Coast Gas Association, n.d. *Data Book of Gas Utilization.* 2 Pine St., San Francisco, CA 94111.

Parker. 1945. The Propulsion of Vehicles by Compressed Methane Gas. Great Britain, Institute of Sewage Purification.

Pauling. 1947. *General Chemistry.* San Francisco: Freeman.

Pickering. 1925. *Compressabilities of Gases.* Miscellaneous Publication of the Bureau of Standards #71. Washington, D.C.

Poincelot. 1972. The Biochemistry and Methodology of Composting. n.p.: Connecticut Agricultural Experiment Station Bulletin #727.

Pomeroy. 1941. Hydrogen Sulfide in Sewage. *Sewage Works J.* 13:498.

Popvich and Herring. 1959. *Fuels and Lubricants.* New York: Wiley.

Porges, Jasewicz, and Hoover. 1952. Measurement of Carbon Dioxide Evolution from Aerated Sludge. *Sewage Ind.* 24:1091.

Pretorius. 1969. Anaerobic Digestion III. Kinetics of Anerobic Fermentation. *Water Research* 3:549.

______. 1973. Principles of Anaerobic Digestion. In *International Conference on Water Pollution,* p. 202. New York: Pergamon.

Pringsheim. 1964. *Pure Cultures of Algae—Their Preparation and Maintenance.* New York: Hafner.

Pyle and Fraenkel, eds. 1975. *Methane.* London: Intermediate Technology Publications.

Quasim. n.d. *SI Units in Engineering and Technology.* New York: Pergamon.

Queer. 1933. Heat Transmission in Sludge. *Sewage Works J.* 5:937.

Rainbow and Rose, eds. 1963. *Biochemistry of Industrial Microorganisms.* New York: Academic.

Rankin. 1947. Scum Control in Digesters. *Sewage Works J.* 18:407.

______. 1948. Digester Capacity Requirements. *Sewage Works J.* 20:478.

Ransome. 1944. Straw-Sludge Manure (Reviews and Abstracts). *Sewage Works J.* 17:395.

Rawn. 1931. Automatic Sludge Stirring Mechanisms. *Sewage Works J.* 3:24.

Rawn, Banta, and Pomeroy. 1937. *Multiple-Stage Sewage Sludge Digestion.* New York: American Society of Civil Engineers.

Reed and Lerner. 1973. Methanol: A Versatile Fuel for Immediate Use. *Science* 182:1299.

Reeves. 1958. Sanitary Aspects of Composted Sewage Sludge and Sawdust. *J. Water Pollu. Control Fed.* 5:557.

Richter. 1935. Relative Permanence of Papers Exposed to Sunlight. *Ind. Eng. Chem.* 27:177.

Riehl. 1938. Gas Service Manual. Robbins Publishing Co.

Riesenfeld and Kohl. n.d. *Gas Purification*. Houston: Gulf.

Rizk, Farag, El-Mofty, and El-Fadl. 1968. Studies on the Production of Methane Gas from Organic Wastes Under Anaerobic Conditions: I. Important Conditions Influencing Formation of Combustible Gases. *Agricultural Research Review,* vol. 46. (Cairo, Egypt).

Romashkevich and Karelina. 1961. Methane and Organic Fertilizers from Wood Waste and Manure Fermentation. *Mikrobiologiya* 30:146.

Rosenburg. 1952. Methane Production from Farm Wastes as a Source of Tractor Fuel. *Journal of the Ministry of Agriculture* 58:487.

Rubins and Bear. 1942. Carbon-Nitrogen Ratios in Organic Fertilizer Materials in Relation to the Availability of their Nitrogen. *Soil Science* 54:411.

Rudolfs. 1927. Effect of Temperature on Sewage Sludge Digestion. *Ind. Eng. Chem.* 19:241.

______. 1932. Gas Yield from Sewage Sludge. *Sewage Works J.* 4:445.

Rudolfs and Heinemann. 1939. Growth Promoting Substances in Sewage and Sludge. *Sewage Works J.* 11:587.

Rudolfs and Miles. 1933. Heat Losses from Sludge Digestion Tanks: II. Field Observation. *Sewage Works J.* 5:17.

Rudolfs and Nemerow. 1952. Rag, Rope and Jute Wastes from Specialty Paper Mills: IV. Anaerobic Digestion. *Sewage Ind.* 24:882.

Rudolfs and Stahl. 1942. Phosphine and Sludge Digestion. *Ind. Eng. Chem.* 34:982.

Rudolfs, Baumgartner, and Setter. 1932. Effects of Coagulants on Sewage Digestion. *Sewage Works J.* 4:628.

Rudolfs, Setter, and Baumgartner. 1929. Effect of Iron Compounds on Sedimentation, Digestion and Ripe Sludge Conditioning. *Sewage Works J.* 1:398.

Sanders and Bloodgood. 1965. The Effect of Nitrogen to Carbon Ratio on Anaerobic Decomposition. *J. Water Pollu. Control Fed.* 37:1714.

Sanitary Engineering Research Laboratory, University of California. 1969. *Comprehensive Studies of Solid Wastes Management: Second Annual Report.* SERL 69–1. Berkeley.

______. 1970*a*. *Comprehensive Studies of Solid Wastes Management: Third Annual Report.* SERL 70–2. Berkeley.

______. 1970*b*. *Hydrolysis Rate of Cellulose in Anaerobic Fermentation.* SERL 70–3. Berkeley.

______. 1972. *Comprehensive Studies of Solid Wastes Management: Final Report.* SERL 72–3. Berkeley.

Savery and Cruzan. 1972. Methane Recovery from Chicken Manure Digestion. *J. Water Pollu. Control Fed.* 44:2347.

Sawyer. 1960. Chemistry for Sanitary Engineers. New York: McGraw-Hill.

Sawyer and Grumbling. 1960. Fundamental Considerations in High Rate Digestion. *J. of the Sanitary Engineering Division of the American Society of Civil Engineers,* . 86.

Sawyer, Howard, and Perhse. 1954. Scientific Basis for Liming of Digesters. *Sewage Ind.* 26:935.

Schatz, et al. 1975. *Transition: A Book on Future Energy*. n.p.: Prometheus Unbound.

Scherfig and Pearson. 1968. Bioenergetics of Acetate and Acetate Precursors in Methane Formation. Report 68–9. Berkeley: Sanitary Engineering Research Laboratory, University of California.

Schlenz. 1936. Experiences with Digester Foaming. *Sewage Works J.* 8:609.

______. 1944. Controlled Digestion. *Sewage Works J.* 16:504.

______. 1947. Scum Control in Digesters—a Discussion. *Sewage Works J.* 18:409.

Schofield and Tark. 1948. New Heating System for Sludge Digestion. *Water and Sewage Works* 1948:172.

Schorger. 1926. *Chemistry of Cellulose and Wood.* New York: McGraw-Hill.

Schroepfer and Ziemke. 1959. Development of the Anaerobic Contact Process: I. Pilot Plant Investigations and Economics. *Sewage Ind.* 31:164.

Schulte, Krocker, Sparling, and Sapp. 1976. Methane Production through Bioconversion of Agricultural

Residues. In *Sharing the Sun: Agriculture, Biomass, Wind—New Developments.* v. 7, p. 119.

Schulze. 1958. Studies on Sludge Digesters and Methane Fermentation I. Sludge Digestion at Increased Solids Concentrations. *Sewage Ind* 30:28.

Scipio. 1931. *Development of a Direct Contact Water Heater.* Purdue University Engineering Experiment Station Research Series #38. Lafayette, IN.

Scott. 1962. Disposal of High Organic Matter Content Wastes on Land. *J. Water Pollu. Control Fed.* 34:932.

Searls. 1944. Operation and Maintenance of Sewage Pumps. *Sewage Works J.* 16:629.

Seil. n.d. *Gas Chemists Manual of Dry Box Purification of Gas: An Annotated Bibliography.* n.p.: American Gas Association.

Shadduck and Moore. 1975. *The Anaerobic Digestion of Livestock Wastes to Produce Methane: 1946–June 1975* (A Bibliography with Abstracts). St. Paul: Agricultural Engineering Dept., University of Minnesota 55108.

Sheehy. 1960. Stream Pollution: Rapid Methods of Solving First Order Equations. *J. Water Pollu. Control Fed.* 32:646.

Sierp. 1930. Specific Heat, Conduction, and Generation of Heat During the Digestion of Sewage Sludge. *Sewage Works J.* 2:191.

______. 1931. The Effect of Sludge Circulation in Digestion Tanks on Gas Production, Sludge Capacity and Scum (abstract). *Sewage Works J.* 3:154.

Singh. 1971. *Biogas Plant—Generating Methane from Organic Wastes.* Ajitmal, Etawah, India: Gobar Gas Research Station.

______. 1972*a*. The Biogas Plant—Generating Methane from Organic Wastes. *Compost Science.* 13:20.

______. 1972*b*. Building a Biogas Plant. *Compost Science* 13:12.

______. 1975. *Biogas Plant: Generating Methane from Organic Wastes and Designs with Specifications.* n.p.: Mother's Print Shop.

Siu. 1951. *Microbial Decomposition of Cellulose.* New York: Reinhold.

Smiser and Gettings. 1977. New Opportunities in Fish Culture. *Organic Gardening and Farming* 24:62.

Smith and Brown. 1975. Process Feasibility Study: The Anaerobic Digestion of Dairy Cow Manuare at the State Reformatory Honor Farm, Monroe, Washington. Seattle: Ecotope Group, Box 5599, University Station 98105.

Smith and Montanaro. n.d. Horizontal Drum Bio-Gas Plant. Austin: Maximum Potential Systems Project, University of Texas.

Smith and Morris. 1953. Use of the Heat Pump for Digester Heating. *Sewage Ind.* 25:1369.

Snell. 1942. Anaerobic Digestion: I. Correction of Errors During the Measurement and Analysis of the Gas. *Sewage Works J.* 14:1304.

______. 1943. Anaerobic Digestion: III. Anaerobic Digestion of Undiluted Human Excreta. *Sewage Works J.* 15:679.

Sobel. 1966. Physical Properties of Animal Manures Associated with Handling. In *Management of Farm Animal Wastes.* Publication SP–0366. n.p.: American Society of Agricultural Engineers.

Society of American Bacteriologists. 1975. *Manual of Microbiological Methods.* New York: McGraw-Hill.

Sparr. 1953. Initial Digestion Problems in a New Plant. *Sewage Ind.* 25:610.

Spencer, Shuckrow, and Ferguson. 1976. *Addition of Powdered Activated Carbon to Anaerobic Digesters: Effects on Methane Production.* Battelle Pacific Northwest Labs Publication #CONF–760738–1. Seattle.

Sperry. 1953. Laboratory Digester and Gasometer. *Sewage Ind.* 25:858.

______. 1959. Gas Recirculation at Aurora, Illinois. *J. Water Pollu. Control Fed.* 31:733.

Steel and Ehlers. 1958. *Municipal and Rural Sanitation.* New York: McGraw-Hill.

Straub. 1943. Digestion Studies on Pure Vegetables. *Sewage Works J.* 15:658.

______. 1944. Effect of Addition of Nitrogen on Digestion of Paper Pulp. *Sewage Works J.* 16:30.

Strock. 1948. *Heating and Ventilating's Engineering Databook.* n.p.: Industrial Press.

Taras, Greenberg, Hoak, and Rand, eds. 1974. *Standard Methods for the Examination of Water and Wastewater.* Washington, D.C.: American Public Health Association.

Thimann. 1955. *The Life of Bacteria—Their Growth, Metabolism, and Relationships.* New York: Macmillan.

Tietjen. 1976. From Biodung to Biogas—Historical Review of European Experience. In *Energy Agriculture and Waste Management,* p. 247. Ann Arbor: Ann Arbor Science Publishers.

Toth. 1960. Using Organic Wastes in Agriculture. *Compost Science* 1:10.

Trasoo. 1972. *Natural Gas.* Beaconsfield, England: Scientific Press.

Travin and Buswell. 1934. The Methane Fermentation of Organic Acids and Carbohydrates. *J. of the American Chemical Society* 56:1751.

United Aircraft Research Laboratories. 1974. *Technology for the Conversion of Solar Energy to Fuel Gas, Annual Report.* 5285 Port Royal Rd., Springfield, VA 22151.

University of Manitoba. 1974. Methane Recovery Project Receives Money from Shell. *University of Manitoba Bulletin.* Aug. 14, 1974.

University of Pennsylvania. 1974. *Technology for the Conversion of Solar Energy to Fuel Gas, Annual Report.*

Van Kleeck. 1943. Use of Sewage Sludge as Fertilizer. *Sewage Works J.* 15:788.

Viets. 1971. Cattle Feedlot Pollution. *Animal Waste Management* Sept.:97.

Visser. 1963. Gas Production in the Decomposition of Cyperus Papyrus. *J. Water Pollu. Control Fed.* 35:973.

Volunteers in Technical Assistance. n.d. *Methane Gas Plant.* Publication 11300. 3706 Rhode Island Ave., Mt. Rainier, MD.

Wade and Hartmann. 1975. *Homegrown Energy.* New York: Scribner.

Walraven. 1932. Development of Power from Sewage Sludge Gas. *Sewage Works J.* 4:614.

______. 1935. Separate Sludge Digestion Tanks. *Sewage Works J.* 7:73.

Ward. 1978. personal communication.

Webber. 1907. *Town Gas and Its Uses.* n.p.: Archibald Constable and Co.

Wilcox. 1948. Agricultural Uses of Reclaimed Sewage Effluent. *Sewage Works J.* 20:24.

Wiley. 1957. *Progress Report on High Rate Composting Studies.* Engineering Bulletin, Proceedings of the 12th Industrial Wastes Conference Series #94, 13–15 May 1957, p. 590.

Wills, et al. 1934. *Gas Engineers Handbook.* New York: McGraw-Hill.

Wilson, et al. 1975. *Energy: Global Prospects 1985–2000.* New York: McGraw-Hill.

Wisely. 1941. Experiences in Digestion Tank Scum Control. *Sewage Works J.* 13:787.

Wong-Chong. 1976. Dry Anaerobic Digestion. In *Energy Agriculture and Waste Management,* p. 366. Ann Arbor: Ann Arbor Science Publishers.

Young. 1948. Corrosion Test of Screen Wire at Sewage Works. *Sewage Works J.* 20:550.

Zeikus and Henning. 1975. Methanobacterium arbophilicum sp.nov.: An Obligate Anaerobe Isolated from Wetwood of Living Trees. *Antonie Van Leewenhoek* 41:543.

Zeikus and Ward. 1974. Methane Formation in Living Trees: A Microbial Origin. *Science* 184:1181.

Zoltek and Gram. 1975. High-Rate Digestor Mixing Study Using Radioisotope Tracer. *J. Water Pollu. Control Fed.* 47:79.

Author's Afterword:
My Beliefs on Saving the World

This note may be read by few, but here the author has a chance to speak unrestrained. One word that shows up infrequently in the rest of the book is "I." Here, however, I hope that you'll grant me the use of that word. Among the many reasons I had for writing this book, one blossoms here. Many people who turn to books like this are interested in helping, in making society and civilization better and more humane. They are interested in solutions. So am I, and I would like to speak about large problems, civilization, history, and a real solution.

The cause of civilization is human unity. And as unity becomes higher and more perfect—economic, then political, then ethical or religious unity—the civilization which is the reflection of that unity also becomes more perfected.

All around us, we see problems. It's difficult to get a perspective on how vast or serious the problems are because we have no reference point. Another way of saying the same thing is: we can't see the forest for the trees. Many of us, especially the young, have grown up in this forest—we don't know what world peace or human unity or the larger love of a true community is, because we have never seen these things ourselves. Yet the things we take for granted, the violence in countries whose names we forget, the huge disparity of education, wealth, opportunity, and physical well-being, between, for example, America and Biafra or Ethiopia—these problems have not always been with us.

Of course, the present time is different. Never before could it be said that the earth is one country and humankind her citizens. Never before in history have we been able to see a picture of that beautiful blue-green jewel as it hung in the nothing blackness of space.

Yet there have been times and places in our past history, humankind's history, when justice was the benchmark of society, and when the good of all was not a social ideal, but a nearly universal personal motive. For example, the civilization of Ashoka of India was founded on the principle of compassion planted in the world by Buddha. Or the civilization of Constantine, Christian king, four centuries after Christ. Consider too the fruits of Mohammed; the university is an Islamic idea, and the Moslems gave us tremendous advances in medicine, astronomy, mathematics, and architecture. In the tenth century, Cordova had a cobbled street, brightly lit along its ten miles, while there was not a paved street in all of Paris. In turning back and seeing these lamps, we may then be instructed as to how to better invent the future.

"Let us examine the two civilizations," wrote Seignobos in his *Histoire de la Civilization au Moyen Age,* "which, in the eleventh century divided the ancient world. In the west—miserable little cities, peasants' huts and great fortresses—a country always troubled by a war, where one could not travel ten leagues without running the risk of being robbed; and in the Orient—Constantinople, Cairo, Damascus, Baghdad—with their marble palaces, their workshops, their schools, their bazaars, their villages, and the incessant movement of merchants who travelled in peace from Spain to Persia. There is no doubt that the musselman and Byzantine worlds were richer, better policed and better lighted than the western world. In the eleventh century these two worlds began to become acquainted; the barberous Christians came into contact with the civilized misselmans in two ways—by war and by commerce. And by contact with the orientals, the occidentals became civilized." (As quoted in *Christ and Baha'u'llah,* by George Townsend, p. 49.)

Other examples can be brought forward, but the point can be made another way as well. Not technology, not law, not politics, not economics, not even the name which we call ourselves, be it Christian or Buddhist or Moslem, but rather ethics, the inner spirit and life of each of those great names, is the basis and fountainhead of true civilization.

Without charity, trust, moderation, and faith, no economy can possibly survive. Without conscience, no laws will be obeyed. Without brotherhood and sisterhood, no society will long exist. Without love this world will not be at peace. My thesis is that these ideals can again be nurtured and grow at the heart of society.

Of course, some will think this is merely a pleasant dream, or a hopeless goal, but this is not the case. What has happened before, in the rise of past civilizations, built on the ashes of previous and failed civilizations, is even now beginning again.

Time and again, as history clearly shows, people have built societies out of a shared sense of good, and time and again those societies have crumbled—as ours is crumbling—when that shared sense of good becomes lost, degenerating into meaningless words and rituals. And what caused these bonds to be strong and what gave them the power to transform men was the same in every case—that is, Authority, given by God.

Constantine, Ashoka, the brilliant star of the Imam Ali, each was a follower of a great Teacher, and a leader of a people who also followed. Buddha, Moses, Christ, Mohammed, and now Baha'u'llah, each of them has given humankind the same priceless Gift. The followers of Baha'u'llah are known as Baha'is, and millions of Baha'is, around the world, affirm of these true Teachers, as Baha'u'llah says: "These sanctified Mirrors, these Daysprings of ancient glory are one and all Exponents on earth of Him Who is the central Orb of the universe, its Essence and ultimate Purpose . . . through them is transmitted a grace that is infinite, and by them is revealed the light that can never fade." (From *Selected Writings of Baha'u'llah,* copyright by the National Spiritual Assembly of the Baha'is of the United States. Reprinted with permission.

Thus Baha'is believe that the Light which shone from each of these Teachers, these Manifestations of God, was the same. The Light is brought to the world, gradually its influence becomes known and takes effect among humanity, and then, gradually, people turn away from it. The outward and visible indicators of this inner spiritual journey are seen in the rise of great civilizations and religions, and their gradual decline and decay, as truth fades to cliché, and inner law degenerates into outer form.

Thus, in this most central realm of human affairs, we have springtime, when the Message is given and begins to dawn in the hearts of people; summer, when it bears fruit; fall, when gradually people turn toward the form and forget the spirit; and winter when the old forms and ways, born in such glory, now impede humanity in its progress.

It is then that the Message is renewed, given again to humanity, through a new Teacher, and like the phoenix, it is reborn.

From age to age, however, the needs of humanity change, their understanding grows, and their capacities enlarge. Thus, the eternal Message is each time revealed more fully, and it is accompanied by laws which bring order to the social realm as it then exists, and according to the needs of that time.

Mosaic truth is Christian truth, but Mosaic law must give way to Christian law, for thus God wills. Now, in this age, the renewed Law will again replace the old and man-changed laws of the past.

This, then is that of which I would like to speak. Not to impel or command, but simply to present. If you wish, investigate. Let those who have an ear . . .

If you would like more information about the Baha'i Faith, please contact the National Spiritual Assembly of your country. In the United States it is located near Chicago, at 112 Linden Avenue, Wilmette, Illinois, 60091.

Index

Subject Index

Author Index